Soils
and
Geomorphology

Peter W. Birkeland

Department of Geological Sciences
University of Colorado

New York   Oxford
OXFORD UNIVERSITY PRESS
1984

Copyright © 1984 by Oxford University Press, Inc.

Library of Congress Cataloging in Publication Data

Birkeland, Peter W.
  Soils and geomorphology.

  Rev. ed. of: Pedology, weathering, and geomorphological
research. 1974.
Includes bibliographies and index.
  1. Soil formation. 2. Weathering. 3. Soil science.
4. Geology, Stratigraphic—Quaternary. I. Birkeland,
Peter W. Pedology, weathering, and geomorphological
research. II. Title.
S592.2.B57 1984   551.3'05   83-13345
ISBN 0-19-503398-1
ISBN 0-19-503435-X (pbk.)

This book is a revised edition of *Pedology, Weathering, and
Geomorphological Research* by Peter W. Birkeland.
Copyright © 1974 by Oxford University Press, Inc.

Printing (last digit): 9 8

Printed in the United States of America

To
my family
and
thesis students

Preface

Although there are many textbooks on soils, there are few that also serve the needs of geomorphologists, sedimentary petrologists, and archeologists working in Quaternary research. This book is an attempt to fill that gap. The emphasis is on the study of soils in their natural setting—the field—since field studies are most significant to Quaternary studies. Such studies are commonly called "pedology." Much of the research that geomorphologists undertake involves the use of soils to date deposits on the basis of soil development and to reconstruct the environment during soil formation. This book focuses on these problems, but other related problems are also discussed. I feel that one cannot adequately use soils for any purpose without understanding the processes and factors that control their formation. Hence, the overall organization of the book is, first, a discussion of soil morphology, weathering, and soil-forming processes and, then, variation in soils with variation in the soil-forming factors (climate, organisms, topography, parent material, and time). My discussion of soil classification is brief and generalized because I feel that it is more important to understand the genesis of a soil than it is to classify a soil. Only the new U.S. soil classification system is used. The books ends with a short discussion of applications of soils in Quaternary studies. For those readers beginning work in soils, I want to stress the importance of sound field work, because pedology is a field science. Laboratory studies are necessary, but they are only as good as the field work and sampling upon which they are based.

This book is a revision of *Pedology, weathering and geomorphological research;* the title has been shortened at the request of some of my colleagues. Many people helped me with the first edition of this book, and they are duly acknowledged there. A revision was considered necessary because much new data are available, and I have seen many more soils since then in the United States as well as overseas. In particular, I use the new horizon nomenclature of the U.S. Department of Agriculture, stress the importance of airborne materials on pedogenesis, use examples from more places, and in the last chapter discuss using soils in Quaternary stratigraphic studies with the new U.S. stratigraphic code, and demonstrate applications to both archeology and the dating of faults.

As with the first edition, I have relied heavily on my own experiences for examples of different aspects of pedology in writing this book. I did this because I feel that one has to work with something or see it in its field setting before one really understands it and can explain it to others. The literature cited is not too extensive, but I have tried to include both the relevant geological and pedological literature in attacking problems common to both disciplines. A more encyclopedic treatment of the literature would have to have been at the cost of deleting materials I consider more important in conveying important principles to the reader.

Two appendices appear at the end of the book. Appendix 1 is designed to give the reader information on the soil data that have to be collected in order to describe a soil profile adequately. Appendix 2 presents climatic maps for the United States. These are included because the examples used in the text generally are taken from many different climatic regions in the United States. For each of these examples one can get a general idea of the local or regional climate by referring to these maps.

I am grateful to several individuals for reading parts of the manuscript, and helping with the index and other chores. J. G. Bockheim and V. T. Holliday reviewed the first edition and made valuable suggestions on changes and additions for this edition. Other valuable suggestions or reviews of small portions have been made by colleagues in Quaternary soil studies: B. L. Allen, J. Boardman, M. N. Machette, and F. C. Ugolini. Most of the reviewing, proofreading, and indexing, however, fell on my present thesis students: Kathy Albino, Margaret Berry, Susan Cannon, Mary Gillam, Vance Holliday, Michael Litaor, Marith Reheis, Dave Swanson, and Emily Taylor: Thanks, friends, and remember that all errors and omissions are mine. Some colleagues sent me theses or unpublished manuscripts that were much appreciated and very helpful; these are M. N. Machette, L. D. McFadden, and D. J. Ponti. Edith Ellis typed the manuscript, and Robin Birkeland and Susan Rose helped get the manuscript ready for the publisher.

Since the first edition, many people have taken the time to help me better understand certain soil topics, to work on soil projects with me, or to take me into new field areas. For these courtesies and for sharing thoughts, I would like to thank B. L. Allen, J. B. Benedict, J. Boardman, J. G. Bockheim, R. M. Burke, W. B. Bull, J. G. Bruce, C. J. Burrows, A. S. Campbell, E. J. B. Cutler, J. Dan, R. Gerson, L. H. Gile, J. W. Harden, J. B. J. Harrison, J. W. Hawley, M. N. Machette, the late D. E. Marchand, L. D. McFadden, C. D. Miller, V. E. Neall, R. R. Shroba, P. T. Tonkin, P. H. Walker, W. T. Ward, A. W. Young, J. C. Yount, and D. H. Yaalon.

In addition to these, I thank all of my thesis students, for I have learned much from them by being involved in their projects.

Finally, I would like to thank the students in my various courses here for being the sounding board for many of these ideas, for challenging the ones they thought were weak, and for providing me with their ideas, and citations in the literature that I had overlooked.

My family has been especially helpful in all of my scientific endeavors. We always go into the field as a group, and each of them, Suzanne, Karl, and Robin, in their own way, has pitched in and helped make each trip a success.

Boulder, Colorado   P.W.B.
August 1983

Contents

Appendix 1   Data necessary for describing a soil profile, 353

Appendix 2   Climatic conditions in the United States, 362

Soils and Geomorphology

1

The soil profile, horizon nomenclature, and soil characteristics

The term "soil" has many definitions, depending upon who is using the term. For example, to engineers "soil" is unconsolidated surficial material, whereas to many soil scientists it is mainly the medium for plant growth. A definition of soil that serves our purpose well is a slight modification of that given by Joffe[34]: a soil is described as a natural body consisting of layers or horizons of mineral and/or organic constituents of variable thicknesses, which differ from the parent material in their morphological, physical, chemical, and mineralogical properties and their biological characteristics; at least some of these properties are pedogenic (Fig. 1-1). Soil horizons generally are unconsolidated, but some contain sufficient amounts of silica, carbonates, or iron oxides to be cemented.

Soils differ from geological deposits generally, but in some places the two are so similar in appearance that they are difficult to tell apart. The focus here is on those aspects of soils of interest to an interdisciplinary group that includes geomorphologists, soil scientists (especially pedologists), sedimentary petrologists, engineering geologists, archeologists, and botanists. Soils will be studied in their natural setting, the field, a part of soil science sometimes called pedology. One cannot stress too much the importance of sound field work in any soil project, for no amount of laboratory work or statistical treatment can correct improper site selection, soil description, or sampling. Hence, the stress will be on field relations, but we will also use data from other specialty fields of soil science (for example, soil chemistry) to help understand the soil in its natural setting. Finally, because most soils have formed during the Quaternary, one needs an adequate understanding of the geological, climatological, and botanical history of the last 2 million years.[12,22]

A horizon

Bt horizon

Cox horizon

Fig. 1-1 Soil formed on Quaternary marine terrace deposits near San Diego, California.

SOIL PROFILE

A soil profile consists of the vertical arrangement of all the soil horizons down to the parent material. In studying a soil, therefore, the investigator must be able to identify the parent material from which the soil formed. This is no easy task and requires a good deal of experience in geology and pedology. However, once the parent material is recognized, and its original properties estimated, one can begin to determine departures in the properties of the original material and identify these materials as soil horizons.

Some geologists distinguish between a soil profile and a weathering profile.[37,50] Where this is done, the soil profile is generally considered to make up the upper part of the much thicker weathering profile. However, because it is difficult to separate these two profiles on the basis of the processes involved, I will not make this distinction. What some would call the weathering profile beneath the soil profile probably would qualify as a Cox soil horizon under the horizon nomenclature used here. A problem with thick weathered zones (say thicker than 30 m) is the separation of the products of soil formation from those of diagenesis. The boundary between the two is difficult to define, and no doubt all gradations exist. Fortunately, most soils are not thick enough for this to be a major problem.

SOIL HORIZON NOMENCLATURE

Two sets of soil horizon nomenclature are in use in the United States, one, diagnostic horizons, for classification purposes, and one for field descriptions. The diagnostic horizons require laboratory analyses, and will be described in Chapter 2. The set used for field descriptions is that introduced by the Soil Survey Staff in 1981,[26] much of which still is unpublished, with some modifications. A comparison between the old and the new field description systems, as well as properties necessary for an adequate soil description, are given in Appendix 1. Soil profile depth functions of various key properties are helpful in visualizing the more common horizons (Fig. 1-2). Many properties form by the vertical movement and accumulation of some material (for example, clay, iron), and these are

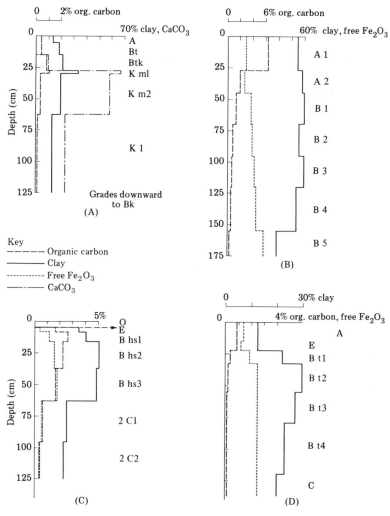

Fig. 1-2 Laboratory data on soil profiles that illustrate properties commonly associated with various soil horizons. (A) Parent material is gravelly alluvium. Percent clay is for the noncarbonate fraction, and both percent clay and carbonate are for the nongravel fraction. (Terino soil from Gile and others,[25] Table 2, © 1966, The Williams & Wilkins Co., Baltimore.) (B) Parent material is serpentine. (Profile No. 27.[57]) (C) Parent material is glacial outwash. (Profile No. 20.[57]) (D) Parent material is loess. (Profile No. 11.[57])

termed illuvial. The overlying horizon(s) from which the illuvial materials were derived is termed eluvial. Many properties have an eluvial-illuvial relationship within the profile.

Three kinds of symbols are used in soil horizon nomenclature; capital letters, lower case letters, and Arabic numerals. Capital letters denote master horizons, lower case letters some specific characteristic or subdivision of the master horizon, and Arabic numerals either a further subdivision of the horizon or parent material layering.

Most soil profiles can be divided into several master or most prominent horizons (Table 1-1). Surface or near-surface horizons relatively high in organic matter are designated O and A horizons, the difference between the two being determined by the amount of organic matter present. Beneath the O or A horizon, in some environments, there is a light-colored horizon relatively leached of iron compounds, the E horizon; this is identical to the A2 horizon of some classifications. The B horizon commonly is beneath the surface horizon or horizons. This horizon encompasses a multitude of soil characteristics relative to those of the assumed parent material. Among the B horizon characteristics are clay accumulation, the production of red color, the accumulation of iron compounds with or without organic matter, the residual concentration of resistant materials following the removal of more soluble constituents under conditions of intensive weathering and leaching, and accumulation of $CaCO_3$ and more soluble salts. The slightly weathered C horizon (Cox) commonly is beneath the B horizon and beneath that is unweathered bedrock, the R horizon, or unconsolidated unweathered material, the Cu horizon. This latter horizon is useful in geomorphological studies. In desert environments, carbonate buildup plays an important role in soil morphology and genesis, and horizons high in carbonate are designated K. Although the K-horizon designation is not used by the Soil Conservation Service, it is used here because most pedologists and geologists working in arid lands find it a very useful term. If a pedogenic carbonate horizon does not meet the criteria for a K horizon, it is designated Bk.

In arid regions, the progressive development of carbonate horizons has been classified into several morphological stages. Gile and others[25] recognized four stages; later, Bachman and Machette[4] further subdivided the higher stages for a total of six. The latter classification is used here (Fig. 1-2; Fig. A-1 and Table A-1, Appendix 1). The stages are identified on particular morphological features that correlate well with age of parent material, and so should be included in any soil description. In general, stage I and II morphologies are associated with Bk horizons, stage III and greater with K horizons.

Under the extreme cold and dry conditions of Antarctica, the morphology of the water-soluble salts follows the ages of the parent material. Six stages of the latter have been defined by Bockheim[10] (Table A-2, Appendix 1). In contrast, in the warm, aridic climatic of Wyoming,

Table 1-1
Soil Horizon Nomenclature

MASTER HORIZONS

O horizon   Surface accumulation of mainly organic matter overlying mineral soil; may or may not be, or has been, saturated with water. Subdivided on the degree of decomposition of organic material as measured by the fiber content after the material is rubbed between the fingers:
Oi horizon Least decomposed organic materials; rubbed fiber content is greater than 40 percent by volume.
Oe horizon Intermediate degree of decomposition; rubbed fiber content is between 17 and 40 percent by volume.
Oa horizon Most decomposed; rubbed fiber content is less than 17 percent by volume.

A horizon   Accumulation of humified organic matter mixed with mineral fraction; and the latter is dominant. Occurs at the surface or below an O horizon; Ap is used for those horizons disturbed by cultivation.

E horizon   Usually underlies an O or A horizon, characterized by less organic matter and/or fewer sesquioxides (compounds of iron and aluminum) and/or less clay than the underlying horizon. Horizon is light colored due mainly to the color of the primary mineral grains because secondary coatings on the grains are absent; relative to the underlying horizon, color value will be higher or chroma lower.

B horizon   Underlies an O, A, or E horizon, shows little or no evidence of the original sediment or rock structure. Several kinds of B horizons are recognized, some based on the kinds of materials illuviated into them, others on residual concentrations of materials. Subdivisions are:
Bh horizon Illuvial accumulation of amorphous organic matter-sesquioxide* complexes that either coat grains, form pellets, or form sufficient coatings and pore fillings to cement the horizon.
Bhs horizon Illuvial accumulation of both organic matter and sesquioxides as organic matter-sesquioxide complexes; both value and chroma are approximately three or less.
Bk horizon Illuvial accumulation of alkaline earth carbonates, mainly calcium carbonate; the properties do not meet those for the K horizon.
Bo horizon Residual concentration of sesquioxides, the more soluble materials having been removed.
Bq horizon Accumulation of secondary silica.
Bs horizon Illuvial accumulation of amorphous organic matter-sesquioxide complexes; both color value and chroma are greater than three.
Bt horizon Accumulation of silicate clay that has either formed *in situ* or is illuvial; hence it will have more clay than the assumed parent material and/or the overlying horizon. Illuvial clay can be recognized as grain coatings; bridges between grains; coatings on ped surfaces or in pores; or thin, single or multiple near-horizontal discrete accumulation layers of pedogenic origin (clay bands or lamellae). In places, subsequent pedogenesis can destroy evidence of illuviation.

Table 1-1   (cont.)
Soil Horizon Nomenclature

MASTER HORIZONS

Bw horizon Development of color (redder hue or higher chroma relative to C) or structure with little or no apparent illuvial accumulation of material.
By horizon Accumulation of gypsum.
Bz horizon Accumulation of salts more soluble than gypsum.

K horizon   A subsurface horizon so impregnated with carbonate that its morphology is determined by the carbonate.[24] Authigenic carbonate coats or engulfs all primary grains in a continuous medium to make up 50 percent or more by volume of the horizon. The uppermost part of a strongly developed horizon commonly is laminated. The cemented horizon corresponds to some caliches and calcretes.

C horizon   A subsurface horizon, excluding R, like or unlike material from which the soil formed or is presumed to have formed. Lacks properties of A and B horizons, but includes materials in various stages of weathering.
Cox and Cu horizons In many unconsolidated Quaternary deposits, the C horizon consists of oxidized C overlying seemingly unweathered C. The oxidized C does not meet the requirements of the Bw horizon. In stratigraphic work, it is important to differentiate between these two kinds of C horizons. It is suggested the Cox be used for oxidized C horizons and Cu for unweathered C horizons. Cn has been used for unweathered horizons, but n now has another meaning. Cu is taken from the nomenclature of England and Wales.[29]
Cr horizon In soils formed on bedrock, there commonly will be a zone of weathered rock between the soil and the underlying rock. If it can be shown that the weathered rock has formed in place, and has not been transported, it is designated Cr. Such material is the saprolite of geologists; *in situ* formation is demonstrated by preservation of the original rock features, such as grain-to-grain textures, layering, or dikes. If such material has been moved, however, the original structural features of the rock are lost, and the transported material may be the C horizon for the overlying soil.

R horizon   Consolidated bedrock underlying soil. It is not unusual for this and the Cr horizon to have illuvial clay in cracks; the latter would be designated Crt.

SELECTED SUBORDINATE DEPARTURES

Lower-case letters follow the master horizon designation. Those that are mainly specific to a particular master horizon are given above. Some can be found in a variety of horizons; they are listed below:

b   Buried soil horizon. May be deeply buried and not affected by subsequent pedogenesis; if shallow, they can be part of a younger soil profile.

c   Concretions or nodules cemented by iron, aluminum, manganese, or titanium.

Table 1-1   (cont.)
Soil Horizon Nomenclature

SELECTED SUBORDINATE DEPARTURES

f   Horizon cemented by permanent ice. Seasonally frozen horizons are not included, nor is dry permafrost material, that is, material that lacks ice but is colder than 0°C.

g   Horizon in which gleying is a dominant process, that is, either iron has been removed during soil formation or saturation with stagnant water has preserved a reduced state. Common to these soils are neutral colors, with or without mottling. Strong gleying is indicated by chromas of one or less, and hues bluer than 10Y. Bg is used for horizon with pedogenic features in addition to gleying; however, if gleying is the only pedogenic feature, it is designated Cg.

j   Used in combination with other horizon designation (Btj, Ej) to denote incipient development of that particular feature or property.[41]

k   Accumulation of alkaline earth carbonates, commonly $CaCO_3$.

m   Horizon that is more than 90 percent cemented. Denote the cementing material (km, carbonate; qm, silica; kqm, carbonate and silica; etc.).

n   Accumulation of exchangeable sodium.

v   Horizon characterized by iron-rich, humus-poor, reddish material that hardens irreversibly when dried. Called plinthite in *Soil Taxonomy*.[58]

x   Subsurface horizon characterized by a bulk density greater than that of the overlying soil, hard to very hard consistence, brittleness, and seemingly cemented when dry (fragipan character).

y   Accumulation of gypsum.

z   Accumulation of salts more soluble than gypsum (for example, NaCl).

*Compounds of iron and aluminum.
(Modified from Guthrie and Witty[26] and 1981 unpublished manuscript of revised U.S. Dept. Agri. Soil Survey Manual.)

Reheis[43] recognizes four stages of gypsum accumulation morphology that are similar to the first four stages of the six-stage carbonate morphology scheme mentioned above.

Within the C horizon, numbers have been used to denote variation with depth, but specific numbers carry no specific meaning (C1, C2, ... C8, C9). In the western United States, many workers find the Cox-Cu subdivision useful. In contrast, workers in Illinois have proposed specific meaning for numbers within the C horizon (Table 1-2), because variation in horizon properties occurs in a predictable order with depth.

In the field, many horizons are transitional rather than sharp. In places the transitional material is not described, but is included in the description of the distinctness of the horizon boundary (Appendix 1). However there may be soils in which the transitional material should be described

Table 1-2
Subdivisions of the C Horizon Used in Illinois

HORIZON	MINERALOGY	CARBONATES	COLOR	STRUCTURE
C1	Strongly altered	Leached	Uniform, mottled, or stained	Some soil structure, peds with clay films; structure of parent material—blocky, layered, or massive—common; often porous
C2	Altered	Unleached	Uniform, mottled, or stained	Less soil structure, clay films along joints; structure of parent material—blocky, layered, or massive—dominant; often porous
C3	Partly altered	Unleached	Uniform, rare stains	Massive, layered, or very large blocky; conchoidal fractures; dense
C4	Unaltered	Unleached	Uniform	Massive or layered, conchoidal fractures, dense

(Taken from Follmer and others,[23] Table A, Appendix 3.)

and sampled. There are two kinds of transitional horizons. In one, the properties of both horizons are mixed, and those of one are dominant; in this case, both capital letters are used and the first letter means that properties of that horizon dominate (AB, BA, AC). The other kind of transitional horizon has distinct parts of both horizons; here the two capital letters are separated by a virgule (A/B, E/B, B/R).

Master horizons are further subdivided by use of both lower-case letters (Table 1-1) and Arabic numerals. The lower-case letters follow the capital letter (Bt), and certain rules pertaining to the order of lower-case letters, if more than one suffix is used, are given in Appendix 1. If the horizon is buried, b is usually written last (Btkb); however, if some soil properties are imparted after burial, the symbol denoting the property follows b (Btbk). Arabic numerals are used to further subdivide horizons identified by a unique set of letters. Such subdivision can be based on slight changes in color, structure, or any other property. The numbering starts at 1 at whatever horizon any element of the letter symbol changes. Horizons in a particular profile could be A, AB, Bt1, Bt2, Btk, Bk1, Bk2, Bk3, and Btb.

This numbering system creates a problem with the K horizon because the numbers (1, 2, and 3) were defined on specific amounts of carbonate materials enclosing grains.[25] One could continue to use that system, or just number consecutively, with depth, each subdivision of the K horizon. To be consistent with the present use of numbers, consecutive numbers are used here.

Many unconsolidated deposits of Quaternary age consist of depositional layers of contrasting texture and/or lithology, and the soil profile extends through more than one layer. Examples are loess/till, floodplain silt/gravelly outwash, and colluvium/outwash. Such primary differences in texture and/or lithology are important in any soil-profile description. Each different geological layer is so noted by an Arabic numeral, counting from the top down. The numerals precede the master horizon designation, and the numeral for the uppermost layer (1) is omitted (Fig. 1-3).

Horizon nomenclature for cumulative and non-cumulative profiles has not received much attention. These are profiles in which, because of hillslope erosion and deposition associated with topographic position, a particular horizon alters to another horizon (Fig. 7-5). For example, a B horizon can convert to a A horizon or vice versa. If the properties of the previous horizon are still detectable, it might be useful to so indicate this with the horizon nomenclature. One suggestion is to use an arrow; hence, A $\rightarrow$ B would indicate a former A horizon converting to a B horizon in a cumulative soil, and B $\rightarrow$ A a former B horizon converting to an A horizon in a non-cumulative soil. In contrast, if the A horizon is only overthickened due to parent material additions (eolian or hillslope deposits), Harrison[28] suggests using the notation Acum to denote the cumulative nature of the horizon; this could be shortened to Ac.

Soils can be classified by their position in a stratigraphic section and in the landscape.[40,47] Three soils are recognized (Fig. 1-3). Relict soils are those that have remained at the land surface since the time of initial formation; they may or may not have acquired most of their properties some time in the past. Buried soils are those that were formed on some ancient land surface, were subsequently buried by a younger deposit, and generally are far enough below the present land surface not to be affected by present pedogenic processes. Exhumed soils were formerly buried but subsequently exposed to current pedogenesis with erosional removal of the overlying material. The letter "P" preceding the master horizon designation denotes exhumed soil horizons.[51]

Soil horizons are more or less parallel to the land surface below which they formed, and indeed this is one criterion for differentiating soil layers from depositional layering (Fig. 1-3). Various properties of the horizons may change laterally, however, due to changing environmental conditions, and this results in lateral change in the overall soil profile. In most cases, the lateral change from one kind of profile to another is gradational; in contrast, lateral changes in parent material can be quite abrupt.

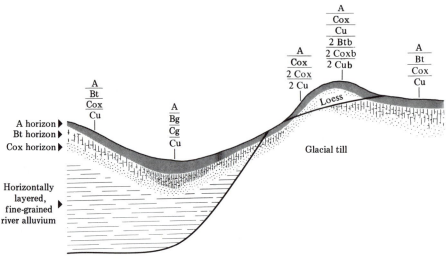

Fig. 1-3 Lateral variation in soil profiles due to lateral variation in environmental conditions and lithology and age of parent materials. Soil horizons parallel the land surface. Those horizons that are developed from river alluvium truncate depositional layering; hence, soil properties are, for the most part, independent of properties of the depositional layers. The soil formed on glacial till was partly truncated by erosion before burial by loess; it grades laterally to the right from a buried to an exhumed soil, because the loess cover has been removed by erosion in that direction.

SOIL CHARACTERISTICS

An undisturbed soil sample consists of a matrix of inorganic and organic solid particles in association with interconnected voids. Depending on local conditions, varying amounts of soil water and gases occupy the voids. I will discuss the main physical and chemical characteristics of the soil, along with an outline of water movement. Only those properties most important to field studies will be dealt with here. These and other properties are treated in more detail elsewhere.[5,8,13,45] Laboratory quantification of some of these properties is desirable, and one should consult Black[9] and Jackson[32] for appropriate analytical methods.

Color

Color is a valuable aid, if used with caution, in qualitatively recognizing processes that are or have been operating in a soil. Indeed, with buried soils color is the property that first catches one's attention. Dark brown to black colors in near-surface horizons reflect an accumulation of humified and/or nonhumified organic matter. Dark colors may also result from the accumulation of MnO_2, but these usually have a bluish cast and are not always close to the surface. Grayish colors (chromas* near 1) and/or hues* bluer than 10Y indicate reducing conditions (gleying), and the color is due mostly to ferrous iron compounds. Yellow-brown to red colors result from the presence of iron oxides and hydroxides and are characteristic of B and C horizons. White or light-gray colors above the B horizon characterize an E horizon and suggest enough leaching by vertically or laterally moving water so that most of the grains are free of colloidal coatings of oxides and hydroxides of aluminum and iron. The same colors below a B horizon are usually due to concentrations of $CaCO_3$.

Intensity of color gives a measure of the amount of pigmenting material present, but not a very accurate one. The reason for this is that the texture of the material (particle size) greatly influences the amount of surface area that has to be coated to impart a certain color on the material. For example, coarse-grained material has a much lower surface area per unit volume than does fine-grained material (Table 1-3), and so it would take much less pigment to impart a certain color to a coarse-grained soil that it would to a fine-grained soil.

Table 1-3
Particle size classes used in pedology and some of their properties

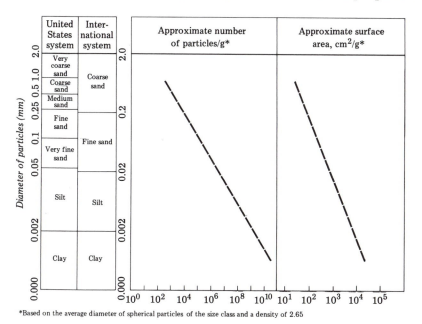

*Based on the average diameter of spherical particles of the size class and a density of 2.65

(Modified from Black,[8] © 1957, John Wiley and Sons.)

*Color terms follow Munsell notation (see Appendix 1).

Texture

Soil texture depends upon the proportion of sand, silt, and clay sizes, as based on the inorganic soil fraction that is less than 2 mm in diameter. Particle-size classes used in pedology are given in Table 1-3; the U.S. system will be used here. Specific combinations of sand, silt, and clay define soil textural classes (Fig. 1-4). Mechanical analysis aids in textural classification, but an approximate classification can be made by the use of simple field tests (Appendix 1).

The variation in particle size originates several ways. Sand and most silt are made up of minerals released by initial weathering or inherited from the parent material, although they may have been reduced in size by weathering. Some of the sand and silt in Bo horizons, however, occurs as aggregates of clay-size material. The clay-size fraction consists of both layer-lattice clay minerals and other crystalline and amorphous materials; this fraction originates in the parent material, is inherited later (for example, eolian influx), or is formed within the soil. Because clay refers to both a size fraction and a suite of minerals, the term "clay" will be used here for all material that is less than 0.002 mm in diameter, and "clay mineral" will be used for the layer-lattice clay minerals.

Texture is one of the more important characteristics of a soil profile. The variation in texture from horizon to horizon can be used to decipher the pedogenic and geological history of a soil. The fine-grained fraction

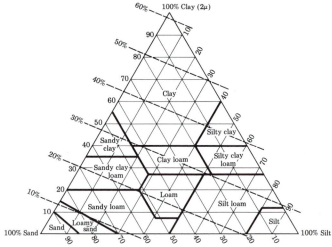

Fig. 1-4 Soil textural classes plotted on a triangular diagram.[58] The dashed lines are values for moisture equivalent calculated from the equation, moisture equivalent = 0.023 sand + 0.25 silt + 0.61 clay. (Taken from Bodman and Mahmud,[11] Fig. 3, © 1932, The Williams & Wilkins Co., Baltimore, and Jenny and Raychaudhuri,[33] Fig. 5.) The moisture equivalent approximates field capacity; thus the plot gives a general relationship between soil-texture classes and field capacity for soils low in organic matter.

also affects many processes operating within a soil because the surface area per unit volume increases markedly as particle size decreases (Table 1-3). Soils with large internal surface areas are more chemically active than are soils with low surface areas because of their greater charge per unit volume and their capacity to hold greater amounts of water by adsorption. For these and other reasons, they weather more rapidly. Many other soil properties, such as organic matter, nutrient content, and degree of aeration, are closely related to soil texture.

Organic Matter

Organic matter is found in varying amounts in mineral soils and is almost always most concentrated near the surface. A wide spectrum of material makes up the soil organic matter, which ranges from undecomposed plant and animal tissue to humus, the latter being defined as "a complex and rather resistant mixture of brown and dark brown amorphous and colloidal substances modified from the original tissues or synthesized by the various soil organisms."[15] Humus commonly makes up the bulk of the soil organic matter, and its chemistry as well as the processes involved in its formation are exceedingly complex.[35] Large amounts of CO_2 are evolved during its formation. Carbon makes up over one-half of the organic matter, and carbon content is commonly used to characterize the amount of organic matter in soils. Generally, the percent of organic matter in a soil is considered to be approximately 1.724 times the percent of organic carbon. The C:N ratio is a rough measure of the amount of decomposition of the original organic material and is related to environmental conditions. The ratio is high (>20) in plant tissue and low (<10) in humus.

Soil organic matter is important to many soil properties, especially to the formation of surface soil structures and to reactions that go on during pedogenesis. It considerably increases both the water-holding capacity of mineral soils and the cation exchange capacity. The organic acids that are produced promote weathering and form chelating compounds that increase the solubility of some ions in the soil environment. The CO_2 that is evolved builds up to reach concentrations higher than those in the atmosphere[14]; this results in the formation of abundant carbonic acid, which lowers the soil pH and thus promotes weathering.

Structure

Structure involves a bonding together into aggregates of individual soil particles. Individual aggregates (peds) are classified into several types on the basis of shape (Table 1-4). Although structure type is associated with soil horizon, details of the origins of many types are rather poorly known. Organic matter is important in the formation of spheroidally shaped structures, as is clay content in the formation of blocky, prismatic, and columnar structures. Some soils in tropical regions, however, are low in

Table 1-4
Description and probable origin of soil structure

TYPE	SKETCH* AND DESCRIPTION	PROBABLE ORIGIN[5,8,45,65]	USUAL ASSOCIATED SOIL HORIZON
Granular	Spheroidally shaped aggregates with faces that do not accommodate adjoining ped faces	Colloids, mainly organic, bind the particles together; clay and Fe and Al hydroxides may be responsible for some binding, and flocculating capacity of some ions, such as Ca^{2+}, may be helpful; periodic dehydration helps form more stable aggregates	A
Angular blocky	Approximately equidimensional blocks with planar faces that are accommodated to adjoining ped faces; face intersections are sharp with angular blocky, rounded with subangular blocky	Many faces may be intersecting shear planes developed during swelling and shrinkage that accompany changes in soil moisture	Bt
Subangular blocky			
Prismatic	Particles are arranged about a vertical line, and ped is bounded by planar, vertical faces that accommodate adjoining faces; prismatic has a flat top, and columnar a rounded top	Faces develop as a result of tensional forces during times of dehydration; rounded column tops may be due to some combination of erosion by percolating water and greater amounts of upward swelling of column centers upon wetting	Bn
Columnar			
Platy	Particles are arranged about a horizontal plane	May be related to particle size orientation inherited from parent material or induced by freeze-thaw processes	E, or those with fragipan
		May be related to layering in cementing material, induced during its precipitation (carbonate, silica, Fe hydroxides)	Km, Bqm, Bs

*Taken from Soil Survey Staff[32]

organic matter, yet they are well aggregated; this appears to result from cementation by iron hydroxides. Finally, high gravel content seems to impede the development of structure.

Structure is important to the movement of water through the soil and to surface erosion. A-horizon structures, although they vary from soil to soil, tend to produce larger-sized pores than would be the case for a structureless surface soil. These larger pores allow the soil to take up large amounts of rainwater over a short period of time, and thus the possibility of runoff and surface erosion is reduced. Remove the A horizon, however, and erosion due to greater runoff may ensue if the infiltration rate in the exposed B horizon is less than that of the A horizon. The fact that many structural aggregates are water stable is important because it means that the percolating waters are fairly free of clay particles. However, aggregates unstable in water can break down and contribute clay to the percolating soil water. This tends to plug some of the pores, with a concomitant decrease in infiltration rate. B-horizon structures provide avenues for the translocation of water and any contained solids along the ped interfaces. Indeed, this is where most clay films are located.

Vesicular structure is not shown in Table 1-4, although it is common to fine-grained desert and arctic A horizons. This structure is characterized by a high volume of near-round small voids that are not unlike the vesicles of some basalts in appearance. In deserts, they appear to form when air is entrapped in the soil as the latter is wetted[21,59]; but a different origin is postulated for those in arctic regions.[17]

Bulk Density

Bulk density is a measure of the weight of the soil per unit volume (g/cc), usually given on an oven-dry (110 °C) basis (Fig. 1-5). Variation in bulk density is attributable to the relative proportion and specific gravity of solid organic and inorganic particles and to the porosity of the soil. Most mineral soils have bulk densities between 1.0 and 2.0. Although bulk densities are seldom measured, they are important in quantitative soil studies, and measurement should be encouraged. Such data are necessary, for example, in calculating soil moisture movement within a profile and rates of clay formation and carbonate accumulation. Even when two soils are compared qualitatively on the basis of their development for purposes of stratigraphic correlation, more accurate comparisons can be made on the basis of total weight of clay formed from 100 g of parent material than on percent of clay alone (Fig. 1-6). To convert percent to weight per unit volume, multiply by bulk density.

Soil Moisture Retention and Movement

Various amounts of water occupy the pore spaces in a soil. This water is held in the soil by adhesive forces between organic and inorganic particles and water molecules and cohesive forces between adjacent water

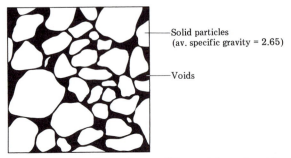

Fig. 1-5 Sketch of soil sample to show solid particle and void space distribution. The mineral grains in many soils are mainly quartz and feldspar, so 2.65 is an adequate average mineral specific gravity for the sand fraction. Bulk density and porosity are calculated as follows:

Weight of oven-dry soil: 63 g
Volume of soil in field: 35 cc

$$\text{Bulk density} = \frac{\text{weight}}{\text{volume}} = \frac{63}{35} = 1.8 \text{ g/cc}$$

$$\text{Porosity } (\%) = \left(1 - \frac{\text{bulk density}}{\text{particle density}}\right) \times 100$$

$$= \left(1 - \frac{1.8}{2.65}\right) \times 100 = 32$$

molecules. Thin films of water are held tightly to the particle surfaces and are relatively immobile, whereas thick films are more mobile and water can migrate from particle to particle both laterally and vertically.

Several soil moisture states are recognized (Fig. 1-7). One can start with a soil devoid of water and begin to add water from the top. Initially, the soil may have its pores saturated with water, but because the outer edge of the water film is under low surface tension, this more loosely held water migrates downward, under the influence of gravity, as a wetting front that more or less parallels the ground surface. After two or three days, redistribution of water ceases for the most part, and the forces that hold water films on the particle surfaces equal the force of downward gravitational pull. At this point, the soil is said to have a water content at field capacity (Fig. 1-7). At field capacity, water can be removed from the soil by evaporation from the surface and transpiration through vegetation, the latter probably being the more important mechanism for water removal from the soil profile. As roots remove water from the pores, the water film becomes thinner and is held by ever stronger forces of attraction until a point is reached at which the water is held so tightly by the particles that the roots can no longer extract it. The water content under these conditions is the permanent wilting point. Many soils under field conditions seldom obtain a water content less than that of permanent wilting point. Available water-holding capacity is the difference between

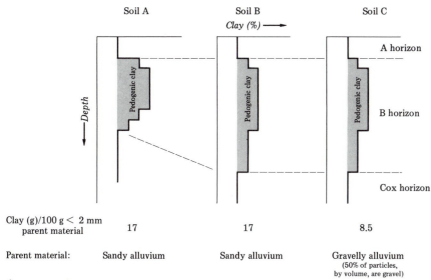

Fig. 1-6 Comparison of three soil profiles based on both percent clay and weight of clay formed. Assume no clay present in all three profiles at the time of deposition. Soils A and B have identical parent materials, but clay is concentrated in a thinner horizon in soil A than in soil B. Soil A, therefore, probably would be classified as more strongly developed than soil B, although the same amount of clay has formed in each. Soils B and C have identical profiles of percent clay, but soil C has formed from a 50 percent gravel parent material (textural data usually are given for the <2 mm fraction). Since gravel contributes little to clay formation, and makes up one-half of the soil solids, the amount of clay formed in soil C is one-half that in soil B.

field capacity and permanent wilting point; if given in water depth units, it is the amount of water required to wet a given thickness of soil from permanent wilting point to field capacity.

It is difficult to determine the moisture content under the above conditions in the field. Joint laboratory and field studies, however, suggest that water held at 15-atmos. tension approximates permanent wilting point for a variety of broad-leaved plants and that held at ⅓-atmos. tension approximates field capacity, although this is not always the case.[54] These are the usual moisture contents reported in the soils literature, and the values are given as percent (P_w)

$$P_w = \frac{H_2O \ (g)}{\text{Oven-dry soil (g)}} \times 100$$

Using moisture percent and bulk density data, one can estimate the depth to which water will wet a soil (Fig. 1-8). Arkley[1] has developed a method for calculating annual water movement within a soil, taking into account the water-holding properties of the soil and the seasonal distribution and amount of precipitation and potential evapotranspiration (see

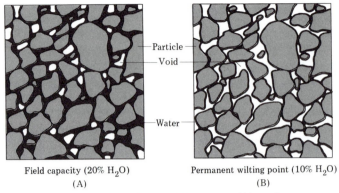

Field capacity (20% H_2O)   Permanent wilting point (10% H_2O)

(A)   (B)

Fig. 1-7 Diagram of water in soil at field capacity (A) and permanent wilting point (B) and an example of how to calculate the amount of water present at these two moisture states. The calculation of amount of water held in 100 cm of soil at the two moisture states in the figure is as follows. (Bulk density = 1.5; P_W = moisture percentage; D = soil horizon thickness; d = amount of water held in soil (cm).

At field capacity (20 g H_2O/100 g soil or 20% H_2O)

$$d = \frac{P_W}{100} \times \text{bulk density} \times D = \frac{20}{100} \times 1.5 \times 100 = 30 \text{ cm } H_2O$$

(30 cm of water are held in 100 cm of this soil at field capacity)

At permanent wilting point (10 g H_2O/100 g soil or 10% H_2O)

$$d = \frac{10}{100} \times 1.5 \times 100 = 15 \text{ cm } H_2O$$

(15 cm of H_2O are held in 100 cm of this soil at permanent wilting point)

Available water-holding capacity is the difference between the above two water contents, or 15 cm of H_2O.

Fig. 11-3). These are important data because percolating water redistributes clay and more soluble constituents downward in the soil or through the entire soil. Thus, many soil properties relate to the depth of wetting.

Soil-moisture retention and movement are strongly related to the surface area per unit volume of the soil mass, and this, in turn, is related to the clay and organic matter content. Buckman and Brady[15] give the following approximate available water-holding capacities for a 10-cm-thick layer of soil: 1.4 cm for clay, 1.7 cm for silt loam, and 1 cm for sandy loam. Thus, for a given rainfall, sandy soils are wetted to greater depths than are more heavily textured soils. Gravelly sands wet even more deeply than sands because gravel-sized particles have a low surface area per unit volume. Because of this close association of moisture content with content of colloid-size material, many workers have tried to correlate the two with varying success[33,42,55]; it might be possible to use some of these data to approximate soil-moisture conditions, provided textural data are available (Fig. 1-4).

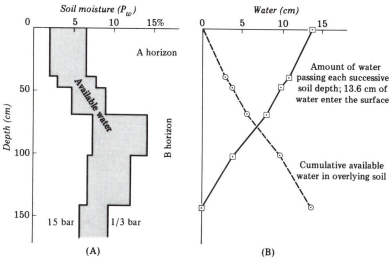

Fig. 1-8 (A) Variation in soil-moisture data with depth and horizon in a Snelling sandy clay loam, California. (Data from Arkley.[2]) (B) Plot of cumulative available water and the amount of water passing through the soil for a 13.6 cm rainfall, calculated by the method presented in the legend for Fig. 1-7. It can be shown that if the soil is at permanent wilting point and begins to receive water from the surface, 13.6 cm of water is necessary to bring the top 142 cm of soil to field capacity, or an influx of 13.6 cm of water (assuming negligible losses) would percolate downward to 142 cm depth. Downward migration would be as a diffuse, near-horizontal wetting front, with the soil above near field capacity and the soil below near permanent wilting point.

Abrupt changes in texture between sedimentary layers have a marked effect on the vertical movement of water. Obviously, water movement is impeded where coarse material overlies fine material. However, if fine material overlies coarse material (for example, loam over sandy gravel), such movement also is impeded and the water content must increase to greater than field capacity before it moves into the coarser material.[3,20,61] Water hanging up in this manner has important implications for soil morphology and the recognition of buried soils. For example, carbonate can accumulate at these boundaries, or gleying features can be produced.

Capillary rise from a shallow water table may deliver water and soluble salts to the overlying horizons. In theory, the smaller the pores the higher the water will rise, but irregularly shaped soil pores hardly make ideal capillary tubes. Rode[45] suggests such rise may be 1 m or less in sands and 3 to 4 m in clays. Parts of the soil profile at distances above the water table greater than these distances should not be influenced by capillary rise. Soluble salts or iron or manganese compounds might accumulate at the top of either the water table or the capillary fringe. Few criteria have been developed, however, to determine if such accumulations are produced by upward- or downward-moving water. If the accumulations are parallel with the ground surface and bear some consistent relationship with the other soil horizons, downward-moving water is suggested. If, however, the accumulations cut across soil horizons, and there is no reasonable depth relationship with the ground surface, derivation from a water table is suggested. If salts of different solubilities are present, the positions of their maximum concentrations may be used to determine the direction of water movement. For example, in a system characterized by downward-moving water, the maximum concentrations of the more soluble salts occur at progressively greater depths; a Bk horizon would overlie a By, which, in turn, would overlie a Bz. These horizon positions might be reversed if the waters were moving upward by capillary action.

It is important to note that field capacity and permanent-wilting-point moisture conditions describe the movement of water and not necessarily the amount of water involved in weathering reactions within the soil. Weathering occurs at all water contents, because a thin water film is always in contact with mineral grains. If field capacity is often reached, the ions in the water film released by weathering may be constantly flushed from the soil. If, however, the film is thin and field capacity is seldom reached, ionic concentrations in the water film may approach saturation, and this would inhibit further weathering unless periodic flushing occurs to lower ionic concentrations.

Cation Exchange Capacity, Exchangeable Cations, and Percent Base Saturation

Most soil colloids, both inorganic and organic, have a net negative surface charge, the origin of which is discussed in Chapter 4. Cations are attracted to these charged surfaces. The strength of cation attraction varies with the colloid and the particular cation, and some cations may exchange for others. The total negative charge on the surface is called the cation exchange capacity, and it is expressed in milliequivalents per 100 g oven-dry material. The exchangeable cations are those that are attracted to the negatively charged surfaces. Base saturation is expressed as the percent of base ions (non-hydrogen) that make up the total exchangeable cations. Thus, there generally is a close relationship between base saturation and pH. Exchangeable sodium percentage is the amount of sodium relative to the total exchangeable cations, and this is important in defining a natric horizon.

Soil pH

Soil pHs have an extreme range of 2 to 11, but most soil pHs range from 5 to 9. Soil pH is dependent on the ionic content and concentration in both the soil solution and the exchangeable cation complex adsorbed to the surfaces of colloids. In general, the ionic concentration increases from a low in the soil solution to a high at the colloid surface (Fig. 5-6). Furthermore, there is an equilibrium between the ions in solution and the exchangeable ions, and thus the ions are present in about the same proportions in both environments. Ions commonly present are Ca^{2+}, Mg^{2+}, K^+, Na^+, H^+, Cl^-, NO_3^-, SO_4^{2-}, HCO_3^-, CO_3^{2-}, and OH^-.

The relative proportion of these ions determines the soil pH (Fig. 1-9). Hydrogen ions are derived from rainfall and from organic and inorganic acids produced within the soil. Rainfall pH varies from 3.0 to 9.8.[19] Pure water, in equilibrium with atmospheric CO_2 at 25 °C, should have a pH of 5.7. Values lower than this are thought to be due partly to atmospheric pollution, whereas higher values are attributed to salts derived both from windblown seawater along coasts or from windblown dust. Carbonic acid is formed within the soil by the combination of CO_2 and water. A pH below 5.7 is possible because CO_2 content up to 10 or more times greater than atmospheric is possible because of respiration by plant roots and by microorganisms.[14] An important H^+ source is the wide variety of organic acids produced within the soil.[35] Still other sources are the exchangeable Al^{3+} and Al-hydroxy ions, which release H^+ during hydrolysis

$$Al^{3+} + H_2O \rightarrow Al(OH)^{2+} + H^+ \text{ (under very acid conditions)}$$

$$Al(OH)_2^+ + H_2O \rightarrow Al(OH)_3 + H^+ \text{ (under less acid conditions)}$$

The exchange complex in acidic soils is dominated by H^+, Al^{3+}, and Al-hydroxy ions, and base content is low. As cation content increases, however, they replace H^+ and Al-hydroxy ions, and OH^- concentration and pH increase. The alkalinity will depend on the strength of the base formed. For example, $Ca(OH)_2$ is formed in the presence of $CaCO_3$ and the resulting pH can approach 8.5. In contrast, $NaCO_3$ and $NaHCO_3$ form NaOH, a stronger base, and this result in pHs over 8.5.

One final point on pH is that the pH of a given soil is not uniform throughout. There may be slight variations from place to place due to

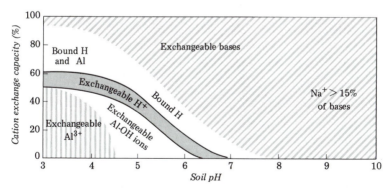

Fig. 1-9 General relationship between pH and exchangeable cations. Precise values vary from soil to soil for many reasons. Bound H^+ is that H^+ held so tightly to colloid surfaces that little of it is exchangeable. (Taken mostly from Buckman and Brady,[15] Fig. 14.1, © 1969, The Macmillan Co.)

slight variations in CO_2 or organic acid concentrations, the content and composition of the exchangeable bases, or the presence of nearby roots, since these commonly contain adsorbed H^+. Thus, although most work on weathering and mineral stabilities uses soil pH values, it should be remembered that these values reflect average conditions and may not reflect conditions where minerals are being weathered or synthesized.

COMPARING THE DEVELOPMENT OF SOILS: USE OF SOIL INDICES

Degree of soil-profile development is used as a qualitative measure of the amount of pedologic change that has taken place in the parent material. It is commonly used in Quaternary stratigraphy where soils are used to correlate unconsolidated deposits.[39,44] The ranking is generally on a relative scale, based on the properties of a sequence of soils in an area. A more quantitative scale would be useful, however, because soil of the same age may vary in its development from place to place due to variations in soil-forming factors. The following qualitative scheme is modified from Birkeland.[7]

A *weakly developed* soil profile is one with an A-Cox and/or Bk or an A-Bw-Cox and/or Bk soil horizon sequence. If carbonate horizons are present, morphology is probably stage I (see Appendix 1).

A *moderately developed* soil profile is one with an A- or A and E-B-Cox and/or a Bk soil horizon sequence. The B horizon may be Bo, Bs, or Bt. If carbonates are present, they probably display stage II morphology.

A *strongly developed* soil profile is similar to a moderately developed one with these exceptions: the B horizon in the strongly developed profile is generally thicker and redder, contains more clay, or other diagnostic components, and has a more strongly developed structure; if carbonates are present, they probably would form a K horizon (stage III and higher morphology).

The distinctions between a moderately and a strongly developed profile are qualitative, but they can be quantified to a degree on color and texture. However, it is difficult to compare soil development on a loess with 20 percent primary clay with that on a gravelly outwash with 5 percent primary clay. Clay content ratios might be used here, and, although soils workers seem to prefer A horizon:B horizon ratios, geologists might prefer B horizon:C horizon ratios. Or increases in the amount of clay (weight or percent) from one horizon to the other (C to B) could be used to distinguish between moderately developed and strongly developed soils. As regards color, again there is no fast rule. Each development rank might be accompanied by a different hue. Thus, if a moderately developed soil has a 10YR hue, a strongly developed soil would have stronger color, perhaps a 7.5YR hue.

Soil development can be better quantified by use of color and profile indices.

Color Indices

Three color indices have been proposed and all are similar to the extent that the Munsell notation is recalculated to a single number. One is the Buntley-Westin color index,[18] in which hue is converted to a number (7.5YR = 4; 10YR = 3; 2.5Y = 2; 5Y = 1), which is multiplied by the chroma. The second index, the Hurst index,[31] also recalculates hue to a number (5R = 5, 7.5R = 7.5, 10R = 10, 2.5R = 12.5, 5YR = 15, 7.5YR = 17.5, and 10YR = 20), and this is multiplied by the product of the fraction value/chroma. Finally, the rubification index of Harden[27] compares the color of each horizon with that of the parent material, and a shift in hue (change 1 hue) and of chroma are each worth 10 points. With all indices, values are calculated for each horizon, and these are multiplied by the horizon thickness.

For the profile color index value, one has several choices. One is to sum the horizon values. However, most soils in a developmental sequence vary in thickness, and this should be taken into account. One way is to divide the profile sum by the thickness of the profile. Another way is to increase the thickness of the lowest horizon in all soils so that the total thickness of all soils is uniform. These latter two choices are preferred over the first. Whatever method is used, the Buntley-Westin and Harden indices increase with soil age, whereas the Hurst index decreases (Fig. 1-10).

Profile Development Indices

Two indices have been developed to assess profile development based on field properties—the Bilzi-Ciolkosz[6] and Harden[27] indices. The properties are compared to those of the parent material or of the C horizon for

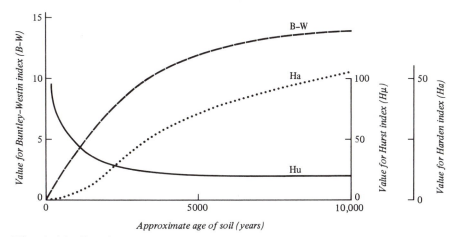

Approximate age of soil (years)

Fig. 1-10 Development of the weighted color index values with time for a sequence of Holocene soils in the alpine region of the Wind River Mountains, Wyoming. For weighted values, the value for each horizon was multiplied by horizon thickness, summed for a profile thickness of 52 cm thickness, and divided by 52. (Data from Miller and Birkeland.[38])

both indices, and points are assigned for the buildup of particular pedologic properties (for example, with Bilzi-Ciolkosz, a change in textural class is worth 1 point, a change in structure grade 1 point, and an abrupt boundary 3 points). Although the Bilzi-Ciolkosz index compares the properties of a horizon with those of the deepest C horizon, the Harden index compares horizon properties with those of the assumed parent material. Because parent material properties might not be represented at depth in the profile, one might have to seek them elsewhere. For example, in working with a river-terrace soil chronosequence, one could use the properties of sediments of the present active floodplain as parent material.

Each of the above profile indices present the data differently. Under the Bilzi-Ciolkosz scheme, the values of each horizon are plotted versus depth and the results graphically presented. One could also sum the horizons or repeat the calculations given above for the color indices to obtain a meaningful number to represent the entire profile. Calculations for the Harden index are more involved. After obtaining a number for each property for each horizon, property values are normalized by dividing the latter by the maximum possible value for that property. This provides a scale ranging from 0 to 1. All the individual property values per horizon are then summed, the total is divided by the number of properties used, and the latter is multiplied by the horizon thickness. One can then sum the data for each profile, or because sequences of profiles are usually of different thicknesses, adjust for this in the ways suggested for the color indices.

Whichever profile index is used, higher values denote greater pedologic development (Table 1-5). These data also can be used to quantitatively define stages of profile development. Finally, if one wishes to use soils to correlate between deposits in two areas using these indices, one could normalize the totals in each area (divide the profile totals by the maximum value in each area). This may produce similar values for soils of roughly similar ages.

INDEX OF PROFILE ANISOTROPY

A problem with presenting laboratory data in, say, pedologic studies (for example, to demonstrate change with time) is that the data themselves change with depth (Fig. 1-2), and without bulk density values there is usually no single number that can be used to represent the variation for a profile so that it can be compared with other profiles. One could use, for instance, a maximum value for material that accumulates in a profile, the numberical difference between the maximum value and that of the parent material, or even a ratio of the latter. Walker and Green,[63] however, have developed an index *(IPA)* that provides a single number for the anisotropy of a profile for laboratory data. The concept is that, ideally, at time = 0 yr, soil properties are isotropic, that is, values are the same irrespec-

Table 1-5
Profile Indices and mIPA Data for Holocene Soils, Ben Ohau Range, New Zealand

Approximate age (years)	Harden profile index		mIPA values***				
			pH	Organic carbon	Al	Fe	P
100	0.02*	0.01**	0.10	0.1	0.8	0.3	0.04
3000	0.28	18.3	0.16	13.8	2.0	0.3	0.47
4000 (range of three profiles)	0.24–0.37	17.7–27.5	0.06–0.17	12.9–21.5	2.9–4.9	0.6–1.6	0.55–0.93
9000 (range of two profiles)	0.28–0.41	20.7–30.6	0.17–0.19	16.8–49.5	5.2–16.0	1.3–6.8	0.42–0.79

*Profile sum divided by profile thickness.
**Profile sum for a fixed thickness.
***Organic carbon is by the Walkley-Black method; Al and Fe are for the oxalate extract; and P is the acid-extractable fraction.
(Unpub. data of author.)

tive of direction (here, however, we are concerned with the variation in properties with depth). In time, however, soil properties change with depth, and the degree of such anisotropy increases. Walker and Green[63] defined the *IPA* as

$$IPA = D\frac{100}{M}$$

where D is the mean deviation of the sampled horizon from the overall weighted mean value (M) for a particular property for the profile. Horizon totals are summed for a profile value. I suggest using a modification of this *(mIPA)*, which takes into account variation from the parent material:

$$mIPA = \frac{D}{PM}$$

where D is the numerical deviation of the particular property from that of the parent material *(PM)*. The *PM* values are obtained as mentioned for the Harden index. Values are calculated for each horizon and multiplied by the horizon thickness; the data for each horizon are summed to give a profile value and further recalculated as outlined for the color indices. Because anisotropy should increase with time, so should the value of the index (Table 1-5). However, with alluvial soils that are made up of contrasting depositional layers of mineral and organic particles, it is possible to have a relatively high m*IPA*, at $t = 0$ yr, for m*IPA* to initially decrease as pedologic mixing occurs, followed by an increase as the usual pedologic processes alter the soil.

SOIL MICROMORPHOLOGY

The soil properties listed above and in Appendix 1 are mostly macroscopic; however, much can be learned of soil development and processes by the study of soils in thin section, in much the same way that geologists study rocks (Fig. 1-11). This branch of pedology, soil micromorphology, was pioneered by Kubiëna,[36] and the standard text and terminology are those of Brewer.[13] The proceedings of the international working-meetings on the topic also provide useful information on the topic.[53] This aspect of pedology is not routinely followed in the United States; perhaps more emphasis should be placed on it for one could only imagine the state of geology today if routine study with a petrographic microscope had not been done.

Brewer recognizes four main pedologic units in thin section. One is voids, which occur in a variety of sizes, shapes, and arrangements. They owe their origin to the original sediment characteristics, faunal activity, freezing and thawing, and the entrapment of air bubbles in the surface of desert soils during rainfall.[21] A second unit is cutans, which are concen-

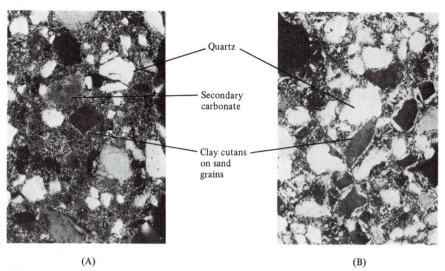

Quartz

Secondary carbonate

Clay cutans on sand grains

(A)   (B)

Fig. 1-11 Thin sections of soil samples from the Lubbock Lake area, Texas.[30] (A) B horizon of soil with 100 years of development, demonstrating rapid development of clay films; (B) Btk horizon of soil with 3500 years of development, showing much greater development of clay films than in (A). Vertical dimension is 86 μ. (Photomicrographs by V. T. Holliday.)

trations of soil constituents by deposition on, or *in situ* modification of, natural soil surfaces. Such surfaces can be grains, peds, channels, and planes. The soil constituents can be clay minerals, sesquioxides, manganese oxides or hydroxides, salts, or silica. Cutans originate mainly by illuviation (as in Bt horizons), diffusion, and stress (for example, due to shrink-swell cycles in Vertisols). A third unit is pedotubules, which are tubular in shape, and are formed, perhaps, by faunal activity or roots and later filled with soil material. The fourth unit is glaebules, which are more or less rounded bodies of constituents such as those listed above for cutans; concretions would be a familiar example.

Such features can be inherited from rock or soils in the source areas or formed during sedimentation, or they may be pedological in origin. Still more features are described by Brewer, and the importance of recognizing them cannot be overemphasized because they can be used to differentiate geological features and processes from pedologic ones, not an easy task. Such study may also help to solve problems of parent material variation and the time and space relationship between several translocated constituents (clay minerals vs. carbonate), or to determine whether clay cutans have been formed by illuvial processes or by stress. Perhaps such studies would help in paleoclimatic studies; for example, during some climatic intervals, clay cutans might form in specific places in the soil, only to be disturbed during a subsequent climatic interval. These latter features could be so small that identification could only come from microscopic examination.

A good illustration of the application of micromorphology to soil genesis studies was done by Bullock and Murphy[16] on interglacial soils in England. They recognized certain features that formed during soil formation under interglacial climatic conditions. Periglacial conditions during the following glacial conditions disrupted some of these features. Although some disruption features can be seen in the field, in places it might be that the thin-section evidence holds the best clues for changing pedogenic conditions.

RECOGNITION OF BURIED SOILS

Some buried soils are so obvious that few people would argue as to their pedogenic origin (Fig. 1-12); others, however, are quite difficult to differentiate from geological deposits. Many people have had the experience in field conferences that one person's soil becomes another's geological deposit. Part of the problem might be that some people have not spent enough time studying surface soils to really understand their properties and profile characteristics.

The same criteria used to recognize and describe surface soils can be used for buried soils.[62,66] The first test is to trace the material laterally in outcrop to be sure it is a soil and not a deposit. The relationship between the soil horizons and bedding should be deciphered because soil horizons can truncate geological bedding. Also, there are predictable lateral changes in soils related to topographic position (Ch. 9). A problem comes up, however, in places where a soil forms from a surface that parallels geological bedding. A common example is the soils formed on alluvial fan deposits in the semiarid southwestern United States. Whenever deposition stopped for a long enough time, a soil could form, but it could be buried later during renewed deposition. In this case, the soil horizons would parallel the bedding in the fan deposit. Some depositional layers in fan deposits can be poorly sorted, however, and resemble soils in their thickness, color, and texture. Thus, one has to look for pedologic features in the zone to be sure soil formation has taken place. Soil features used to recognize buried soils should persist after burial.[67] In general, the organic matter in the A horizon does not persist after burial,[60] but the mineral part of the A horizon may still be present and recognizable by a slightly lower clay content than that of the B horizon. Generally, the buried B horizon is the most important horizon for recognizing buried soils. If it is a Bt, it will have a greater clay content, redder or browner colors, and have a better developed structure than the C horizon. In drier regions, the presence of a carbonate-enriched horizon beneath the Bw or Bt horizon is helpful in identifying a buried soil. If a K horizon is present, it may be

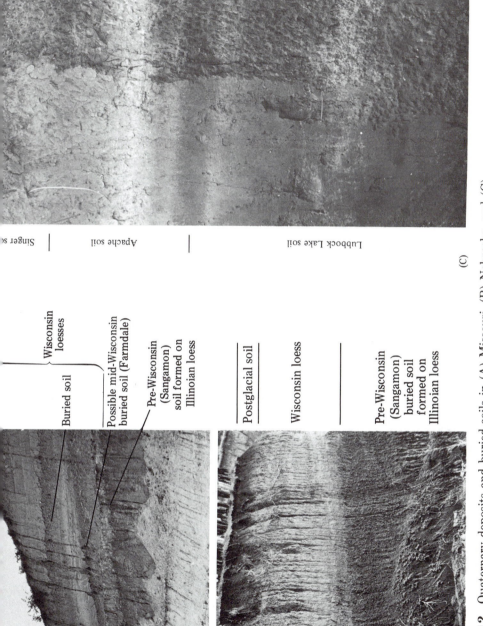

Fig. 1-12 Quaternary deposits and buried soils in (A) Missouri, (B) Nebraska, and (C) Texas. (Courtesy of V. T. Holliday.)

(A)

Wisconsin loesses

Buried soil

Possible mid-Wisconsin buried soil (Farmdale)

Pre-Wisconsin (Sangamon) soil formed on Illinoian loess

(B)

Postglacial soil

Wisconsin loess

Pre-Wisconsin (Sangamon) buried soil formed on Illinoian loess

(C)

A
B
Ab1
Btb1
Bkb1
A1b2
A2b2
Ab3
Btkb3
Btb3
Btkb3
Cgb3

Singer soil

Apache soil

Lubbock Lake soil

laminated in its uppermost part. Moreover, carbonate horizons should have a distinct relationship to the buried land surface so that ground-water origin can be ruled out.[48]

One other criterion that is helpful in the recognition of a buried soil is the abruptness of the horizon boundaries (Fig. 1-2). Quite commonly, the upper boundary of a soil horizon is sharper than the lower boundary, and this need not be the case in geological layering. For example, with depth in a well-developed soil profile the transition from a low-clay-content A horizon to a Bt horizon may take place over a few centimeters, whereas with greater depth in the B horizon there is a more gradual decrease in clay content toward the C horizon. The same criterion appears to hold for pedogenic carbonate horizons, that is, the upper horizon boundary is sharp and marked by a thin transition zone between the overlying non-carbonate material and the carbonate-enriched horizon. If a Bw horizon is present, the color is redder or browner in the upper part of the horizon, and chroma gradually diminishes with depth. In contrast, post-burial diagenetic alteration within a deposit may result in a gradual decrease in color both upward and downward from the zone of maximum alteration.

Another criterion that is helpful in the recognition of buried soils is the mineralogical characteristics of the profile. Some nonclay silicate minerals become weathered and/or etched upon weathering (Fig. 8-6). If the zone studied is a B horizon, weathering and etching may be greater there than in the underlying C horizon and overlying deposit. Mineral depletion during weathering may be reflected by resistant-to-less-resistant mineral ratios that have a consistent relationship with depth (Figs. 8-4 and 8-5). Clay minerals may also give a clue on the pedogenic origin of a horizon. Quite commonly, the clay minerals that form during soil formation vary in type with depth in a profile (Ch. 4), provided such variation is not due to variations in the parent material. If clay minerals were originally in the deposit and later underwent weathering within a soil, they may have been selectively altered with depth to other clay minerals (Fig. 8-15).

A more difficult problem in the recognition of buried soils comes when the upper part of the soil has been removed by erosion, leaving only that part of the profile that was below the B horizon. Here, oxidation colors help in identifying the material as part of a buried soil, as long as post-burial, groundwater alteration can be ruled out. In the midcontinent, evidence for such a history is shown by the carbonate content of superimposed loesses. The older loess may have been leached of carbonate during an interval of soil formation, the soil B horizon may have been removed during a subsequent period of erosion, and the leached loess may then have been buried by carbonate-bearing loess.[49,64] Thus, the major remaining evidence for an unconformity and an interval of soil formation is the presence of carbonate-bearing loess overlying loess that had been leached of its carbonate prior to burial. Recognition of the C-horizon nomenclature of Table 1-2 could help in this regard.

Stone lines may help in the location of buried soils. These are thin, buried, more or less planar layers of stones.[46] Although several origins have been suggested for them, including biological activity, some are distinctly of geological origin—such as deposition by running water when the associated erosional surface was cut or concentration at the base of a creep mantle. Whatever the geological origin, a stone line could be a field indicator of a hiatus in deposition, and if so, have a soil or truncated soil associated with it.

A problem with soils in some places is that the buried soil can merge laterally with a surface soil to form a fairly complex surface soil morphology. If these latter situations are not recognized, all properties of the complex surface soil are incorrectly ascribed to one set of pedogenic processes operating since the uppermost sediment was deposited. Ruhe and Olson[52] propose the term soil welding for such occurrences and point out that physical and mineralogical properties can be used to help identify welded soils. Such soil occurrences are also called composite soils.[40]

Soil chemical data can help with the identification and interpretation of buried soils. Constituents measured, however, should be those that persist after burial.[67] Iron, aluminum, and phosphorus trends probably would persist, but pH and exchangeable cations could be altered soon after burial and give no information of the preexisting soil values. In arid regions, calcium carbonate could be subsequently translocated into a buried Bt horizon to give the common Btbk horizon, but in most places, the preburial carbonate morphology below the Bt horizon should be recognized.

Finally, thin sections of the buried-soil horizons should be studied. In places, it may be difficult to differentiate parent material features from pedologic ones, and the only real clues that the materials are soils could come from thin-section analysis.

It is difficult to classify buried soils to the same degree of accuracy as surface soils. This is because during burial changes take place in properties critical to such classification. For example, A horizons are critical to classification, yet are rare in buried soils. Upon burial, other changes take place in pH and base saturation, and these are important to classification. One should try, however, to classify buried soils on as many of the same criteria as are used to classify surface soils.

Finally, if many buried soils are stacked vertically in a sequence, the present nomenclature is not adequate to differentiate one buried soil from the others. One solution is to give them names, but this could result in unnecessary proliferation of names or the introduction of formal names prematurely. It is suggested that we follow B.L. Allen (personal communication 1982) and number the buried soils from the surface downward, and that the number follow the b (for example, 2Btkb2) (Fig. 1-12 c).

REFERENCES

1. Arkley, R.J., 1963, Calculations of carbonate and water movement in soil from climatic data: Soil Sci., v. 96, p. 239–248.
2. ———, 1964, Soil survey of the eastern Stanislaus area, California: U.S. Dept. Agric., Soil Surv. Series 1957, no. 20. 160 p.
3. Aylor, D.E., and Parlange, J., 1973, Vertical infiltration into a layered soil: Soil Sci. Soc. Amer. Proc., v. 37, p. 673–676.
4. Bachman, G.O., and Machette, M.N., 1977, Calcic soils and calcretes in the southwestern United States: U.S. Geol. Surv. Open-file Rept. 77–794, 163 p.
5. Baver, L.D., 1956, Soil physics: John Wiley and Sons, New York, 489 p.
6. Bilzi, A.F., and Ciolkosz, E.J., 1977, A field morphology rating scale for evaluating pedological development: Soil Sci., v. 124, p. 45–48.
7. Birkeland, P.W., 1967, Correlation of soils of stratigraphic importance in western Nevada and California, and their relative rates of profile development, p. 71–91 *in* R.B. Morrison and H.E. Wright, Jr., eds., Quaternary soils: Internat. Assoc. Quaternary Research, VII Cong., Proc. v. 9, 338 p.
8. Black, C.A., 1957, Soil-plant relationships: John Wiley and Sons, New York, 332 p.
9. ———, 1965, Methods of soil analysis (parts 1 and 2): Amer. Soc. Agron., Madison, Series in Agron., no. 9. 1572 p.
10. Bockheim, J.G., 1979, Relative age and origin of soils in eastern Wright Valley, Antarctica: Soil Sci., v. 128, p. 142–152.
11. Bodman, G.B., and Mahmud, A.J., 1932, The use of the moisture equivalent in the textural classification of soils, Soil Sci., v. 33, p. 363–374.
12. Bowen, D.Q., 1978, Quaternary Geology: Pergamon Press, Oxford, England, 221 p.
13. Brewer, R., 1964, Fabric and mineral analysis of soils: John Wiley and Sons, New York, 407 p.
14. Brook, G.A., Folkoff, M.E., and Box, E.O., 1983, A world model of soil carbon dioxide: Earth Surface Proc. and Landforms, v. 8, p. 79–88.
15. Buckman, H.O., and Brady, N.C., 1969, The nature and properties of soils: The Macmillan Co., Toronto, 653 p.
16. Bullock, P., and Murphy, C.P., 1979, Evolution of a paleo-argillic brown earth (Paleudalf) from Oxfordshire, England: Geoderma, v. 22, p. 225–252.
17. Bunting, B.T., 1977, The occurrence of vesicular structures in arctic and subarctic soils: Zeitschrift für Geomorph., v. 21, p. 87–95.
18. Buntley, G.J., and Westin, F.C., 1965, A comparative study of developmental color in a Chestnut-Chernozem-Brunizem soil climosequence: Soil Sci. Soc. Amer. Proc., v. 29, p. 579–582.
19. Carroll, D., 1962, Rainwater as a chemical agent of geologic processes—a review: U.S. Geol. Surv. Water-Supply Pap. 1535-G, 18p.
20. Clothier, B.E., Scotter, D.R., and Kerr, J.P., 1977, Water retention in soil underlain by a coarse-textured layer: Theory and field application: Soil Sci., v. 123, p. 392–399.
21. Evenari, M., Yaalon, D.H., and Gutterman, Y., 1974, Note on soils with vesicular structure in deserts: Zeitschrift für Geomorph., v. 18, p. 162–172.
22. Flint, R.F., 1971, Glacial and Quaternary geology: John Wiley and Sons, Inc., New York, 892 p.

23. Follmer, L.R., McKay, E.D., Lineback, J.A., and Gross, D.L., 1979, Wisconsinan, Sangamonian, and Illinoian stratigraphy in central Illinois: Illinois State Geol. Surv. Guidebook 13, 139 p.
24. Gile, L.H., Peterson, F.F., and Grossman, R.B., 1965, The K horizon: A master soil horizon of carbonate accumulation: Soil Sci., v. 99, p. 74–82.
25. ———, 1966, Morphological and genetic sequences of carbonate accumulation in desert soils: Soil Sci., v. 101, p. 347–360.
26. Guthrie, R.L., and Witty, J.E., 1982, New designations for soil horizons and layers and the new *Soil Survey Manual:* Soil Sci. Soc. Amer. Jour., v. 46, p. 443–444.
27. Harden, J.W., 1982, A quantitative index of soil development from field descriptions: Examples from a chronosequence in central California: Geoderma, v. 28, p. 1–28.
28. Harrison, J.B.J., 1982, Soil periodicity in a formerly glaciated drainage basin, Ryton Valley, Craigieburn Range, Canterbury, New Zealand: Unpubl. M.A.S. thesis, Lincoln College, Canterbury, New Zealand, 189 p.
29. Hodson, J.M., 1976, compiler and editor, Soil survey field handbook: Soil Surv. Tech. Mono. No. 5, Rothamsted Experimental Station, Harpenden, Herts., England, 99 p.
30. Holliday, V.T., 1982, Morphological and chemical trends in Holocene soils at the Lubbock Lake archeological site, Texas: Unpubl. Ph.D. thesis, University of Colorado, Boulder, 285 p.
31. Hurst, V.J., 1977, Visual estimation of iron in saprolite: Geol. Soc. Amer. Bull., v. 88, p. 174–176.
32. Jackson, M.L., 1973, Soil chemical analysis—advanced course: Dept. Soil Sci., Univ. of Wisconsin, Madison, Wisc., 894 p.
33. Jenny, H., and Raychaudhuri, S.P., 1960, Effect of climate and cultivation on nitrogen and organic matter reserves in Indian soils: Indian Council of Agricultural Research, New Delhi, 126 p.
34. Joffe, J.S., 1949, Pedology: Pedology Publ., New Brunswick, N.J., 662 p.
35. Kononova, M.M., 1966, Soil organic matter: Pergamon Press, New York, 554 p.
36. Kubiëna, W.L., 1970, Micromorphological features of soil geography: Rutgers Univ. Press, New Brunswick, N.J., 254 p.
37. Leighton, M.M., and MacClintock, P., 1962, The weathered mantle of glacial tills beneath original surfaces in north-central United States: Jour. Geol., v. 70, p. 267–293.
38. Miller, C.D., and Birkeland, P.W., 1974, Probable pre-Neoglacial age of the type Temple Lake moraine, Wyoming: Discussion and additional relative-age data: Arc. Alp. Res., v. 6, p. 301–306.
39. Morrison, R.B., 1964, Lake Lahontan: Geology of the southern Carson Desert, Nevada: U.S. Geol. Surv. Prof. Pap. 401, 165 p.
40. ———, 1978, Quaternary soil stratigraphy—concepts, methods and problems, p. 77–108 *in* W.C. Mahaney, ed., Quaternary soils: Geo Abstracts Ltd., Univ. of East Anglia, Norwich, England.
41. National Soil Survey Committee of Canada, 1974, The system of soil classification for Canada: Canada Dept. Agri. Publ. 1455, 255 p.
42. Nielsen, D.R., and Shaw, R.H., 1958, Estimation of the 15-atmosphere moisture percentage from hydrometer data: Soil Sci., v. 86, p. 103–105.

43. Reheis, M.C., in press, The soil chronosequence on the Kane fans, Bighorn Basin, Wyoming: U.S. Geol. Surv.

44. Richmond, G.M., 1962, Quaternary stratigraphy of the La Sal Mountains, Utah: U.S. Geol. Surv. Prof. Pap. 324, 135 p.

45. Rode, A.A., 1962, Soil science: Israel Program for Scientific Translations, Jerusalem, 517 p.

46. Ruhe, R.V., 1959, Stone lines in soils: Soil Sci., v. 87, p. 223–231.

47. ———, 1965, Quaternary paleopedology, p. 755–764 *in* H.E. Wright, Jr., and D.G. Frey, eds., The Quaternary of the United States: Princeton Univ. Press, Princeton, 922 p.

48. ———, 1967, Geomorphic surfaces and surficial deposits in southern New Mexico: New Mex. Bur. Mines and Min. Resources, Memoir 18, 65 p.

49. ———, 1968, Identification of paleosols in loess deposits in the United States, p. 49–65 *in* C.B. Schultz and J.C. Frye, eds., Loess and related eolian deposits of the world: Internat. Assoc. Quaternary Res., VII Cong., Proc. v. 12.

50. ———, 1969, Quaternary landscapes in Iowa: Iowa State Univ. Press, Ames, 255 p.

51. Ruhe, R.V., and Daniels, R.B., 1958, Soils, paleosols, and soil-horizon nomenclature: Soil Sci. Soc. Amer. Proc., v. 22, p. 66–69.

52. Ruhe, R.V., and Olson, C.G., 1980, Soil welding: Soil Sci., v. 130, p. 132–139.

53. Rutherford, G.K., ed., 1974, Soil microscopy: The Limestone Press, Kingston, Ontario, 857 p.

54. Salter, P.J., and Williams, J.B., 1965, The influence of texture on the moisture characteristics of soils: I. A critical comparison of techniques for determining the available-water capacity and moisture characteristic curve of a soil: Jour. Soil Sci., v. 16, p. 1–15.

55. ———, 1967, The influence of texture on the moisture characteristics of soils: IV. A method of estimating the available-water capacities of profiles in the field: Jour. Soil Sci., v. 18, p. 174–181.

56. Soil Survey Staff, 1951, Soil survey manual: U.S. Dept. Agri. Handbook no. 18, 503 p.

57. ———, 1960, Soil Classification, a comprehensive system (7th approximation): U.S. Dept. of Agri., Soil Cons. Service, 265 p.

58. ———, 1975, Soil taxonomy: U.S. Dept. Agri. Handbook No. 436, 754 p.

59. Springer, M.E., 1958, Desert pavement and vesicular layer of some soils of the desert of the Lahontan Basin, Nevada: Soil Sci. Soc. Amer. Proc., v. 22, p. 63–66.

60. Stevenson, F.J., 1969, Pedohumus: Accumulation and diagenesis during the Quaternary: Soil Sci., v. 107, p. 470–479.

61. Stuart, D.M., and Dixon, R.M., 1973, Water movement and caliche formation in layered arid and semiarid soils: Soil Sci. Soc. Amer. Proc., v. 37, p. 323–324.

62. Valentine, K.W.G., and Dalrymple, J.B., 1976, Quaternary buried paleosols: A critical review: Quaternary Res., v. 6, p. 209–222.

63. Walker, P.H., and Green, P., 1976, Soil trends in two valley fill sequences: Aust. Jour. Soil Res., v. 14, p. 291–303.

64. Willman, H.B., AND Frye, J.C., 1970, Pleistocene stratigraphy of Illinois: Illinois State Geol. Surv. Bull. 94, 204 p.

65. White, E.M., 1966, Subsoil structure genesis: Theoretical consideration: Soil Sci., v. 101, p. 135–141.
66. Yaalon, D.H. (chairman), 1971a, Criteria for the recognition and classification of Paleosols, p. 153–158 *in* D.H. Yaalon, ed., Paleopedology: Israel Univ. Press, Jerusalem, 350 p.
67. ———, 1971b, Soil-forming processes in time and space, p. 29–39 in D.H. Yaalon, ed., Paleopedology: Israel Univ. Press, Jerusalem, 350 p.

2

Soil classification

We classify soil by common properties for the purposes of systematizing knowledge about soils and determining the processes that control similarity within a group and dissimilarities among groups. The numbers of individual soils in a group are a function of the limits one allows in the defining properties. Thus, there are many soil series and types at the lowest category of classification because many restrictions are imposed by the limits of the diagnostic properties. Different series can be grouped together at higher levels of classification, and these, in turn, can be regrouped until one ends up at the highest category with a few orders, each with wide limits on allowable variations in differentiating properties. In spite of the large number of members, soils in each order should share many properties in common because they have been formed under a somewhat similar set of pedogenic processes. To be really useful, however, classification has to be based on soil properties and not geological, climatic, or vegetational properties. Maps of soils at any category of classification are useful for geomorphic interpretation.

SOIL CLASSIFICATION AT LOWER LEVELS

Soil maps are prepared by the study of many soil profiles. With time, it becomes apparent that the soils can be grouped on the basis of similar profile characteristics and that these characteristics can change with changes in the soil-forming factors. In general, boundaries between mapping units, although commonly gradational, will be governed by one or more of the factors.[6,7] The mapping of soils is thus complex, because as mapping proceeds one must keep the multitude of factors in mind when predicting and drawing boundaries around the mapping units. Most mapping nowadays is done with the aid of aerial photographs; these increase mapping effectiveness because subtle tonal changes are many times the clue to lateral changes in soil properties and therefore the mapping unit.

The soil series is the basic classification unit used in soil mapping,[11,12] and in a sense it is similar in concept to geological formations. A soil series is a group of soil profiles with somewhat similar profile characteristics, such as the kind, thickness, arrangement, and properties of soil horizons. The properties used to differentiate series must be observable in the field and must have pedogenic significance. Series, therefore, are conceptual and are defined according to permissible ranges in the properties. Series names are derived from a local place name, such as a town or a county. Many times, changes in parent material lithology and texture are reflected in basic soil profile differences, and these have been used to differentiate series. This is not always the case, and present practice bases most of the differentiating criteria on observable soil properties. The texture of the A horizon is allowed to vary within a series; this variation is shown by mapping units that are a subdivision of the series, called soil types.

Soil types are subdivided into soil phases, which are the mapping units used in detailed soil surveys. Phase differentiation is based on slope steepness, physiographic position, thickness of soil profile or individual horizons, amount of erosion, stoniness, and salinity within the series limits.

In the United States, soil maps are published on a vertical air photograph base, and each phase is represented by a number (Fig. 2-1). Although the map gives the impression that each mapping unit is pure, and includes just that phase, in reality phases of other series commonly are included in the mapping unit, and the percentage of these other units usually is specified.

Soil maps in the United States are made with many purposes in mind. Agriculture, of course, is the main reason for such maps. Other uses, however, are planning for urban areas, highways, and recreational uses, but for these, derivative maps are made.[1] To make such maps more readable, the soil phase names are deleted and a new, much simplified explanation written for that specific map. For example, from the soil map of Malibu, California, derivative maps of several kinds were published.[10] One, a soil erosion map, depicts areas of slight, moderate, and high soil-erosion hazard. Another outlines areas of soil shrink-swell behavior (low, moderate, and high). Such maps are obviously useful to regional or local land-use considerations, and in places I have noticed that a person knowledgeable in soils and soil patterns in the landscape is much better at predicting troublesome areas than are geologists or engineers not trained in soils. A detailed soil map at the series, type, and phase level is useful in interpretating the geology of an area. In Iowa, for example, the soils mapped at the surface are a complex mosaic that can best, and perhaps only, be understood by mapping at the series and phase levels and by knowing the local geology (Fig. 2-2). There, the different series and phases elucidate parent material differentiation and the presence of buried soils and their

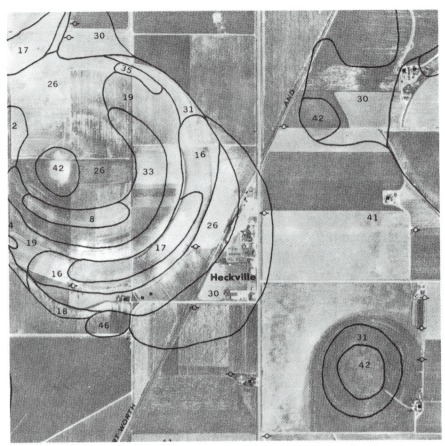

Fig. 2-1 Standard USDA-SCS soil map of Heckville, Texas.[2] The numbered soil phases are as follows: The flat High Plains surface is underlain by Pleistocene eolian sediments in which the Pullman (41), Olton (30, 31), Acuff (2), and Posey (35) soils have formed. The High Plains surface is pockmarked with numerous circular depressions (playas), formed by wind deflation. The rest of the soils are formed in lacustrine clays and calcareous loamy material on the playa floors or in calcareous eolian sediments derived from the lacustrine materials by prevailing westerly winds; the eolian sediments occur on the lee side of the playas. The Randall soil (42) has formed in the lacustrine clays, whereas the soils formed in calcareous lacustrine loams include the Midessa (26), Arch (8), and Portales (33) soils. Soils formed in the calcareous eolian sediments include the Drake (16,17), Midessa (26), and Zita (46) soils. The Midessa is mapped in both lacustrine and eolian sediments; however, for geological studies, different names would probably be used for soils formed from the different sediments. Finally, the Pleistocene eolian sediments of the High Plains crop out along the gently sloping margins of the playas, and the Estacado (18,19) soil is formed in them. This latter soil is probably the uppermost buried soil in the Blackwater Draw Formation. (Photograph from Blackstock and others.)

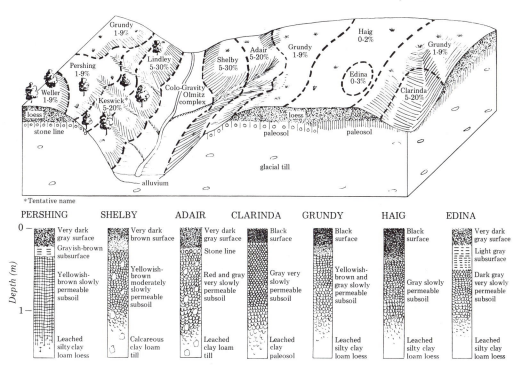

Fig. 2-2 Relationship of slope (%), vegetation, and parent material to soil series mapped in southern Iowa. Upland soils have formed from Wisconsin loess, and valley side soils have formed either from re-exposed, pre-Wisconsin soils or from unweathered Kansan Till. Morphology of some of the diagnostic soil series is diagrammatic. (Taken from Oschwald and others,[8] Fig. 15.)

effect on overlying soils, and could provide a qualitative measure of the rate of hillslope erosion. In other places, geological maps can be made directly from soil maps. This is especially so in areas of widespread Quaternary surficial units (river-terrace deposits, loesses, sand dune deposits) that might vary in age. Because of the difference in the parent material textures of the above materials, each will be assigned a different series. Furthermore, the ages of each kind of deposit can be approximated by the degree of soil development (see Ch. 8). Ideally, one can make such maps directly from the soil maps and follow it up with some field checking. Soil maps and accompanying text, however, should be detailed and include all the pertinent soil and soil-forming-factor data; only in this way can accurate derivative maps be made without additional mapping at considerable cost. In contrast, if a map were made for only one purpose, say soil erosion, it would be very difficult to derive other useful maps from it.

SOIL CLASSIFICATION AT THE HIGHER CATEGORIES

Many classification schemes have been proposed over the years, and the history of these is reviewed elsewhere.[4,9,12] There is still no worldwide agreement on soil classification; in general, each country has its own, although many of the systems now in use have some features in common. At present, an effort is being made to put together a soil classification scheme (FAO) for use with a soils map of the world.[5] Workers from many parts of the world have had a part in this study, which is continuing, and perhaps it will gain international recognition and use. Until that time, one must either know several classifications or have access to conversion tables.[12]

Soil Taxonomy[12] is the classification scheme in use in the United States, and one has no alternative but to learn at least some of it. The classification is based mainly on observable properties and not some genetic concept, is quantitative, runs over 700 pages, and carries such exotic combinations of Latin and Greek as Cryaqueptic Haplaquoll, Aquic Ustochrept, and Natraqualfic Mazaquerts. To use the classification in detail requires a good working knowledge of it and long experience. In many instances, field observations are not enough to classify a soil; these must be supplemented by laboratory quantification of some properties, as well as by such information as the number of wet or dry days per year or the mean annual soil temperature. In places, it is not at all clear why certain decisions were made in the classification, but some of these are explained in a recent series of informative articles in *New Zealand Soil News,* reprinted in *Soil Survey Horizons.* The articles are called "Conversations in taxonomy" and are just that—a series of published conversations between M. Leamy of New Zealand and G.D. Smith, the person who was most responsible for Soil Taxonomy. Finally, to aid in classifying soils, workers in New Zealand are publishing a set of flow-diagram keys.[13] These are quite useful aids, especially for the person beginning to use Soil Taxonomy.

The new classification does have some appeal for geomorphologists and ecologists, at least down to the suborder level. The reason for this is that soil profile development is included in the classification, as well as base saturation, amount of organic matter, and properties indicative of relative wetness and dryness. Most geomorphological soils studies deal with climatic and time factors, and the above properties will later be shown to be related to both factors.

Ten orders are recognized in the new classification, and these are subdivided into 47 suborders. The orders are basically differentiated by the horizon or horizon combinations that occur in the soil profile. These usually can be recognized in the field without recourse to laboratory analysis. One criticism of the new classification, from a geomorphological point of view, is the overemphasis on soil classification by surface horizon. In contrast, horizons beneath the A horizon probably are more important to geomorphologists. Classification into suborders requires an increasingly quantitative knowledge of soil properties and soil-moisture and soil-temperature regimes. Many times, however, it is not necessary to take these measurements because, with experience, soil classification can be estimated from properties recognizable in the field.

In order to classify a soil at the order and suborder level, one must be able to identify the diagnostic horizons as well as the soil-moisture and temperature regimes. Epipedons are the surface diagnostic horizons. The diagnostic horizons are somewhat similar to the field-designated soil horizons, although in places they can encompass several different field soil horizons. For example, the mollic epipedon can include both the A and B horizons. For some diagnostic horizons, the criteria are so complex that one has to read the defining criteria in detail with both the field and laboratory data. Only the main discriminating criteria can be given here (Table 2-1); Soil Taxonomy must be consulted for the details, however. One detail of Soil Taxonomy that has bothered geologists and some pedologists is the restriction of the cambic horizon to textures of very fine sand, loamy very fine sand, or finer texture. This restriction excludes some B horizons from the cambic, influences the classification of the soil in an important way, and might be disregarded when soils are studied for geological and other non-agronomic purposes.

Plinthite is material important to soil classification, but it is not given diagnostic horizon status. This material is usually found in humid tropical regions and is what some people would call laterite. Plinthite is mottled, iron-rich material that can harden irreversibly on exposure to repeated wetting and drying. Saturation with water at some season seems to be common for its formation. It can be found in epipedons, or in cambic, argillic or oxic horizons.

Soil-moisture regimes are difficult to summarize because they are based on estimates of actual soil moisture at depths that vary with texture and on soil temperatures at a depth of 50 cm. Only the basic concepts can be given here. Six moisture regimes are recognized, based on the length of time a soil is either moist (water content more than that held at 15 atmos.) or dry (water content is that held at 15 atmos.). From drier to wetter, the regimes are

Aridic moisture regime: Usually soils in arid climates; never moist for as long as 90 consecutive days when temperature is above 8 °C, and dry more than one-half the time (cumulative) temperature is above 5 °C

Ustic moisture regime: Soils dry for 90 or more cumulative days; depending upon the temperature, soil (a) moist for at least 180 cumulative and 90 consecutive days or (b) not dry in all parts over one-half the time, and do not meet xeric moisture regime requirements

Table 2-1
Diagnostic Horizons Used in Soil Taxonomy

DIAGNOSTIC HORIZON	DEFINING CRITERIA	PROBABLE FIELD HORIZON EQUIVALENT
EPIPEDONS		
Mollic epipedon	Must be 10 cm thick if on bedrock, otherwise a minimum of 18 or 25 cm thick depending on subhorizon properties and thicknesses; color value darker than 3.5 (moist) and 5.5 (dry); chroma less than 3.5 (moist); organic carbon content at least 0.6%; structure developed and horizon not both massive and hard; base saturation $\geq$50%	A, A + E + B, A + B
Umbric epipedon	Meets all criteria for mollic epipedon, except base saturation <50%.	A
Ochric epipedon	Epipedon that does not meet requirements of either mollic or umbric epipedons	A
Histic epipedon	Complex thickness requirements, but >20 cm thick; >12% organic carbon, with some adjustment for percent clay; saturated with water for 30 consecutive days or more per year, or artificially drained	O
SUBSURFACE HORIZONS		
Albic horizon	Light colored with few to no coatings on grains—light color is that of grains; if color value (dry) is 7 or more, or color value (moist) is 6 or more, chroma is 3 or less; if color value (dry) is 5 or 6, or color value (moist) is 4 or 5, chroma is closer to 2 than to 3	E
Argillic horizon	Complex thickness requirements, but at least 7.5 or 15 cm thick depending on texture and thickness of overlying horizons; must have these greater amounts of clay relative to overlying eluvial horizon(s) or underlying parent material: (a) if the latter horizons have < 15% clay, argillic horizon must	Bt

Table 2-1 (cont.)
Diagnostic Horizons Used in Soil Taxonomy

DIAGNOSTIC HORIZON	DEFINING CRITERIA	PROBABLE FIELD HORIZON EQUIVALENT
	have a 3% absolute increase (10 vs. 13%); (b) if the latter horizons have 15 to 40% clay, the ratio of clay in argillic horizon relative to them must be 1.2 or more, and (c) if the latter horizons have >40% clay, the argillic horizon must have an 8% absolute increase (42 vs. 50%); in most cases, evidence for translocated clay should be present (clay as bridges between grains or clay films in pores or on ped faces)	
Natric horizon	In addition to properties of argillic horizon: prismatic or columnar structure; 15% or more exchangeable sodium; exchangeable magnesium and sodium exceed exchangeable calcium and exchange acidity	Btn
Spodic horizon	One or more of the following: (a) a subhorizon >2.5 cm thick cemented by some combination of organic matter with aluminum or iron or both, (b) a sandy or coarse loamy particle-size class, and sand grains covered with cracked coatings or dark pellets present or both, and (c) complex combinations of the amounts of iron, aluminum, clay and organic matter are required, as well as a particular value for an accumulation index of amorphous material	Bh, Bs, Bhs
Cambic horizon	Base usually at least 25 cm deep; stronger chroma or redder hue relative to underlying horizon; soil structure or absence of rock or sediment structure; weatherable minerals present; carbonates removed if originally present; no cementation or brittle consistence	Bw

Table 2-1 (cont.)
Diagnostic Horizons Used in Soil Taxonomy

DIAGNOSTIC HORIZON	DEFINING CRITERIA	PROBABLE FIELD HORIZON EQUIVALENT
Oxic horizon	At least 30 cm thick; $>15\%$ clay and sandy loam or finer; cation exchange capacity ≤ 16 meq/100 g soil; few weatherable minerals	Bo
Calcic horizon	At least 15 cm thick; 15% $CaCO_3$; relative to underlying horizon, has at least 5% more $CaCO_3$, or at least 5% by volume secondary carbonate	Bk
Petrocalcic horizon	Horizon continuously cemented with $CaCO_3$	Km
Gypsic horizon	At least 15 cm thick; at least 5% more gypsum than underlying horizon; product of thickness (cm) times content ($\%$) is 150 or more	By
Petrogypsic horizon	Strongly cemented gypsic horizon, commonly with greater than 60% gypsum	Bym
Salic horizon	At least 15 cm thick; at least 2% salts more soluble than gypsum; product of thickness (cm) times content ($\%$) is 60 or more	Bz
Duripan	Silica cementation is strong enough that fragments do not slack in water	Bqm
Fragipan	Horizon of high bulk density relative to overlying horizons; formed in loamy material; although seemingly cemented with a brittle appearance, slacks in water; slowly permeable to water, so usually mottled; very coarse prismatic structure, usually with some bleached faces	Bx, Cx

(Taken from Soil Survey Staff.[12])

Xeric moisture regime: Soils in Mediterranean climates with moist winters and dry summers

Udic moisture regime: Associated with humid climates with enough summer precipitation that latter plus stored moisture exceed evapotranspiration; not dry for as long as 90 cumulative days

Perudic moisture regime: In all months, water moves through the soil because precipitation exceeds evapotranspiration

Aquic moisture regime: Soil saturated with water for at least a few days per year; level of water can fluctuate with season; water is essentially free of dissolved oxygen, conditions are reducing

Soil-temperature regimes are based on the mean annual temperature at 50 cm depth, and they can be estimated by the mean annual air temperature of the site. Six regimes are recognized: *Pergelic,* below 0 °C; *Cryic,* 0–8 °C and summer temperature restrictions depending on whether or not an O horizon is present and if the soil is saturated with water; *Frigid,* 0–8 °C and warmer in summer than cryic regime; *Mesic,* 8–15 °C; *Thermic,* 15–22 °C; *Hyperthermic,* more than 22 °C. In the frigid, mesic, thermic, and hyperthermic regimes, the difference between mean summer and mean winter temperatures is more than 5 °C. In contrast, if the differences in the latter temperatures are less than 5 °C, the prefix *iso* is added to the name (for example, isofrigid).

Although at times in Soil Taxonomy it seems as though we are classifying soils more on climatic parameters than on pedologic features, many of the latter features are related to climate (Ch. 11). Take, for example, the calcic horizon. It is common in the aridic regime; common, but at greater depth in the ustic regime; rare and only present at still greater depth in the udic regime; and absent in the perudic regime.

I have tried, in Table 2-2, to list the horizons most diagnostic for classification of soils at the order and suborder level. Only those suborders thought to be most useful to Quaternary research are included. It should be stressed that because of the extremely complex nature of the new classification any such simplification is bound to contain some errors, especially at the suborder level. It is an attempt, anyway, to simplify the system so that it can be used by workers who have neither the necessary time nor desire to learn the new system in detail.

Entisols are soils in which pedologic processes have left only a faint imprint. Thus, only a weakly developed surface horizon is present, and there may be salt or silica accumulations at depth.

Inceptisols are better devloped than Entisols in that a well-developed epipedon can be present, and oxidation extends below the base of the epipedon to form a cambic horizon (Fig. 2-3, A). Almost all other orders, exclusive of the Vertisols and the Histisols, differ from the Inceptisols in having better developed B horizons and/or $CaCO_3$ accumulations.

Table 2-2
Generalized Key to Soil Orders and Some Suborders Based on One or More Diagnostic Horizons

DIAGNOSTIC SOIL HORIZONS	ENTISOL	INCEPTISOL	ARIDISOL	MOLLISOL	ALFISOL
Histic		Aquept[1]*		Aquoll	Aqualf[1]*
Mollic		Andept[2]* Aquept[1]* Tropept*		Rendoll[3]* Boroll[4]* Udoll[5]* Ustoll[6]* Xeroll[7]*	
Umbric		Aquept[1]* Umbrept* Tropept*			—*
Ochric	—*	Aquept[1]* Ochrept* Tropept*	—*		—*
Albic	If at the surface			Alboll[8]*	Boralf[9]* Udalf[5]*
Cambic		—*	Orthid*		
Argillic			Argid*	Base saturation > 50% in some part*	Base saturation > 35%*
Natric			—*		Xeralf[7]*
Spodic					
Oxic					
Petrocalcic			—*		
Calcic	—*	—*	—*	Ustoll[6]* Xeroll[7]*	Ustalf[6]* Xeralf[7]*
Gypsic, Saliz	—*	—*	—*		

1. Histic epipedon may not always be present; soils usually are saturated with water at some time of year, and so display evidence of gleying
2. Contains high content of allphane, volcanic ash, or both
3. Calcareous parent material; cambic horizon may be present
4. Mean annual soil temperature is below 8°C; mositure regime is usually ustic
5. Udic moisture regime; soluble constituents not common at depth
6. Ustic moisture regime; soluble constituents common at depth

ORDER

SPODOSOL	ULTISOL	OXISOL	VERTISOL	HISTOSOL
Aquod[1]*	Aquult[1]*	Aquox[1]*	May or may not have diagnostic horizons; has > 30% clay and shrinks and swells with moisture variation to form cracks that extend to the surface; can have slickensides and gilgai microrelief	Organic soils, many of which are water-saturated; suborders defined partly or degree of decomposition of organic matter
	—*	—*		
	—*	—*		
	—*	Torrox[15]*		
—*	—*			
	Base saturation < 35%* Humult[13]* Udult[14]* Ustult[14]* Xerult[14]*			
Ferrod[10]* Orthod[11]* Humod[11]*				
		Humox[16]* Orthox[17]* Ustox[18]*		

7. Xeric moisture regime; soluble constituents common at depth

8. If argillic, or natric, horizon underlies albic horizon; some evidence of gleying

9. Frigid or cryic temperature regime; udic moisture regimen common; Bk horizon may be present in areas of least rainfall

10. Free Fe:C > 6

11. Free Fe:C < 6, > 0.2

12. Free Fe:C > 0.2

13. High organic matter content (>12 kg) in uppermost 1 m^3 of soil, exclusive of O horizon

14. Less organic matter than Humults, and indicated moisture regime

15. Have present-day aridic moisture regime, so marked climatic change from humid to arid is required

16. High organic matter content (>16 kg) in uppermost 1 m^3 of soil, exclusive of O horizon

17. Can have organic matter content of Humox if certain temperature or base saturation conditions are met

18. Less organic matter content than Humox; Ustic moisture regime

(A) Inceptisol, Searles Lake, California   (B) Argid, Las Cruces, New Mexico

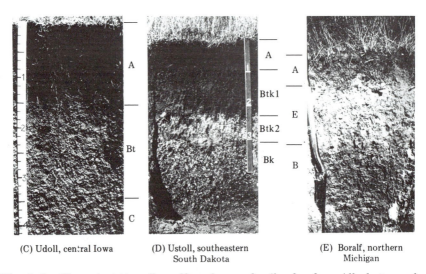

(C) Udoll, central Iowa   (D) Ustoll, southeastern South Dakota   (E) Boralf, northern Michigan

Fig. 2-3 Characteristic soil profiles of several soil suborders. All photographs but A and B are from the Marbut Memorial Slide Collection, prepared and published by the Soil Science Society of America (Madison, Wisconsin) in 1968. Originals are in color; scale, feet and inches.

Aridisols are usually characterized by an aridic moisture regime, an ochric epipedon, and salt or silica accumulations at depth. The two suborders are based on the presence or absence of an argillic horizon (Fig. 2-3, B). This is a significant differentiation for geomorphic studies because clay buildup in many soils generally is related to duration of soil formation.

Mollisols have a mollic epipedon as a major distinguishing property, and the base saturation throughout is usually higher than 50 percent. At the suborder level, Mollisols can be differentiated on properties indicative of wetness or dryness such as the presence of an E horizon, and the presence or absence of salts or silica at depth (Fig. 2-3, C and D).

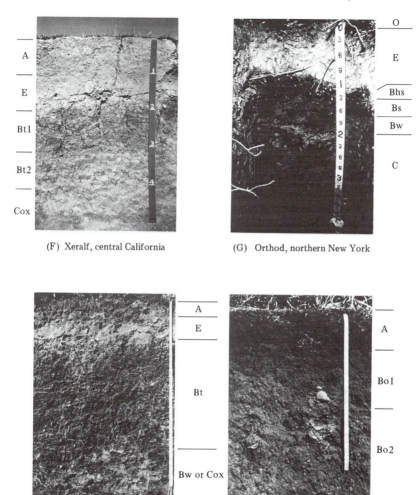

(F) Xeralf, central California

(G) Orthod, northern New York

(H) Udult, western Arkansas

(I) Orthox, Hawaii

Alfisols may have a content of organic matter similar to that of the Mollisols, but the epipedon properties do not meet the requirements of a mollic epipedon, and base saturation in some parts of the argillic horizon is lower than 50 but higher than 35 percent (Fig. 2-3, E and F). Like the Mollisols, soils that are dry for various parts of the year can have $CaCO_3$ or silica accumulations at depth.

Spodosols differ from all other orders in the presence of a spodic horizon. An E horizon is not mandatory, but it is usually present in uncultivated sites, along with an O horizon (Fig. 2-3, G). Various combinations of iron and organic matter in the spodic horizon serve to differentiate most of the suborders.

Ultisols may have any surface A horizon, but they differ from the other orders in having less than 35 percent base saturation in the argillic horizon (Fig. 2-3, H). Organic matter content in the uppermost cubic meter of the mineral soil and properties associated with dryness are some criteria that serve to separate Ultisols at the suborder level.

Oxisols differ from the other orders by the presence of an oxic horizon (Fig. 2-3, I). Plinthite may be present. Again, organic matter content and properties associated with dryness serve to separate Oxisols at the suborder level.

Vertisols differ from some other orders in that a diagnostic horizon or set of such horizons is not required. These are high-clay content soils that markedly shrink on drying and swell on wetting. Open cracks extend from the surface to variable depths during the dry season. Cracking and shrinking and swelling can be so extensive that man-made structures are broken. Suborders are generally defined on the basis of the number of times per year the cracks open and close and how long they remain open.

In places, exclusions are made in Soil Taxonomy for little apparent reason. For example, soils with pergelic temperature regime are excluded from Aridisols. This seems strange because Bockheim[3] reports that the arid climate is the main soil-forming factor at high latitudes. He suggests ways to change the definitions of both the cambic horizon and Aridisols so that soils of the polar deserts can be classed as Aridisols. In essence, delete the textural restrictions of the cambic horizon and the temperature restrictions of the Aridisols.

Suborder names are formed by combining two syllables to indicate the order and some distinguishing characteristic of the suborder (Table 2-3). The last syllable of each suborder name is the clue to the order in which it is grouped. The first syllable indicates some diagnostic property or groups of properties common to the suborder. Thus, Aquolls are gleyed Mollisols. Argids are Aridisols with argillic horizons, and Humox are Oxisols characterized by large amounts of organic matter to fairly great depth.

Suborders can be subdivided into great groups, of which there are about 185 in the United States. It would be too lengthy to go into these taxonomic units in this book, but once one gets acquainted with Soil Taxonomy, it is not too difficult to classify soils approximately to the great group level. Much information can also be gained by classifying to that level, as shown by the following examples.

Several trends can be deciphered from the great groups of the Ustolls (Table 2-4). An age sequence would be as follows:

Haplustoll → Argiustoll → Paleustoll, or Calciustoll → Paleustoll

with the age-related soil features better expressed in the older soils to the right of each sequence. A similar age sequence appears at the suborder level of the Aridisols, with Orthids (cambic horizon) developing into Argids (Argillic horizon) with time.

Table 2-3
Suborder Nomenclature Key for Soil Taxonomy

PREFIX INDICATES SOME DIAGNOSTIC PROPERTY OF THE SUBORDER		SUFFIX INDICATES THE ORDER
alb-	presence of albic horizon	-ent (Entisol)
aqu-	associated with prolonged wetness	-ept (Inceptisol)
arg-	presence of argillic horizon	-id (Aridisol)
bor-	associated with cool environment; relatively high organic matter content	-oll (Mollisol)
		-od (Spodosol)
ferr-	high iron content	-alf (Alfisol)
fluv-	recent river deposit	-ult (Ultisol)
hum-	high organic matter content	-ox (Oxisol)
ochr-	presence of an ochric epipedon	-ert (Vertisol)
orth-	the group of soils within the suborder that best typifies the order	-ist (Histosol)
pale-	old or excessive development	
psamm-	sandy material	
rend-	calcareous parent material	
torr- ⎫	associated with dryness lasting various lengths	
ust- ⎬	of time; usually low organic matter	
xer- ⎭	content	
trop-	associated with continually warm climate	
ud-	associated with humid climates; moderate to low organic matter content	
umbr-	presence of umbric epipedon	

Table 2-4
Some Great Groups of Ustolls and Boralfs

ORDER	SUBORDER	GREAT GROUP*
Mollisol	Ustoll	Durustoll (duripan)
		Natrustoll (natric horizon)
		Paleustoll (high clay content, red, petrocalcic)
		Calciustoll (calcic horizon)
		Argiustoll (argillic horizon)
		Haplustoll (the other Ustolls; cambic horizon)
Alfisol	Boralf	Paleboralf (upper boundary of argillic horizon deeper than 60 cm)
		Fragiboralf (fragipan)
		Natriboralf (natric horizon)
		Cryoboralf (cryic temperature regime)
		Eutroboralf ($>60\%$ base saturation in argillic horizon)
		Glossoboralf ($<60\%$ base saturation in some part of argillic horizon)

*In Soil Taxonomy, one always reads down lists such as these, so that if the first name encountered does not apply to the soil being classified, one goes down to the next name, and so forth.

In contrast to the Mollisols, within the Boralfs there is generally no such obvious age sequence because Boralfs must have an argillic horizon. One might suspect that Paleboralf would denote characteristics of great age, but the manner in which it is defined does not (depth to top of argillic horizon); the same is true for the Paleborolls. Leaching conditions, important to agronomists and botanists, can be read in some of the Boralfs for

$$\text{Glossoboralf} \rightarrow \text{Eutroboralf} \rightarrow \text{Natriboralf}$$

are in the direction of increasing base saturation.

DISTRIBUTION OF SOIL ORDERS AND SUBORDERS IN THE UNITED STATES

A soils map of the United States has been prepared, based on the new classification (Fig. 2-4). It shows some regional trends that can be roughly related to the soil-forming factors, mainly climatic and vegetation patterns and geology. One major trend is that, generally, the soil distribution east of the Rocky Mountains seems less complex than that to the west.

The soil pattern east of the Rocky Mountains generally follows the gradual regional climatic gradient, although there is some variation probably due in part to erosion in mountainous areas and to age of the landscape. Just east of the Rockies is a wide expanse of Ustolls. Entisols are interspersed with the Ustolls and are related to the erosive shales of eastern Montana and to the dune sands of Nebraska. Udolls lie east of the Ustolls, but in part of the glaciated region to the north these give way to Borolls and Aquolls, the latter being mainly associated with the relatively impermeable sediments of glacial Lake Agassiz. Proceeding eastward, there is a large area of Udults formed on fairly old landscapes south of the glacial boundary that extend almost to the Aquults of the East Coast. Florida does not follow this regional trend and is covered mainly with Aquods, Entisols, and Histosols. Inceptisols are associated with the Ultisols, as shown by Aquepts on the Mississippi River floodplain deposits and Ochrepts in the Appalachian and Ouachita Mountains. Just east of the Mississippi River is a belt of Udalfs that seems to be associated with loess deposits of glacial age. In the glaciated region of the central and northeastern United States, Udalfs are common to the south, and these grade into Orthods to the north. The only major exceptions are Udolls over much of Illinois and Boralfs in northern Minnesota.

West of the eastern front of the Rocky Mountains there is an intricate mosaic of climatic, vegetation, topographic, and geological patterns. The topography consists of a multitude of mountain ranges separated by intermontane valleys. Bedrock makes up most of the mountains, whereas unconsolidated alluvium of various ages underlie the valleys. In almost all places, the climate and vegetation are closely associated with the topography, with the valleys being the driest and the mountain slopes receiving

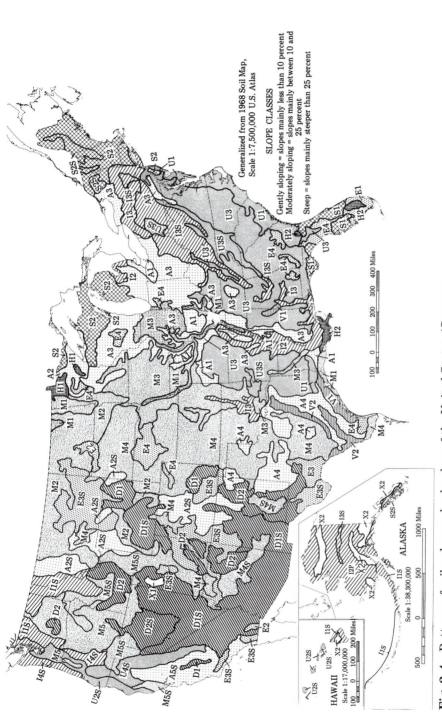

Fig. 2-4 Patterns of soil orders and suborders of the United States. (Courtesy of the U.S. Dept. Agriculture, Soil Conservation Service.)

Generalized from 1968 Soil Map,
Scale 1:7,500,000 U.S. Atlas

SLOPE CLASSES

Gently sloping = slopes mainly less than 10 percent
Moderately sloping = slopes mainly between 10 and 25 percent
Steep = slopes mainly steeper than 25 percent

LEGEND

Only the dominant orders and suborders are shown. Each delineation has many inclusions of other kinds of soil. General definitions for the orders and suborders follow. For complete definitions, see Soil Survey Staff.[12] Approximate equivalents in the modified 1938 soil classification system are indicated for each suborder.

 ALFISOLS ... Soils with gray to brown surface horizons, medium to high base supply, and subsurface horizons of clay accumulation; usually moist but may be dry during warm season

A1 AQUALFS (seasonally saturated with water) gently sloping; general crops if drained, pasture and woodland if undrained (Some Low-Humic Gley soils and Planosols)

A2 BORALFS (cool or cold) gently sloping; mostly woodland, pasture, and some small grain (Gray Wooded soils)

A2S BORALFS steep; mostly woodland

A3 UDALFS (temperate, or warm, and moist) gently or moderately sloping; mostly farmed, corn, soybeans, small grain, and pasture (Gray-Brown Podzolic soils)

A4 USTALFS (warm and intermittently dry for long periods) gently or moderately sloping; range, small grain, and irrigated crops (Some Reddish Chestnut and Red-Yellow Podzolic soils)

A5S XERALFS (warm and continuously dry in summer for long periods, moist in winter) gently sloping to steep; mostly range, small grain, and irrigated crops (Noncalcic Brown soils)

 ARIDISOLS ... Soils with pedogenic horizons, low in organic matter, and dry more than 6 months of the year in all horizons

D1 ARGIDS (with horizon of clay accumulation) gently or moderately sloping; mostly range, some irrigated crops (Some Desert, Reddish Desert, Reddish Brown, and Brown soils and associated Solonetz soils)

D1S ARGIDS gently sloping to steep

D2 ORTHIDS (without horizon of clay accumulation) gently or moderately sloping; mostly range and some irrigated crops (Some Desert, Reddish Desert, Sierozem, and Brown soils, and some Calcisols and Solonchak soils)

D2S ORTHIDS gently sloping to steep

 ENTISOLS ... Soils without pedogenic horizons

E1 AQUENTS (seasonally saturated with water) gently sloping; some grazing

E2 ORTHENTS (loamy or clayey textures) deep to hard rock; gently to moderately sloping; range or irrigated farming (Regosols)

E3 ORTHENTS shallow to hard rock; gently to moderately sloping; mostly range (Lithosols)

E3S ORTHENTS shallow to rock; steep; mostly range

E4 PSAMMENTS (sand or loamy sand textures) gently to moderately sloping; mostly range in dry climates, woodland or cropland in humid climates (Regosols)

HISTOSOLS ... Organic soils

H1 FIBRISTS (fibrous or woody peats, largely undecomposed) mostly wooded or idle (Peats)

H2 SAPRISTS (decomposed mucks) truck crops if drained, idle if undrained (Mucks)

INCEPTISOLS ... Soils that are usually moist, with pedogenic horizons of alteration of parent materials but not of accumulation

I1S ANDEPTS (with amorphous clay or vitric volcanic ash and pumice) gently sloping to steep; mostly woodland; in Hawaii mostly sugar cane, pineapple, and range (Ando soils, some Tundra soils)

I2 AQUEPTS (seasonally saturated with water) gently sloping; if drained, mostly row crops, corn, soybeans, and cotton; if undrained, mostly woodland or pasture (Some Low-Humic Gley soils and Alluvial soils)

(continued)

I2P AQUEPTS (with continuous or sporadic permafrost) gently sloping to steep; woodland or idle (Tundra soils)

I3 OCHREPTS (with thin or light-colored surface horizons and little organic matter) gently to moderately sloping; mostly pasture, small grain, and hay (Sols Bruns Acides and some Alluvial soils)

I3S OCHREPTS gently sloping to steep; woodland, pasture, small grains

I4S UMBREPTS (with thick dark-colored surface horizons rich in organic matter) moderately sloping to steep; mostly woodland (Some Regosols)

MOLLISOLS . . . Soils with nearly black, organic-rich surface horizons and high base supply

M1 AQUOLLS (seasonally saturated with water) gently sloping; mostly drained and farmed (Humic Gley soils)

M2 BOROLLS (cool or cold) gently or moderately sloping, some steep slopes in Utah; mostly small grain in North Central States, range and woodland in Western States (Some Chernozems)

M3 UDOLLS (temperate or warm, and moist) gently or moderately sloping; mostly corn, soybeans, and small grains (Some Brunizems)

M4 USTOLLS (intermittently dry for long periods during summer) gently to moderately sloping; mostly wheat and range in western part, wheat and corn or sorghum in eastern part, some irrigated crops (Chestnut soils and some Chernozems and Brown soils)

M4S USTOLLS moderately sloping to steep; mostly range or woodland

M5 XEROLLS (continuously dry in summer for long periods, moist in winter) gently to moderately sloping; mostly wheat, range, and irrigated crops (Some Brunizems, Chestnut, and Brown soils)

M5S XEROLLS moderately sloping to steep; mostly range

SPODOSOLS . . .
Soils with accumulations of amorphous materials in subsurface horizons

S1 AQUODS (seasonally saturated with water) gently sloping; mostly range or woodland; where drained in Florida, citrus and special crops (Ground - Water Podzols)

S2 ORTHODS (with subsurface accumulations of iron, aluminum, and organic matter) gently to moderately sloping; woodland, posture, small grains, special crops (Podzols, Brown Podzolic soils)

S2S ORTHODS steep; mostly woodland

ULTISOLS . . . Soils that are usually moist, with horizon of clay accumulation and a low base supply

U1 AQUULTS (seasonally saturated with water) gently sloping; woodland and pasture if undrained, feed and truck crops if drained (Some Low - Humic Gley soils)

U2S HUMULTS (with high or very high organic matter content) moderately sloping to steep; woodland and pasture if steep, sugar cane and pineapple in Hawaii, truck and seed crops in Western States (Some Reddish-Brown Lateritic soils)

U3 UDULTS (with low organic-matter content; temperate or warm, and moist) gently to moderately sloping; woodland, pasture, feed crops, tobacco, and cotton (Red-Yellow Podzolic soils, some Reddish-Brown Lateritic soils)

U3S UDULTS moderately sloping to steep, woodland, pasture

U4S XERULTS (with low to moderate organic-matter content, continuously dry for long periods in summer) range and woodland (Some Reddish-Brown Lateritic soils)

VERTISOLS. . . Soils with high content of swelling clays and wide deep cracks at some season

V1 UDERTS (cracks open for only short periods, less than 3 months in a year) gently sloping; cotton, corn, pasture, and some rice (Some Grumusols)

V2 USTERTS (cracks open and close twice a year and remain open more than 3 months); general crops, range, and some irrigated crops (Some Grumusols)

☐ AREAS with little soil . . .

X1 Salt flats X2

X2 Rock land (plus ice fields in Alaska)

NOMENCLATURE

The nomenclature is systematic. Names of soil orders end in *sol* (L. *solum*, soil), e.g., ALFISOL, and contain a formative element used as the final syllable in names of taxa in suborders, great groups, and subgroups.

Names of suborders consist of two syllables, e.g., AQUALF. Formative elements in the legend for this map and their connotations are as follows:

and — Modified from Ando soils; soils from vitreous parent materials

aqu — L. *aqua*, water; soils that are wet for long periods

arg — Modified from L. *argilla*, clay; soils with a horizon of clay accumulation

bor — Gr. *boreas*, northern; cool

fibr — L. *fibra*, fiber; least decomposed

hum — L. *humus*, earth; presence of organic matter

ochr — Gr. base of ochros, pale; soils with little organic matter

orth — Gr. *orthos*, true; the common or typical

psamm — Gr. *psammos*, sand; sandy soils

sapr — Gr. *sapros*, rotten; most decomposed

ud — L. *udus*, humid; of humid climates

umbr — L. *umbra*, shade; dark colors reflecting much organic matter

ust — L. *ustus*, burnt; of dry climates with summer rains

xer — Gr. *xeros*, dry; of dry climates with winter rains

an increasing amount of moisture relative to altitude; vegetation follows these trends. The one major exception to these climatic trends is the lowland regions west of the crests of the northern Sierra Nevada and the Cascade Range; these areas receive abundant moisture from air masses moving inland from the Pacific Ocean. Still, even here the general climate-altitude relationship holds.

Soil patterns in the Cordilleran mountain ranges follow the overall climatic trends. Boralfs are common in the eastern parts of the Rocky Mountains, whereas the western parts are dominated by Ustolls to the south, as well as in the mountains of Arizona; Xerolls dominate to the north. Andepts are found in parts of the northern Rockies and in the Cascade Range, where they are most often associated with volcanic rock. The Sierra Nevada are dominated by Xerults, although soils south of Lake Tahoe might be closer to Xeralfs. Entisols occur in the ranges south of the Sierra Nevada. The ranges along the West Coast grade from Umbrepts in the north to Humults in the central sector to Xeralfs in the south.

The basins show a soil variation from north to south. Those in eastern Oregon and Washington are dominated by Xerolls, with Orthids in the drier parts of the Columbia Plateau. Orthids are common also in the Snake River Plain of southern Idaho and in the northwestern Basin and Range Province, and they grade into Argids to the south in that province. This gradation can be explained, at least in part, by age of landscape, because the widespread alluvial fan and pediment deposits to the south seem to be older than those to the north. Entisols and Orthids are the major soils of the Colorado Plateau.

Humults dominate in the Hawaiian Islands. Oxisols are present there to a limited extent. Most of Alaska has Inceptisols, with Spodosols more common in the southern parts of the state.

REFERENCES

1. Bartelli, L.J., Klingebiel, A.A., Baird, J.V., and Heddleson, M.R., eds., 1966, Soil surveys and land use: Soil Sci. Soc. Amer., 196 p.
2. Blackstock, D.A., 1979, Soil survey of Lubbock County, Texas: Soil Cons. Service, U.S. Dept. Agri., 105 p.
3. Bockheim, J.G., 1980, Properties and classification of some desert soils in coarse-textured glacial drift in the Arctic and Antarctic: Geoderma, v. 24, p. 45–69.
4. Bunting, B.T., 1965, The geography of soil: Aldine Publ. Co., Chicago, 213 p.
5. Dudal, R., 1968, Definitions of soil units for the soil map of the world: World soil resources reports no. 33, World soil resources office, FAO, Rome, 72 p.
6. Harris, S.A., 1968, Comments on the validity of the law of soil zonality: 9th Internat. Cong. Soil Sci., Trans. v. 4, p. 585–593.
7. Jenny, H., 1946, Arrangement of soil series and types according to functions of soil-forming factors: Soil Sci., v. 61, p. 375–391.
8. Oschwald, W.R., Riecken, F.F., Dideriksen, R.I., Scholtes, W.H., and Schaller, F.W., 1965, Principal soils of Iowa—their formation and properties: Iowa State Univ. Coop. Ext. Serv. Spec. Rep. 42, 76 p.
9. Simonson, R.W., 1962, Soil Classification in the United States: Science, v. 137, p. 1027–1034.
10. Soil Conservation Service, 1967, Soils of the Malibu area, California: U.S. Dept. Agri., 89 p.
11. Soil Survey Staff, 1951, Soil survey manual: U.S. Dept. Agri. Handbook no. 18, 503 p.
12. ———, 1975, Soil taxonomy: U.S. Dept. Agri. Handbook No. 436, 754 p.
13. Thomas, R.F., Blakemore, L.C., and Kinloch, D.I., 1979–1982, Flow-diagram keys for "Soil Taxonomy": New Zealand Soil Bureau Scientific Report 39: A. Diagnostic horizons and properties: mineral soils (1979); B. Soil moisture and temperature regimes, and diagnostic horizons and properties for organic soils (1980); C. The key to soil orders (1980); D. Histosols and Spodosols (1981); E. Oxisols and Vertisols (1981); F. Aridisols (1981); G. Ultisols (1982).

3

Weathering processes

Weathering is the physical and chemical alteration of rock or minerals at or near the earth's surface. Most rocks and minerals exposed at and immediately beneath the earth's surface are in an environment quite unlike that under which they formed. This is especially true for igneous or metamorphic rocks that formed under high temperatures and, with the exception of some volcanic rocks, at great confining pressures. Weathering can be defined as the process of rock and mineral alteration to more stable forms under the variable conditions of moisture, temperature, and biological activity that prevail at the surface.

Two main types of weathering are recognized.[38,42,60] One is physical weathering, in which the original rock disintegrates to smaller-sized material, with no appreciable change in chemical or mineralogical composition. The other is chemical weathering, in which the chemical and/or mineralogical composition of the original rock and minerals is changed. In nature, physical and chemical weathering occur together, and it may be difficult to separate the effects of one from the effects of the other. Winkler[65] provides an excellent quantitative review of this topic.

PHYSICAL WEATHERING

The mechanism common to all processes of physical weathering is the establishment of sufficient stress within the rock so that it breaks. If the rock is ruptured along fracture planes, blocks or sheets of varying size are produced. If, however, the lines of weakness are along mineral grain boundaries, physical weathering can produce materials whose size is determined by the size of the grains in the original rock; smaller sizes are possible if the minerals are cross-cut by small-scale fractures. The processes that are reported to be most common to physical weathering are unloading by erosion, expansion in cracks or along grain boundaries by freezing water or crystallizing salts, fire, and possibly thermal expansion and contraction of the constituent minerals.[42]

Rock bodies that are either homogeneous or layered can have numerous fracture planes or joints, nearly parallel to the ground surface, that divide the rock into a series of layers or sheets (Fig. 3-1). The spacing between the joints generally increases with depth, and they can be observed for several tens of meters below the surface. The origin of the fractures seems to be the release of stresses contained within the rock.[10] While buried, the rock is under high confining pressures. With erosion of the overlying rock mass, however, the rock has less overburden pressure, so it can expand. If it is close enough to the surface, expansion can only be upward or toward the valley wall—in any direction in which the rock body is not confined. This expansion can lead to rupture of the rock along fracture planes oriented at right angles to the direction of the pressure release and, thus, to development of sheeting parallel to the surface.

The role of freeze-thaw in physical weathering has been debated a long time and the end is not in sight. The concept is as follows. Water, upon freezing, can set up pressures sufficient to disintegrate most rocks. At 0 °C, the increase in volume with the conversion from water to ice is 9 percent. At localities in which there is a sufficient supply of moisture and a low enough temperature, the moisture contained in the rock can freeze, and the accompanying internal pressures are sufficiently great to exceed the strength of the rock, and the rock ruptures. Even though the water in

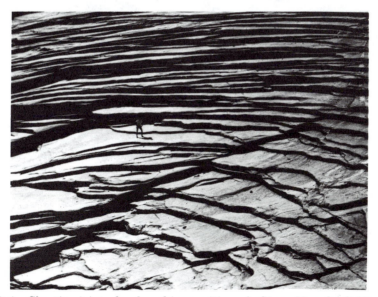

Fig. 3-1 Sheeting joints developed in granitic rock, Sierra Nevada, California. (Photograph by W. C. Bradley.)

the system might not be confined, pressures in an unconfined system are probably great enough to rupture most rock types. The direction of the fractures produced could be determined by minute, preexisting planes of weakness, such as joints, or along fractures produced by unloading. In this case, the result could be a large field of angular blocks. In some places this process might be responsible for granular disintegration, provided water has access to voids or cracks between the grains. This process could be most effective in environments in which surface temperatures fluctuate across 0 °C many times each year.

Recent reviews cast doubt on the primary role of freeze-thaw in physical weathering; instead, more workers seem to favor a little understood process in which thin films of adsorbed water might be the agent that does the weathering.[8,58,63,64] These films can be adsorbed so tightly that they cannot freeze. Pressures exerted during the migration of these films along microfractures could pry the rock apart. There still may be a connection with freezing, as such migration could take place during times of freezing.

Saline solutions, if they gain access to fractures in the rock or to the boundaries between grains, can bring about disintegration of rocks either into blocks or individual grains.[15,18,23,25,65] Several processes are recognized as important to the breakup of the rock by salts. One is the internal pressures set up during the growth of crystals from solution (Fig. 3-2). A com-

Fig. 3-2 Weathered granitic stones in Baja California. In many places, the undersides of these stones are virtually unweathered. Salt crystallization in small cracks may be a major factor in the surface weathering of the stones.

mon pedologic example demonstrating this effect is the development of K horizons. In many places, the K horizon contains over 70 percent $CaCO_3$, and the original silicate grains are no longer in contact with each other. Because there is little evidence that the original grains were dissolved, it appears that they were pushed aside during the crystallization of the $CaCO_3$ from the soil solution. Precipitation of gypsum from a solution has similar effects. Another important process is the thermal expansion of salts upon heating, which occurs because many common salts have thermal expansion coefficients higher than those of some common rocks.[18] This process might be important in many of the hot deserts that are characterized by large changes in daily temperatures. A final process that is important to rock disintegration is the stresses set up by volume increase that accompanies hydration of the various salts. Winkler,[64] for example, feels that the main cause of exfoliation of Cleopatra's Needle is hydration of salts that formed in the monument in Egypt; these salts were hydrated and expanded once the monument was moved to the humid climate of New York City. Hydration of clay minerals may have similar disruptive effects on rocks.

The salts responsible for the weathering can come from several sources. The obvious one is the sea. Other sources are salt deposits in deserts (from bedrock or playa deposits), which are transported as dust, dissolved in rain, or both. Still another, more nebulous source of salt is that contained in fluid inclusions in rock minerals. This latter source was invoked by Bradley and others[11] to explain what seemed to be a good case for salt weathering in Australia when no other process seemed plausible.

Fire is probably an important, but often overlooked, factor in the physical weathering of rock. Because rock is a poor conductor of heat, a thin surface layer attains a high temperature during a fire, and there is a rapid decrease in temperature with depth. The heated surface layer will expand more rapidly than will rock at greater depth, and this expansion can lead to rupture of the rock into thin sheets that eventually fall to the ground. An example of the weathering effects of fire is seen on Mt. Sopris in western Colorado (Fig. 3-3). The rock-glacier deposits on Mt. Sopris do not have a forest cover. The stones on the surfaces of the rock-glacier deposits and of non-vegetated blockfields have thick oxidation weathering rinds, stone surfaces are oxidized to a reddish color, and the corners of some stones, even after several thousands of years of weathering, still are not too rounded.[4] However, stones adjacent to or in the forest that burned several decades ago have more rounded corners and thin or no weathering rinds, and they lack the pronounced surface oxidation. The differences in corner rounding and in weathering are attributed to fire, in both historic and prehistoric times.

These observations on the effect of fire on weathering are important to Quaternary stratigraphic studies, because commonly the weathered condition of the surface of a stone is one criterion used for age differen-

(A)

(B)

Fig. 3-3 Comparison of stone weathering in (A) a forested and (B) an adjacent non-forested environment, western Colorado. Note the abundance of fresh fracture faces in (A), due to expanison on heating by fire, and their absence in (B), which experiences no burning.

tiation (Ch. 8). One finds in parts of the Colorado Front Range and the Sierra Nevada, for example, that stones in young tills above treeline have more highly pitted surfaces than have stones in much older tills within the present forest zone (Fig. 3-4). One suspects that fire may explain the difference in weathering of individual stones because, in the forest zone, the stone surfaces can be continually renewed by spalling during a fire. The evidence for ancient fire is indirect, but surely lightning was an important cause. Blackwelder[6] emphasized that fire, in some places, might be the main weathering process in physical weathering, and I think the stratigraphic and boulder weathering data from some places would support him. If present stone weathering can indeed be related to the presence or absence of fire, and thus to the treeline, stone-weathering studies might be used as one criterion to help estimate the expansion of the upper and lower limit of past forests.

Diurnal fluctuations in surface temperatures, if great enough, are thought by some workers to bring about surface fracturing or granular disintegration of some rocks.[7] This idea is based on repeated observations of weathered rocks in deserts, weathering that seemed best ascribed to

Fig. 3-4 Pitted stone on surface of Holocene till above the treeline in the Colorado Front Range. Weathering features such as these could take about 10,000 years to form.

physical processes. In theory, the surface of a rock, where temperature fluctuations are greatest, should expand and contract the most.[44,55] The effect is not unlike that of fire, that is, the rock is eventually weakened to the point where a part of the surface flakes off or mineral grains are dislodged. The process envisaged is that each different mineral will expand and contract a different amount at a different rate with surface-temperature fluctuations. With time, the stresses produced are sufficient to weaken the bonds along grain boundaries, and thus flaking of rock fragments or dislodging of grains occurs. The importance of this process in rock weathering has been debated over the years. Griggs,[27] in laboratory experiments in which rock samples were heated and cooled over large fluctuations in temperature to simulate 244 years of weathering in the absence of water, showed that little disaggregation of the rock occurred. Ollier[42] argues, however, that the time factor was not taken sufficiently into account in Griggs's study and that small stresses applied over long periods of time might lead to permanent strain. This process might explain some weathering in arid regions. In most desert regions, however, some moisture is present, as are salts, and therefore other weathering processes could be operative.

Plant growth contributes to some physical weathering. Pressures exerted by roots during growth are sometimes able to rupture the rock or to force blocks apart. That such pressures can do this is shown by the common observation of cracked and heaved concrete sidewalks adjacent to tree roots. Lichens growing on rock surfaces also contribute to physical weathering. Loose mineral and rock fragments become attached to the undersides of the lichens and are pulled free of the surface when the lichen contracts during a dry spell. If the lichen is removed from the rock surface, it takes rock material with it. One would suspect, however, that the grains would have had to be loosened by some other process prior to removal from the rock surface with the lichen.

Of the above processes, weathering by freeze-thaw and by insolation have been the most hotly debated. Many workers disregard both as major factors in weathering. One reason for this is that experimental work on both has not yielded the expected results. However, the problem also could be that of performing experiments at a scale and time span different than those in nature. As an example, Rice[49] has cautioned workers against abandoning insolation as a weathering agent because of the importance of both sample size and number of temperature fluctuations in artificial weathering experiments. Time is also difficult to scale down. Perhaps one should still keep an open mind on the efficacy of either one.

CHEMICAL WEATHERING

Chemical weathering occurs because rocks and minerals are seldom in equilibrium with near-surface waters, temperatures, and pressures. The products that form, however, are more stable in near-surface environments. If the soil environment does change with time, so too can the initial products of weathering. That some change occurs during chemical weathering is shown by field evidence for oxidation and for clay formation, by the different chemical and mineralogical composition of the weathered material relative to that of the assumed parent material, and by the chemistry of the waters that move through the soil. The change on weathering can be very slight and involve nothing more than the oxidation state of iron ions, or it can be quite intense and result in a product much different from the assumed parent material, such as the formation of an Oxisol from a mafic rock. There are several processes involved in chemical weathering of common rocks and minerals.

Congruent and Incongruent Dissolution

Chemical weathering can be subdivided into congruent and incongruent dissolution.[9,19,21,24] Congruent dissolution is when the mineral goes into solution completely with no precipitation of other substances. In contrast, with incongruent dissolution, all or some of the ions released by weathering precipitate to form new compounds. Put another way, the amounts of substances in solution do not correspond to the formula of the weathering mineral.

The soluble salt minerals, halite and gypsum, and silica minerals are examples of congruent dissolution. The respective reactions are

$$NaCl = Na^+ + Cl^-$$
$$c \quad\;\; aq \quad\;\; aq$$

$$CaSO_4 \cdot 2H_2O = Ca^+ + SO_4^{2-} + 2H_2O$$
$$c \quad\;\; aq \quad\;\; aq \quad\;\; l$$

$$SiO_2 + 2H_2O = H_4SiO_4$$
$$c \quad\;\;\; l \quad\;\;\; aq$$

where c is the crystalline, l is the liquid, and aq is the aqueous species. The solubilities of these minerals vary markedly, being about 260 to 350 g/liter of water for halite and 1.6 to 2 g/liter for gypsum, whereas silica has a range from 7 ppm for quartz to 120 ppm for silica gel (1000 ppm = 1 g/liter). The solubilities of other forms of silica lie between these latter two values. Higher temperatures slightly increase the solubilities of gypsum and silica species. Furthermore, the solubility of silica is somewhat pH dependent; above pH 9 it increases drammatically (Fig. 4-3) according to the reaction

$$H_4SiO_4 = H_3SiO_4^- + H^+$$
$$aq \quad\quad aq \quad\;\; aq$$

The weathering of calcium carbonate is another example of congruent dissolution. It is quite soluble under surface conditions and dissolves according to the equation

$$CaCO_3 + CO_2 + H_2O \rightleftharpoons Ca^{2+} + 2HCO_3^-$$
$$ c \quad\quad g \quad\quad l \quad\quad aq \quad\quad aq$$

Krauskopf[35] discusses carbonate equilibria and points out that the solubility of $CaCO_3$ varies with differences in CO_2 pressure and in H^+ concentration. Increases in either CO_2 pressure or in H^+ concentration will increase the rate at which $CaCO_3$ dissolves. The partial pressure of CO_2 in the soil atmosphere under vegetation is greater than atmospheric, and therefore $CaCO_3$ solubility is greater under vegetated surfaces than under surfaces that lack vegetation. Rode[50] gives $CaCO_3$-solubility values for various CO_2 pressures. It is also known that CO_2 partial pressure in water is temperature dependent, with colder waters able to contain more CO_2 than warmer waters. Thus, $CaCO_3$ should dissolve more readily in cooler climates than in warmer climates. Arkley[1] gives two graphs showing the relationship of $CaCO_3$ solubility with pH and with temperature (Fig. 5-8), and these may be helpful in estimating the rate of weathering of carbonate rocks. In addition, chelating agents combine with Ca^{2+} and in this way increase the rate of solution of $CaCO_3$ and the mobility of Ca^{2+}.

Soils formed from limestone can consist only of the insoluble residue left behind as Ca^{2+} and HCO_3^- are leached from the system. The properties of the soil, therefore, are strongly dependent on the properties of the insoluble fraction, as long as subsequent eolian additions can be ruled out.

The chemical weathering of the aluminosilicate minerals is a good example of incongruent dissolution. A general reaction is

$$\text{Aluminosilicate} + H_2O + H_2CO_3$$
$$ c \quad\quad l \quad\quad aq$$
$$= \text{clay mineral} + \text{cations} + OH^- + HCO_3^- + H_4SiO_4$$
$$ c \quad\quad aq \quad\quad aq \quad\quad aq \quad\quad aq$$

The usual reaction is that of water and acid on the mineral. The acid shown here is H_2CO_3. Other acids, such as those resulting from the decay of organic matter, also are important H ion sources. The common by-products are H_4SiO_4, HCO_3^-, and OH^-, along with clay minerals if aluminum is present in the decomposing minerals and if certain chemical conditions are met. More detailed weathering equations are given in Table 3-1; these reactions are commonly termed hydrolysis.

The fate of the by-products of weathering varies, and this will be discussed in more detail later. Cations can remain in the soil either in the soil solution, as part of the crystal lattice of the clay mineral, or as exchangeable ions adsorbed to the surfaces of the colloidal particles. Some ions can be cycled through the biosphere from the soil and back again. Some cations can be removed from the system, along with HCO_3^-, with the percolating waters; indeed, one measure of the rate of chemical weathering of a region can be gained from the composition of the waters draining the region. Silica is quite soluble over the normal soil pH range (Fig.

Table 3-1
Equations Representing the Hydrolysis of Orthoclase and Albite with Various Clay Minerals as a By-Product

$$2KAlSi_3O_8 + 2H^+ + 9H_2O = H_4Al_2Si_2O_9 + 4H_4SiO_4 + 2K^+$$

    c   aq   1   c   aq   aq

  (orthoclase)   (kaolinite)

$$3KAlSi_3O_8 + 2H^+ + 12H_2O = KAl_3Si_3O_{10}(OH)_2 + 6H_4SiO_4 + 2K^+$$

    c   aq   1   c   aq   aq

  (orthoclase)   (illite)

$$2NaAlSi_3O_8 + 2H^+ + 9H_2O = H_4Al_2Si_2O_9 + 4H_4SiO_4 + 2Na^+$$

    c   aq   1   c   aq   aq

  (albite)   (kaolinite)

$$8NaAlSi_3O_8 + 6H^+ + 28H_2O$$

    c   aq   1

  (albite)

$$= 3Na_{0.66}Al_{2.66}Si_{3.33}O_{10}(OH)_2 + 14H_4SiO_4 + 6Na^+$$

  (smectite)   c   aq   aq

4-3), and it is almost always present in the parent minerals in higher amounts than are necessary to form most clay minerals; therefore some is removed in solution. Aluminum is not very soluble over the normal soil pH range (Fig. 4-3), and so it generally remains near the site of release by weathering to form clay minerals or hydrous oxides. Iron also remains near the point of release in most soils and gives the soil or weathered rock the commonly observed oxidation colors.

One effect of the hydrolysis reaction is that hydrogen ion is consumed, hydroxide is produced, and the solution becomes more basic. This effect is especially noticeable when the various silicate and aluminosilicate minerals are ground in distilled water, and the pH of the solution, called the abrasion pH, is taken.[56] The pH resulting from this grinding and initial hydrolysis is a function of the rapidity at which cations are released to the solution and the strengths of the bases formed (Table 3-2). In any weathering environment, the leaching of the cations and the production of hydrogen ion offsets this tendency for most reactions to become basic as weathering proceeds. Grant[26] has shown that the abrasion pH of weathered material that includes some clay is less than the pH of the original rock, because some cations have been removed, and abrasion pHs of clay minerals commonly are lower than those of the common rock-forming minerals.

Little is known of the details of the reactions that take place at the crystal surface during hydrolysis. Jenny,[34] however, has presented a useful model. He notes that there are unsatisfied bonds between the cations and the oxygens and hydroxides at the crystal surface. Water, being a polar substance, is attracted to the differently charged sites on the mineral surface (Fig. 3-5). The water dipoles may be attracted by silicon and aluminum with such force that they dissociate; the hydrogen ions then can com-

Table 3-2
Abrasion pH Values for Some Common Minerals

MINERAL	FORMULA	ABRASION pH
Silicates		
Olivine	$(Mg,Fe)_2SiO_4$	10, 11
Augite	$Ca(Mg,Fe,Al)(Al,Si)_2O_6$	10
Hornblende	$Ca_2Na(Mg,Fe)_4(Al,Fe,Ti)_3$ $Si_6O_{22}(O,OH)_2$	10
Albite	$NaAlSi_3O_8$	9, 10
Oligoclase*	$Ab_{90-70}An_{10-30}$	9
Labradorite*	$Ab_{50-30}An_{50-70}$	8, 9
Biotite	$K(Mg,Fe)_3AlSi_3O_{10}(OH)_2$	8, 9
Microcline	$KAlSi_3O_8$	8, 9
Anorthite	$CaAl_2Si_2O_8$	8
Hypersthene	$(Mg,Fe)_2Si_2O_6$	8
Muscovite	$KAl_3Si_3O_{10}(OH)_2$	7, 8
Orthoclase	$KAlSi_3O_8$	8
Montmorillonite	$(Al_2,Mg_3)Si_4O_{10}(OH)_2 \cdot nH_2O$	6, 7
Halloysite	$Al_2Si_2O_5(OH)_4$	6
Kaolinite	$Al_2Si_2O_5(OH)_4$	5–7
Oxides		
Boehmite	$AlO(OH)$	6, 7
Gibbsite	$Al(OH)_3$	6, 7
Quartz	SiO_2	6, 7
Hematite	Fe_2O_3	6
Carbonates		
Dolomite	$CaMg(CO_3)_2$	9, 10
Calcite	$CaCO_3$	8

*Ab = albite; An = anorthite.
(Taken from Stevens and Carron,[56] by permission from the Mineralogical Society of America.)

bine with the oxygen ions of the crystal surface, and the hydroxide ions combine with either silicon or aluminum. Hydrogen ions also may replace cations in the crystal lattice. This exchange of hydrogen ions for cations has a disrupting effect on the crystal surface because of the high charge-to-size ratio of hydrogen. The polyhedra of aluminum and silicon also are no longer held tightly to the mineral, and they are able to move from the crystal surface into the soil solution.

Experimental work by Wollast[67] adds more detail to the above general scheme. He notes that in artificially weathering potassium-feldspar over the normal pH range, the increase of potassium ion in solution is accompanied by a decrease in hydrogen ion. This is expected from the general hydrolysis equation and abrasion pH data. He also found that the aluminum and silicon released by weathering formed a thin surface coating

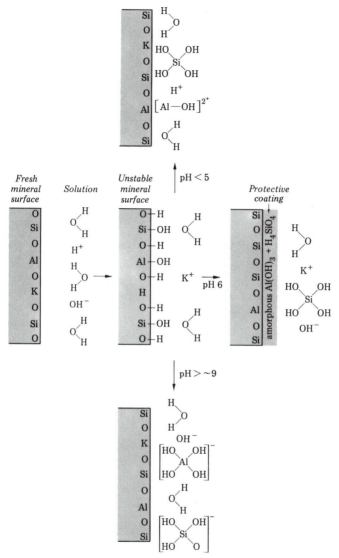

Fig. 3-5 Scheme of an orthoclase surface reacting with water at various pHs. Conditions in acid soils might approximate that depicted for pH less than 5, those for near-neutral soils for pH 6, and those for quite basic soils for pH near 9. (Data partly from Jenny[34] and Wollast.[65])

composed of amorphous $Al(OH)_3$ and SiO_2 or H_4SiO_4 (Fig. 3-5). The solution is saturated with respect to aluminum but not silicon. Silicon continues to be released by weathering but at a lower rate, because it must diffuse through the surface amorphous coating. Thus, the initial weathering products, if they accumulate on the surface of the weathering mineral, can bring about a decrease in weathering rate with time. Below a pH of 5, aluminum is soluble as Al^{3+} or aluminum-hydroxy ion (Fig. 4-3), and the blocking effect is not seen (Fig. 3-5). Perhaps this explains, in part, the observation of more rapid weathering of minerals at low pHs.

Minerals also weather at high pHs, and one explanation for this may be seen in the model of Jenny and Wollast. At high pH, there are abundant hydroxide ions to combine with the aluminum and silicon at the crystal surface, and the compounds thus formed are quite soluble (Fig. 4-3), perhaps as negatively charged hydroxy ions. Aluminum, therefore, would not have a blocking effect on weathering by forming a film on the mineral grain surfaces; thus, ions released by weathering would go into the solution. The rate of weathering, as under acid conditions, is governed somewhat by the rate of removal of cations from the site of weathering so that the mineral and the solution are not in equilibrium, a point well made by Todd.[59]

The idea of an amorphous or crystalline protective layer at the surface of weathering grains has been challenged. Petrović and others[45] used the results of artificial weathering experiments and theoretical considerations to argue that no such layer exists. Later studies on naturally and artificially weathered feldspars by Berner and Holdren[3] have provided scanning electron photomicrographs that show weathered mineral surfaces to be free of any such layer (Fig. 3-6). In contrast, Nixon[41] reports the existence of such a layer on a naturally weathered microcline grain. At present, there seems to be a growing number of workers arguing against the layer; perhaps more work could be done in a variety of different natural environments before any conclusions are drawn.

The weathering of iron-bearing minerals in oxygenated waters also is an example of incongruent dissolution; it is commonly termed oxidation. Oxidation is the process by which an element loses an electron. This loss results in an increase in positive valency for the element. Iron is the element most commonly oxidized in a soil or weathering environment, and the oxidation products give the altered material the characteristic yellowish brown to red colors. In soils and in many other weathering environments, the common oxidizing agent is oxygen dissolved in the water involved in the weathering reactions.

Weathering of iron-bearing minerals commonly releases Fe^{2+} which, if in contact with oxygenated waters, is oxidized to form an oxide or hydrous oxide of iron. An example of this reaction for fayalite is given by Krauskopf[35]

$$Fe_2SiO_4 + 2H_2CO_3 + 2H_2O \rightarrow 2Fe^{2+} + 2HCO_3^- + H_4SiO_4 +$$
$$\text{c}\text{aq}\text{l}\text{aq}\text{aq}\text{aq}$$
$$+ 2OH^- \text{ (hydrolysis)}$$
$$\text{aq}$$

$$2Fe^{2+} + 4HCO_3^- + \tfrac{1}{2}O_2 + 2H_2O \rightarrow Fe_2O_3 + 4H_2CO_3 \text{ (oxidation)}$$
$$\text{aq}\phantom{Fe^{2+} +}\text{aq}\text{g}\phantom{\tfrac{1}{2}O_2 +}\text{l}\text{c}\text{aq}$$

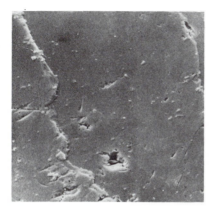 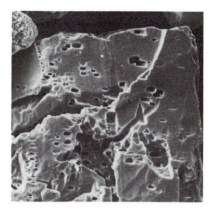

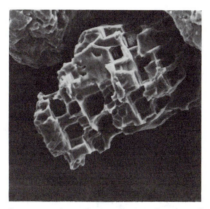

Fig. 3-6 Scanning electron microscope photomicrographs showing the increasing amounts of etching of feldspars (clockwise from left) in the A horizon of a mountain soil in New Mexico. The etching here is considered to be selective along crystal lattice dislocations. (Taken from Berner and Holdren,[3] Figs. 2 and 3.)

where g denotes the gaseous phase. Here the Fe^{2+} is released by weathering, and it enters an oxidizing environment and forms a precipitate, in this case Fe_2O_3. Usually, other more hydrated iron compounds are formed (for example, goethite).

The rate at which oxidation takes place and is noticeable in many soils seems to depend on the rate of release of iron upon weathering. Stratigraphic studies of unconsolidated materials indicate that the depth and intensity of oxidation increase with age, and that this increase is a fairly slow process on a geological time scale. For example, chronosequences of soils in the western United States that span tens of thousands of years commonly show differences in the depths of oxidation that are a function of soil age. Although there are few data on oxygen content with depth in soils, those of Black[5] suggest that porous, well-drained soils can be well oxygenated below the zone of noticeable oxidation (Cox horizon) and into seemingly unoxidized materials (Cu horizon). It would seem, therefore, that the slow extension of oxidation with depth in the individual members of a soil chronosequence is directly related to the rate of release of iron during weathering because oxygenated waters are passing through materials that appear to be unweathered by most field criteria.

A complex alteration involving the iron-bearing minerals is that of hematite and goethite.[53] Although sometimes the conversion of one to the other is considered a hydration-dehydration phenomena,

$$Fe_2O_3 + H_2O \rightleftharpoons 2FeOOH$$

$$\text{c} \qquad \text{l} \qquad \text{c}$$

$$\text{(hematite)} \qquad \text{(goethite)}$$

the reaction is more involved than that because dissolution of one species precedes formation of the other. In an oxidizing environment, the formation of hematite seems to require the formation of a precursor, ferrihydrite, a relatively new iron mineral species (Ch. 5). Hematite can alter to goethite only by dissolution and re-precipitation. Goethite can convert to hematite, but dissolution and formation of the precursor ferrihydrite also seems to be necessary. The reactions are quite complex and can involve reduction/oxidation and the formation of organic complexes.

Oxidation of iron in a mineral can alter the mineral.[2] Iron exists as Fe^{2+} in the common rock-forming minerals. Oxidation to Fe^{3+} disrupts the electrostatic neutrality of the crystal, such that other cations leave the crystal lattice to maintain neutrality. These latter cations leave vacancies in the crystal lattice, vacancies that either bring about the collapse of the lattice or render the mineral more susceptible to attack by other weathering processes. The alteration of biotite to vermiculite is one example of weathering primarily due to oxidation.

Oxidizing and reducing conditions could influence the rate at which iron-bearing minerals weather. It has been suggested that under oxidizing conditions, the Fe that is released precipitates as an oxide on the mineral and that this could protect the mineral from further weathering, or at least slow down the rate of weathering.[54] In contrast, a protective coat would not form in a reducing environment, and could result in relatively higher rates of weathering. Theoretically, reducing conditions during part of the Precambrian, when atmospheric oxygen levels were low, could have accelerated weathering of iron-bearing minerals.

Chelation

Evidence suggests that chelating agents are responsible for a considerable amount of weathering; in fact, in some places the amount of weathering by this process might exceed that brought about by hydrolysis alone. Chelating agents are formed by biological processes in soil and excreted by lichens growing on rock surfaces. Their structure is varied and complex and can be described as "the formation of more than one bond between the metal and a molecule of the complexing agent and resulting in the formation of a ring structure incorporating the metal ion" (Lehman,[36] p. 167; see Fig. 3-7). Hydrogen ion is released from the organic molecule during the reaction and can participate in hydrolysis reactions. Once in solution, the chelate may be stable at pH conditions under which the included cations would ordinarily precipitate out, a topic to be discussed later. The problem here, however, is to assess the effect of chelating agents on rock and mineral weathering.

Recent laboratory and field work has demonstrated that chelating agents in contact with rocks or minerals can bring about a significant amount of weathering. Schalscha and others[51] ground up various minerals and granodiorite and allowed these materials to react with solutions containing chelating agents. Cations were released to the solution at rates greater than would be predicted by a hydrogen ion effect alone. In fact, these workers found little correlation between pH and weathering rate. They concluded that the weathering of these materials is a combination of the effect of chelating agents and hydrogen ion. This is a departure from much past thinking that has ascribed much weathering to the hydrogen ion only.

Lichens growing on rock surfaces or on soils can bring about substantial amounts of weathering. Lichens excrete chelating agents,[52] and, thus, are important to the understanding of weathering of rocks and minerals. Jackson and Keller[32] have studied the weathering of a recent basalt in Hawaii under a lichen cover and in the absence of such a cover (Table 3-3). They found good evidence that weathering rinds are thicker and chemical alteration is more extensive beneath a lichen cover than in lichen-free areas of the same rock. Data for the lichen-free rock show slight enrich-

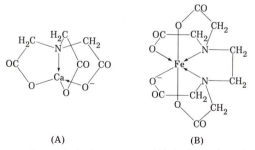

(A)   (B)

Fig. 3-7 Structure of two chelating agents. (A) Aminotriacetic acid binding calcium; (B) ethylenediaminetetracetic acid (EDTA) binding iron. (Taken from Ponomareva.[46] English translation by Israel Program for Scientific Translations, Jerusalem.)

Table 3-3
Comparison of Weathering of Hawaiian Basalt Erupted in 1907 under Two Surface Weathering Environments: Lichen-Covered and Lichen-Free

		LICHEN-COVERED ROCK	LICHEN-FREE ROCK
Mean thickness of weathering rind (mm)		0.142 (*color:* 1OR 3/4–4/6)	<0.002
Concentration ratio of weathered crust: fresh rock for these elements	Fe	6.36	1.21
	Al	0.58	0.47
	Si	0.21	1.20
	Ti	0.27	0.965
	Ca	0.004	1.24

*The lichen is *Stereocaulon volcani.*
(Taken from Jackson and Keller.[32])

ment in iron, silicon, and calcium, little change in titanium, and some depletion of aluminum. In contrast, the lichen-covered rock showed a six-fold enrichment of iron and depletion in varying amounts of all other elements. This study is unique in that it provides a comparison of weathering with and without biological input. These workers concluded that chelating agents in the presence of high hydrogen ion concentration, due to respiratory CO_2 and organic acids, increased weathering beneath lichens.

Little is known of the mechanism by which chelating agents bring about the weathering of a mineral grain surface. The agents might combine directly with exposed cations, much in the manner in which hydroxide ions become attached to aluminum at the grain surface (Fig. 3-5), and this could be followed by the movement of the complex into solution.

Chelating agents render substances more soluble as chelates under certain pH conditions. A common example is aluminum, which may be soluble as a chelate over pHs at which it is insoluble as an ion (Fig. 4-3). Hence, solubilizing aluminum with a chelating agent could offset the blocking action depicted in Figure 3-5 and thus allow more rapid weathering of the mineral surface.

Hydration and Dehydration

Hydration and dehydration are processes by which water molecules are added to or removed from a mineral. The result is the formation of a new mineral. These processes probably are not too important in overall chemical weathering because few minerals are affected, and they are not too common. An example of hydration-dehydration is the formation of gypsum and anhydrite by adding or removing water

$$CaSO_4 \cdot 2H_2O \rightleftharpoons CaSO_4 + 2H_2O$$
$$c \qquad\qquad c \qquad\quad l$$

(gypsum)   (anhydrite)

A marked increase in volume accompanies the reaction anhydrite to gypsum, and, if this takes place within a rock, physical disintegration can occur.

Ion Exchange

Some weathering from one mineral to another can occur through the exchange of ions between the solution and the mineral. The most readily exchangeable cations are those between the layers of the phyllosilicates, such as sodium and calcium. During the exchange, the basic structure of the mineral is unchanged, but interlayer spacing may vary with the specific cation. Because this mechanism is important in the alteration of one clay mineral to another, it is discussed more fully later.

MEASUREMENT OF THE AMOUNT OF CHEMICAL WEATHERING THAT HAS TAKEN PLACE

Total chemical analysis is the most widely used way of determining the amount of chemical weathering that has taken place in a rock. There are several things one must know, or decide, when using such analyses. First, one must decide if the analysis is to be done on the whole material or on fractions thereof (clay or clay plus silt). This decision will be based on the purpose of the study—that is, is it most important to focus on total change or rather on the changes within the more reactive finer-grained fraction? Second, all data are presented as oxides, not because they appear that way but because the main balancing anion usually is oxygen. Third, sometimes one or two water percentages are given. The H_2O^- is merely that loosely held water that is lost at temperatures less than 105 °C and probably should not be reported. In contrast, H_2O^+ is termed crystal lattice water, but it really is the weight loss at higher temperatures due to the conversion of crystal lattice OHs to water vapor and oxygen gas. The amount of weight loss varies with clay mineral species, percent of clay minerals present, and temperature.[31] Fourth, because analyses are given in percentages, we are dealing with relative increases and decreases.

Values for the individual oxides vary as a function of many soil-forming processes (Ch. 5), and whether they are essential to the products formed, so only broad statements can be made, and all are relative to the parent rock. Silicon dioxide is almost always present in the parent material in amounts greater than needed to form clay minerals, so it commonly decreases. And Al_2O_3, which has very low solubility over the usual pH range, is essential to most clay minerals, so it commonly shows a relative increase. Iron in most rock-forming minerals is present in the Fe^{2+} form (reported as FeO), and on weathering in an oxidizing environment converts to various Fe^{3+}-bearing substances or makes up part of the clay minerals; the latter is reported in analyses as Fe_2O_3. Commonly, the Fe_2O_3/FeO ratio increases on weathering. A reducing environment will complicate the interpretation of iron values, however. One of the more insoluble constitutents, TiO_2, should show a relative increase. Of the major remaining elements (MnO_2, CaO, MgO, Na_2O, and K_2O), Mg is an essential part of a weathering product (Mg-chlorite); part of the K is tightly held in interlayer position by illite; Mn can form bluish black mottles; and Mg, Ca, K, and Na are the major exchangeable cations. Most of the latter are depleted in wet environments, but in dry environments, Ca can increase at depth if $CaCO_3$ has accumulated.

An example of chemical losses and gains is a highly weathered volcanic rock in northern California. Table 3-4 compares analytical data for fresh rock with data for various stages of weathered rock (saprolite), rated from 1 (least weathered) to 4 (most weathered). Note that all data are in percent, and therefore all one has are the relative changes that take place upon weathering, not absolute gains and losses. The obvious relative

Table 3-4
Chemical Analyses, Molar Ratios of Oxides, and Abrasion pH Values for Two Andesite Rocks and Their Weathered Products, Southern Cascade Range, California

	Hypersthene andesite	*Saprolite samples (increasing weathering→)*			
		1	*2*	*3*	*4*
SiO_2	57.0	45.7	41.3	38.9	39.9
Al_2O_3	16.7	22.3	29.1	31.6	31.8
Fe_2O_3	2.0	5.2	9.8	11.1	11.3
FeO	4.7	4.2	1.1	0.50	0.40
TiO_2	1.0	1.0	1.2	1.3	1.3
MnO	0.12	0.24	0.26	0.27	0.26
P_2O_5	0.15	0.07	0.03	0.03	0.03
CaO	7.2	4.8	1.2	0.62	0.43
MgO	6.1	7.2	2.7	1.4	0.27
Na_2O	3.1	1.2	0.86	0.79	0.62
K_2O	1.2	0.24	0.28	0.11	0.15
H_2O^+	0.35	7.4	11.8	12.6	12.9
Total	99.6	99.9	99.6	99.2	99.4
Molar sa ratio[†]	5.79	3.47	2.40	2.09	2.13
$\dfrac{\text{Molar sa ratio saprolite}}{\text{Molar sa ratio rock}}$		0.60	0.41	0.36	0.37
Molar ba ratio[‡] ($\times 10^{-3}$)	3.36	2.81	1.03	0.59	0.26
$\dfrac{\text{Molar ba ratio saprolite}}{\text{Molar ba ratio rock}}$		0.84	0.31	0.18	0.08
Abrasion pH	8.9	5.5	5.1	4.9	4.9

*Weight of each oxide (g) assuming that Al_2O_3 content remains constant on weathering.
**Gains and losses by weight (g) of each oxide obtained by subtracting the rock analysis from that of column 4a. Total weight is an approximation of that that remains from the weathering of 100 g of rock.

trends are the loss of silicon and gains of iron and aluminum upon weathering. It is commonly assumed that Al_2O_3 content does not change upon weathering because it is relatively insoluble at normal pHs, and much of it is tied up in the clay minerals that form. Thus, if one assumes a constant Al_2O_3 content, all constituents can be recalculated by multiplying by the factor

$$\% \, Al_2O_3 \text{ in fresh rock}/\% \, Al_2O_3 \text{ in weathered material}$$

This is done in Table 3-4 (columns 4a) for saprolites weathered to stage 4. Gains and losses can be determined by subtracting data in columns 4a from those of the fresh rock, and these are shown in columns 4a-R. The main error in this method is that the gains and losses depend upon Al_2O_3 content remaining constant on weathering, a condition not always attained.

A more accurate way of determining chemical weathering is by gains and losses in weight of material on a volume basis. Many times this cannot be done, but two examples from northern California serve to demonstrate

| | | | Saprolite samples (increasing weathering→) | | | | | |
4a*	4a-R**	Olivine andesite	1	2	3	4	4a*	4a-R**
20.9	−36.1	53.8	42.6	38.3	36.3	36.8	20.2	−33.6
16.7	0	16.9	21.7	30.3	30.5	30.7	16.9	0
5.9	+3.9	2.3	6.7	15.2	16.0	15.6	8.6	+6.3
0.21	−4.49	5.8	2.9	0.36	0.24	0.10	0.06	−5.74
0.68	−0.32	1.2	1.1	1.5	1.5	1.5	0.83	−0.37
0.14	+0.02	0.12	0.15	0.09	0.17	0.10	0.06	−0.06
0.02	−0.13	0.15	0.25	0.04	0.03	0.04	0.02	−0.13
0.23	−6.97	8.4	5.2	0.79	0.20	0.12	0.07	−8.33
0.14	−5.96	7.4	9.7	0.22	0.15	0.15	0.08	−7.32
0.33	−2.77	2.5	0.89	0.19	0.09	0.08	0.04	−2.46
0.08	−1.12	0.60	0.31	0.27	0.15	0.15	0.08	−0.52
6.8	+6.45	0.71	9.1	13.1	13.9	13.7	7.5	+6.79
52.13	−47.48	99.9	100.6	100.3	99.3	99.0	54.44	−45.44
		5.40	2.88	2.15	2.03	2.04		
			0.53	0.40	0.38	0.38		
		3.72	3.44	0.25	0.11	0.09		
			0.92	0.07	0.03	0.02		
		8.6	6.3	5.6	5.4	5.5		

†Molar sa ratio = Sio_2/Al_2O_3
‡Molar ba ratio = $(CaO + MgO + Na_2O + K_2O)/Al_2O_3$.

(Taken from Hendricks and Whittig,[29] Table 1.)

the usefulness of the method (Table 3-5). Because rock structure is retained in saprolite, it could be shown that there is little volume change in going from rock to saprolite. Hence, from data from the chemical analyses (Table 3-4) and the bulk densities for fresh material and for material of all weathering stages, the actual gains and losses in weight can be calculated (Table 3-5). It can be seen from this analysis that Al_2O_3 is depleted, and thus the assumption of a constant Al_2O_3 content must be reevaluated. It does appear fairly certain, however, that by assuming constant Al_2O_3 one can at least show the minimum changes that have occurred upon weathering (compare relative losses in columns 4a-R of Table 3-4 with that in columns 4-R of Table 3-5).

Table 3-5
Weight Change in Oxide Content upon Weathering for Two Andesite Rocks of the Southern Cascade Range, California*

	HYPERSTHENE ANDESITE	SAPROLITE (INCREASING WEATHERING→)				OLIVINE ANDESITE	SAPROLITE (INCREASING WEATHERING→)			
		1	2	3	4		1	2	3	4
SiO_2	157	77	48	43	42	143	82	51	47	46
Al_2O_3	46	38	34	35	34	45	42	40	39	39
Fe_2O_3	5.5	8.8	11	12	12	6.1	13	20	20	20
FeO	13	7.1	1.3	0.6	0.4	15	5.6	0.5	0.3	0.1
TiO_2	2.8	1.7	1.4	1.4	1.4	3.2	2.1	2.0	1.9	1.9
MnO	1.3	0.4	0.3	0.3	0.3	0.3	0.3	0.1	0.2	0.1
P_2O_5	0.4	0.1	0.03	0.03	0.03	0.4	0.5	0.05	0.05	0.05
CaO	20	8.1	1.4	0.7	0.5	22	20	1.1	0.3	0.2
MgO	17	12	3.1	1.6	1.0	20	19	0.2	0.3	0.1
Na_2O	8.5	2.0	1.0	0.9	0.4	6.6	1.7	0.3	0.1	0.1
K_2O	3.3	0.4	0.3	0.1	0.2	1.6	0.6	0.4	0.2	0.2
H_2O	1.0	13	14	14	14	1.6	18	17	18	17
Total Fe as Fe_2O_3	20	17	12	13	13	23	19	21	20	20

DIFFERENCES BETWEEN SAPROLITE (1, 2, 3, 4) AND PARENT ROCK (R)

	Hypersthene Andesite				*Olivine Andesite*			
	1 − R	2 − R	3 − R	4 − R	1 − R	2 − R	3 − R	4 − R
SiO_2	−80	−109	−114	−115	−61	−92	−96	−97
Al_2O_3	−8	−12	−12	−11	−3	−5	−6	−6
Fe_2O_3	+3.3	+5.5	+6.5	+6.5	+7	+14	+14	+14
FeO	−5.9	−11.7	−12.4	−12.6	−9	−15	−15	−15
TiO_2	−1.1	−1.4	−1.4	−1.4	−1.1	−1.2	−1.3	−1.3
CaO	−12	−19	−19	−19	−12	−21	−22	−22
MgO	−5	−14	−15	−16	−1	−20	−20	−20
Na_2O	−6.5	−7.5	−7.6	−8.1	−5	−6	−7	−7
K_2O	−2.9	−3.0	−3.2	−3.1	−1.0	−1.2	−1.4	−1.4
H_2O	+12	+13	+13	+13	+16	+15	+16	+15
Total Fe as Fe_2O_3	+3.2	−7.5	−7.6	−8.1	−3.4	−2.0	−2.3	−2.5

*All values in centigrams per cubic centimeter. The volume of the rock was shown to not change upon weathering to the various stages of saprolite.
(Taken from Hendricks and Whittig,[29] Table 2 and 3.)

In regional comparisons of many soils or rock weathering by chemical data, it is cumbersome to use total chemical analyses. Workers generally recalculate the data to a single number for these comparative studies. Molar ratios (percent oxide divided by molecular weight) provide the best data. Common molar ratios are as follows.[16,33]

Silica:alumina	SiO_2/Al_2O_3
Silica:iron	SiO_2/Fe_2O_3
Silica:sesquioxides	$SiO_2/(Al_2O_3 + Fe_2O_3)$
Silica:R_2O_3	$SiO_2/(Al_2O_3 + Fe_2O_3 + TiO_2)$
Bases:alumina	$(K_2O + Na_2O + CaO + MgO)/Al_2O_3$
Bases:R_2O_3	$(K_2O + Na_2O + CaO + MgO)/(Al_2O_3 + Fe_2O_3 + TiO_2)$
Parkers weathering index	$100\ (K_2O/0.25 + Na_2O/0.35 + CaO/0.7 + MgO/0.9)$
Reiche's weathering potential index	$100\ (\Sigma\ bases - H_2O)/(\Sigma\ bases + SiO_2 + R_2O_3)$
Reiche's product index	$100\ (SiO_2)/(SiO_2 + R_2O_3)$
Iron species ratio	Fe_2O_3/FeO

The first six ratios should decrease on weathering in a leaching environment, and some examples of these are given in Table 3-4. In some of the ratio's R_2O_3 is used as the sum of the most stable components in a weathering environment. One can use the above ratios for the parent material and for weathered materials or soil horizons to calculate ratios that express differences between the two. One such ratio is what Jenny[33] calls the leaching factor

$$\text{Leaching factor} = \frac{(K_2O + Na_2O)/SiO_2 \text{ of weathered horizon}}{(K_2O + Na_2O)/SiO_2 \text{ of parent material}}$$

Such ratios have the advantage of expressing six analytical values as a single number. To illustrate this method, the SiO_2:Al_2O_3 ratio and (CaO + MgO + Na_2O + K_2O):Al_2O_3 ratio for saprolite:rock are presented in Table 3-4.

Another example of data reduction using chemical analyses is the study of Colman[16] on weathering rinds formed on volcanic clasts contained in tills in the western United States. Most of the rinds are up to several millimeters thick, and the clasts were collected from soil depths of 20 to 50 cm. For a rind representing about 140,000 years of weathering, most oxides are depleted; the main increase is in H_2O^+ (Table 3-6). In this study TiO_2 is assumed constant, as it seems to be the most insoluble of all the elements released on weathering. These values and sums in Table 3-6 are equivalent to those remaining from the weathering of 100 g of rock. However, for comparison of many samples of different age, it might be best to use the normalized weights calculated by Colman. The ratios, molecular or normalized, display the expected trends with depletions of

Table 3-6
Weathering Data on a Weathering Rind (0–0.5 mm Depth) Compared to That of Fresh Rock (>2 mm depth) for a Clast in Till Near McCall, Idaho

*Weights assuming TiO_2 constant**

Interval (mm)	SiO_2	Al_2O_3	Fe_2O_3	FeO	MgO	CaO	Na_2O	K_2O	TiO_2	MnO	H_2O^+	Sum
0–0.5	16.3	8.1	7.3	2.0	1.3	1.5	0.1	0.2	2.2	0.5	4.4	43.9
>2	53.7	14.1	1.9	9.8	4.2	8.1	2.8	1.5	2.2	0.1	1.2	99.6

*Normalized weights assuming TiO_2 constant***

Interval (mm)	SiO_2	Al_2O_3	Fe_2O_3	FeO	MgO	CaO	Na_2O	K_2O	TiO_2	MnO	H_2O^+	
0–0.5	0.3	0.6	3.9	0.2	0.3	0.2	0.1	0.2	1.0	0.5	3.6	
>2	1.0	1.0	1.0	1.0	1.0	1.0	1.0	1.0	1.0	1.0	1.0	

Molecular ratios

Interval (mm)	SiO_2/R_2O_3	Bases/R_2O_3	Fe_2O_3/FeO	Parker's index	WPI	PI
0–0.5	1.6	0.4	1.6	11.7	−35.7	62.0
>2	3.6	1.3	0.1	32.3	16.8	78.4

Normalized molecular ratios

Interval (mm)	SiO_2/R_2O_3	Bases/R_2O_3	Fe_2O_3/FeO	Parker's index	WPI	PI
0–0.5	0.5	0.3	18.8	0.4	−2.1	0.8
>2	1.0	1.0	1.0	1.0	1.0	1.0

*Calculated by multiplying weight percentage by the ratio % TiO_2 (fresh rock)/% TiO_2 (weathered sample).
**Calculated by dividing the value for each interval by the value for that oxide in the fresh rock (here taken as the >2-mm-depth material).
(Taken from Colman,[16] Tables 11, 14, 15, and 16.)

SiO_2 and bases and an increase in the iron ratio. Parker's[45] weathering index is a bit different from many indices in that each oxide amount is divided by a number related to the strengths of the cation-oxygen bond. Loss of bases results in a decrease in the index. The weathering potential index and product index of Reiche[47] incorporate all the major elements into two numbers. Colman considers them the most useful indices, especially when they are plotted together on one graph (Fig. 3-8). Both indices decrease on weathering, and the decrease of the former index is more rapid than the decrease of the latter because the former reflects not only the leaching of bases, but also the increase in crystal-lattice water on weathering. Although many of the indices that have been mentioned were initially intended for use in rock weathering studies, they can also be used to depict trends in total chemical analyses of soils. Still more indices are suggested by Wakatsuki and others.[62]

The abrasion pH of weathered materials can be used as a rough measure of the weathering that has taken place. Grant[26] has shown that

$$\text{Abrasion pH} = f \frac{Na + K + Ca + Mg}{\text{Clay minerals}}$$

These are compared with various molar ratios in Table 3-4, and it is seen that as the various molar ratios decrease, so too does the abrasion pH. One could also combine abrasion pHs into ratios of weathered material:rock, in much the same way that various oxides can be combined.

Oxide ratios can be calculated on the basis of either weathered rock:parent material or weathering by-product:parent material. The former ratio gives the overall changes in the parent material. The latter, however, gives the direction in which the weathering reactions are going, as reflected in the composition of the by-product. This is especially true for the $SiO_2:Al_2O_3$ ratio because this ratio differs with the clay mineral that forms.

CHEMICAL WEATHERING DISGUISED AS PHYSICAL WEATHERING

In many field situations, it is difficult to determine quantitatively the amount of weathering due to physical processes relative to that due to chemical processes. Indeed, if visual evidence for chemical weathering is lacking, one usually is tempted to look for evidence supporting some physical weathering process. The weathering may be chemical, however, and the evidence for it is either quite subtle or has been removed from the site of weathering.

One common example is the formation of granitic grus meters or tens of meters thick. In some localities, there is little visual evidence for chem-

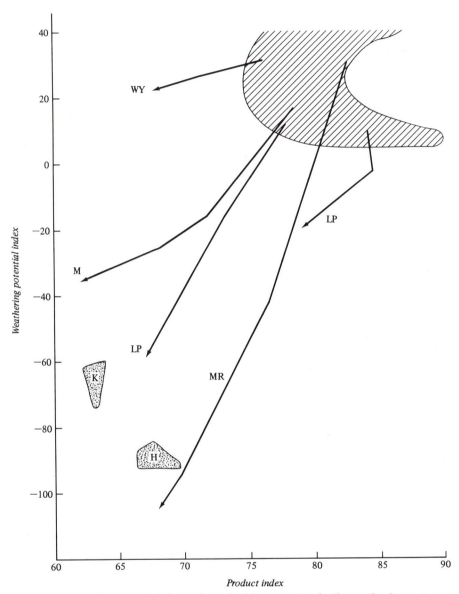

Fig. 3-8 Weathering-rind data plotted with respect to both weathering potential index and product index: IR, field for igneous rocks; K, kaolinite, and H, halloysite (Reiche[47]). The trend from the unaltered rock to the outermost part of the weathering rind is shown by the position of the arrow. All rinds are about 140,000 years old or older. Rocks from West Yellowstone, Montana (WY) and McCall, Idaho (M) are basalt; those from Mt. Rainier, Washington (MR) and Lassen Peak, California (LP) are andesites. (Taken from Colman,[16] Fig. 24.)

ical weathering, such as clay formation or iron oxidation, and the problem confronting many workers is just how is this deep weathering accomplished. The answer lies in the examination of the rock and constitutent minerals under a petrographic microscope or by X-ray. Wahrhafitig[61] studied the origin of grus in the Sierra Nevada and was able to show that the granitic rock was shattered to considerable depth by the expansion of several minerals upon alteration. Some biotite and some plagioclase had altered to clay. Volume increase accompanying the alteration was sufficient to fracture the surrounding minerals, as seen in thin section, and this brought about the disaggregation of a great thickness of rock. Other work confirms this origin for grus.[30] For example, biotite alteration was found to be important in the formation of grus in Wyoming.[22] In this case, however, the biotite was initially altered during Precambrian, high-temperature oxidation. Later weathering in a near-surface environment exploited the already altered biotite grains, causing them to alter further, expand, and internally shatter the rock to grus. In Australia, hydrothermal cracking, followed by volume increase during biotite and plagioclase weathering seems to account for the formation of grus at least 13 m thick.[20] Grus formation in southern California is also attributed to biotite alteration.[40] Although biotite alteration was not always seen with the petrographic microscope, it could be deciphered by X-ray. As an example of the amount of expansion possible, it was pointed out that the complete alteration of biotite to vermiculite can be accompanied by a 40 percent volume increase, and some data were presented to indicate that this had happened in places.

Biotite-induced shattering of rocks opens up a whole list of possibilities for investigation into weathering processes. One can no longer rely on field evidence alone to estimate the weathering process involved in rock breakdown. Thin-section study of the rock is mandatory, and this may lead to X-ray studies as well as polished-section study under high magnification.

Another common example of weathering that might be attributed to physical processes is that which goes on at high altitudes. For example, stones lying on the surface of Holocene deposits above timber line in the Colorado Front Range are deeply pitted (Fig. 3-4). Commonly, the felsic mineral bands of coarse-grained gneisses form depressions, and the mafic mineral bands stand in relief. The climate is very rigorous, with long cold winters and short cool summers. Because many of the rock surfaces appear fresh, and there is little visual evidence for chemical weathering, one might suspect that most of the weathering is due to physical processes. However, fine-grained igneous and metamorphic rocks in the Colorado Front Range and in other parts of the Colorado Rockies, presumably of the same age or older, show considerable development of weathering rinds that result from chemical weathering (Fig. 3-9). Thus,

Fig. 3-9 Weathering rind, 1 cm thick, developed from a granitic rock on the surface of a late-Wisconsin rock glacier, western Colorado.

chemical weathering of exposed rocks goes on under the rigorous climatic conditions of high altitudes. This relationship suggests that the coarse-grained, pitted rocks (Fig. 3-4) also are undergoing chemical weathering, that this weathering loosens the bonds between mineral grains, and that once the grains are loosened, physical processes remove them from the rock surface. The evidence, therefore, for chemical weathering does not remain on the rock surface, which can always have a fresh appearance. In contrast, chemical weathering of fine-grained rocks results in a weathering rind that remains intact on the rock surface and thus provides the basic evidence for chemical weathering.

RATE OF PRESENT-DAY CHEMICAL WEATHERING

It would be helpful to have a measure of regional rates of chemical weathering in near-surface environments. Tombstone weathering studies can provide data for above-ground processes.[39] Experiments can be devised to determine rates.[60] For example, packets of crushed rock can be placed in the ground in different environments and weight loss determined over various time spans.[12] Clay formation in dated soils could also be used, but again we are hindered by the lack of data. Furthermore, clays form so slowly in some environments that thousands of years may have to elapse before clay formation is noticeable. Another complication is that the clays are a by-product of weathering and thus represent only part of the material released by weathering. Finally, some clay can be eolian and not related to weathering.[17]

One approach to estimating the present-day rate of chemical weathering is to examine the chemistry of surface waters because many of the ions in water come from weathering reactions. The system we are dealing with is an exceedingly complex one (Fig. 3-10). The main source of ions

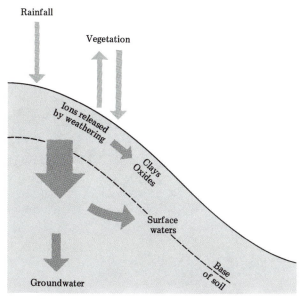

Fig. 3-10 Hypothetical flow of ions in a near-surface environment. Arrow width approximates the ion content involved in the reaction or transfer.

in the soil solution is release upon mineral weathering or organic matter decomposition. Rainfall will add some ions, but this amount can be determined and subtracted from the overall ion content. Some ions will cycle through the biosphere, but it is probable that the amounts gained from and lost to the soil soon reach a steady state. Analyses of waters that have moved through the soil should provide the best data.[28,37,66] Of the water that moves through the soil, part goes to groundwater reservoirs and part becomes surface-water runoff. Assuming that the dissolved load in surface water is proportional to the rate of chemical weathering, one might be able to use stream dissolved-load data to roughly estimate chemical weathering in weight loss per unit area per year.

An example of the above is the mass-balance study of Cleaves and others[14] in a 38.5-hectare watershed in Maryland. Data were collected for a 2.25-year period in the late 1960s. Ideally, the inputs of dissolved solids (precipitation plus that from weathering) should equal the outputs of dissolved solids (runoff, storage, and biomass) at the steady state (Table 3-7). Such equivalence could not be proven because biomass transfers were not measured, but instead taken as the algebraic sum of the other four quantities. Probable sources for dissolved solids in precipitation are pollution (for example, the large amount of SO_4^{2-}) and the sea. Dissolved solids from weathering were estimated from knowledge of the minerals being weathered and the weathering products (see, for example,

Table 3-7
Geochemical Balance for 2.25 Years for a Small Watershed in
Maryland*

	Precipi- tation		Weather- ing		Runoff		Storage		Bio- mass
SiO_2	1.04	+	23.54	=	19.00	+	5.58	+	0
Na^+	2.13	+	2.62	=	3.51	+	1.24	+	0
K^+	1.68	+	2.30	=	1.95	+	0.54	+	1.48
Ca^{2+}	3.16	+	1.28	=	2.99	+	0.35	+	1.11
Mg^{2+}	1.04	+	1.14	=	1.83	+	0.35	+	0
HCO_3^-	3.16	+	28.65	=	16.72	+	3.73	+	11.36
Cl^-	6.30	+	—	=	4.42	+	2.00	−	0.12
SO_4^{2-}	17.83	+	—	=	3.93	+	0.69	+	27.05
H^+	0.27								

*Units, kg/hectare/year.
(Taken from Cleaves and others,[14] Table 8.)

Table 3-1). In addition, Cleaves and others were able to estimate the dissolved solids released by each of the weathering reactions and the amounts of silicate minerals consumed and formed for that period of time (Table 3-8). Finally, although data for the short term indicate that chemical denudation is five times that of mechanical denudation, they suggest that both are equal in the long term. At the calculated rates, the present landscape could have formed in about 1.5 million years.

The above study is an excellent example of the integration of several disciplines to help understand the complex physical, chemical, and biological processes operating in a watershed. Many assumptions had to be made, however, so the results are only approximations. Other studies are reviewed by Drever.[21] Clayton[13] also reviewed these and other methods related to the rate of release of dissolved solids from rock weathering. The reason for the latter study was that data are needed on the rate of nutrient supply from rock weathering so that forests can be harvested without depleting the nutrients required by the trees.

Strakhov[57] has complied dissolved-load data for many rivers of the world, and from these he has calculated chemical denudation. River basins differ markedly in their rate of chemical denudation (Table 3-9), and this variation, no doubt, is influenced by many factors. Some important factors are climate, susceptibility of the rocks to weathering, and length of time the water is in contact with the weathering material. Although Strakhov[57] does not discuss these factors in detail, he does show that significant chemical denudation occurs in many diverse environments, including those characterized by low temperatures, and that in some areas chemical denudation is more rapid than mechanical denudation. The idea of high values for chemical weathering and denudation in

Table 3-8
Amounts of Dissolved Solids Produced, and Minerals Consumed
and Produced by Weathering in a Small Watershed in Maryland

DISSOLVED SOLIDS (kg)

Reaction	SiO_2	Na^+	Ca^{2+}	Mg^{2+}	K^+	HCO_3^-
Oligoclase to kaolinite	1195	229	112	..	..	948
Biotite to vermiculite	408	..	..	61	199	620
Vermiculite to kaolinite	122	..	..	37	..	915
Kaolinite to gibbsite	314	..	..	..	..	..
Total	2039	229	112	98	199	2483

SILICATE MINERALS CONSUMED (MINUS SIGN) OR FORMED (PLUS SIGN) (KMOLES)

Reaction	olig	bio	vermic	ka	gb
Oligoclase to kaolinite	−12.8	..	..	+7.8	..
Biotite to vermiculite	..	−2.8	+1.7	..	..
Vermiculite to kaolinite	..	..	−1.0	+1.5	..
Kaolinite to gibbsite	..	..	..	−2.6	+5.2
Total	−12.8	−2.8	+0.7	+6.7	+5.2

*Data are for the entire watershed for the 2.25-year period.
(Taken from Cleaves and others,[14] Table 9.)

mountainous, glacierized areas is supported by dissolved solid data for a watershed in the northern Cascade Mountains[48] and by extractive chemistry and mineralogy of soils in the Southern Alps, New Zealand (Ch. 8). Finally, on a continental basis, mechanical denudation seems to exceed chemical denudation for most continents (Table 3-10).

The point to this general discussion is to show that one might be able to make rough estimates on regional rates of chemical weathering. The important factors of soil formation and of weathering have to be taken into careful consideration, however, so that one is fairly certain of the influence of each. For example, the influence of rock type could be held constant by studying only those basins underlain with granitic rocks. The time factor could be held reasonably constant by restricting the study to areas characterized only by weakly or moderately developed soils. One could devise ways of keeping the other factors reasonably constant. Dissolved-load data then could be compared with variation in regional climates to determine if regions do differ in their rate of chemical weathering. Of course, this analysis carries with it the assumption that there is a strong correlation between the dissolved load of a stream and the rate of weathering in a near-surface environment, say, in a soil. Unfortunately, this ideal situation might not always obtain. Finally, many studies suffer because soil processes are not adequately considered.

Table 3-9
Data on Annual Mechanical and Chemical Denudation for Various Rivers of the World

RIVER	MECHANICAL DENUDA-ATION (TONS/KM^2)	CHEMICAL DENUDA-ATION (TONS/KM^2)	RIVER	MECHANICAL DENUDA-ATION (TONS/KM^2)	CHEMICAL DENUDA-ATION (TONS/KM^2)
I. Northern rivers of temperate and cold climates			*II. Rivers of temperate-hot, subtropical, and tropical climates*		
MOUNTAIN RIVERS			PLAINS RIVERS		
Kolyma	7	5.5	Dnieper	4.0	17.0
Yana	10	3.9	Southern Bug	11.5	14.0
Pechora	20	17.0	Don	18.3	22.0
Indigirka	24	11.0	Ural	18.6	15.0
Amur	28	10.1	Volga	18.6	32.5
Yukon	103	22.0	Dniester	31.5	3.5
PLAINS RIVERS			RIVERS STARTING IN MOUNTAINS OR LARGE UPLANDS		
Neva	3.9	10.0			
Yenisei	4.0	11.4	Kuma	33	9.0
Luga	4.0	17.0	Kalaus	40	5.0
Narova	4.0	17.7	Amazon	60	13.0
Onega	4.0	20.0	La Plata	75	18.0
Ob	6.0	12.2	Mississippi	118	28.4
Western Dvina	6.0	25.0	Kuban	180	35.0
Mezen	10.0	16.0	MOUNTAIN RIVERS		
Northern Dvina	16.5	48.0	Kura	213	23.4
			Amu Darya	440	78.1
			Rivers of south-eastern Asia (average)	390	93.0
			Terek	587	125
			Rion (Rioni)	2000	209
			Samur	1700	270
			Sulak	2000	290

(Taken from Strakhov.[57])

REFERENCES

1. Arkley, R.J., 1963, Calculation of carbonate and water movement in soil from climatic data: Soil Sci., v. 96, p. 239–248.
2. Barshad, I., 1964, Chemistry of soil development, p. 1–70 *in* F.E. Bear, ed., Chemistry of the soil: Reinhold Publ. Corp., New York, 515p.
3. Berner, R.A., and Holden, G.R., Jr., 1977, Mechanism of feldspar weathering: Some observational evidence: Geology, v. 5, p. 369–372.

Table 3-10
Annual Chemical and Mechanical Denudation of the Continents

CONTINENT	ANNUAL CHEMICAL DENUDATION *(metric tons/km²)*	ANNUAL MECHANICAL DENUDATION *(metric tons/km²)*	RATIO MECHANICAL TO CHEMICAL
North America	33	86	2.6
South America	28	56	2.0
Asia	32	310	9.7
Africa	24	17	0.7
Europe	42	27	0.65
Australia	2	27	>10.0
Total			Overall 4.7

(Taken from Garrels and MacKenzie,[24] Table 5.1.)

4. Birkeland, P.W., 1973, Use of relative age-dating methods in a stratigraphic study of rock glacier deposits, Mt. Sopris, Colorado: Arctic Alp. Res., v. 5, p. 401–416.

5. Black C.A., 1957, Soil-plant relationships: John Wiley and Sons, New York, 332p.

6. Blackwelder, E.B., 1927, Fire as an agent in rock weathering: Jour. Geol., v. 35, p. 134–140.

7. ———, 1933, The insolation hypothesis of rock weathering: Amer. Jour. Sci., v. 226, p. 97–113.

8. Bloom, A.L., 1978, Geomorphology, a systematic analysis of late Cenozoic landforms: Prentice-Hall, Inc., Englewood Cliffs, N.J., 510 p.

9. Bohn, H.I., McNeal, B.L. and O'Connor, G.A., 1979, Soil chemistry: John Wiley and Sons, New York, 329 p.

10. Bradley, W.C., 1963, Large-scale exfoliation in massive sandstones of the Colorado Plateau: Geol. Soc. Amer. Bull., v. 74, p. 519–528.

11. Bradley, W.C., Hutton, J.T., and Twidale, C.R., 1978, Role of salts in development of granitic tafoni, South Australia: Jour. Geol., v. 86, p. 647–654.

12. Caine, N., 1979, Rock weathering rates at the soil surface in an alpine environment: Catena, v. 6, p. 131–144.

13. Clayton, J.C., 1979, Nutrient supply to soil by rock weathering, p. 75–96 *in* Proceedings, Impact of intensive harvesting on forest nutrient cycling: State Univ. of New York, Syracuse, N.Y.

14. Cleaves, E.T., Godfrey, A.E., and Bricker, O.P., 1970, Geochemical balance of a small watershed and its geomorphic implications: Geol. Soc. Amer. Bull., v. 81, p. 3015–3032.

15. Coleman, J.M., Gagliano, S.M., and Smith, W.G., 1966, Chemical and physical weathering on saline tidal flats, Northern Queensland, Australia: Geol. Soc. Amer. Bull., v. 77, p. 205–206.

16. Colman, S.M., 1982a, Chemical weathering of basalts and andesites: Evidence from weathering rinds: U.S. Geol. Surv. Prof. Pap. 1246, 51 p.

17. ———— 1982b, Clay mineralogy of weathering rinds and possible implications concerning the sources of clay minerals in soils: Geology, v. 10, p. 370–375.

18. Cooke, R.U., and Smalley, I.J., 1968, Salt weathering in deserts: Nature, v. 220, p. 1226–1227.

19. Curtis, C.D., 1976, Chemistry of rock weathering: Fundamental reactions and controls, p. 25–57 *in* E. Derbyshire, ed., Geomorphology and climate: John Wiley and Co., New York.

20. Dixon, J.C., and Young, R.W., 1981, Character and origin of deep arenaceous weathering mantles on the Bega batholith, southeastern Australia: Catena, v. 8, p. 97–109.

21. Drever, J.I., 1982, The geochemistry of natural waters: Prentice-Hall, Inc., Englewood Cliffs, N.J. 388 p.

22. Eggler, D.H., Larson, E.E., and Bradley, W.C., 1969, Granites, grusses, and the Sherman erosion surface, southern Laramie Range, Wyoming: Amer. Jour. Sci., v. 267, p. 510–522.

23. Evans, I.S., 1969, Salt crystallization and rock weathering: A review: Rev. Geomorph. Dyn., v. 19, p. 153–177.

24. Garrels, R.M., and MacKenzie, F.T., 1971, Evolution of sedimentary rocks: W.W. Norton and Co., Inc., New York, 397 p.

25. Goudie, A., Cooke, R., and Evans, I., 1970, Experimental investigation of rock weathering by salts: Inst. British Geographers (London) Area no. 4, p. 42–48.

26. Grant, W.H., 1969, Abrasion pH, an index of weathering: Clays and Clay Minerals, v. 17, p. 151–155.

27. Griggs, D.T., 1936, The factor of fatigue in rock exfoliation: Jour. Geol., v. 44, p. 781–796.

28. Hay, R.L., and Jones, B.F., 1972, Weathering of basaltic tephra on the island of Hawaii: Geol. Soc. Amer. Bull., v. 83, p. 317–332.

29. Hendricks, D.M., and Whittig, L.D., 1968, Andesite weathering II. Geochemical changes from andesite to saprolite: Jour. Soil Sci., v., 19, p. 147–153.

30. Isherwood, D., and Street, A., 1976, Biotite-induced grussification of the Boulder Creek Granodiorite, Boulder County, Colorado: Geol. Soc. Amer. Bull., v. 87, p. 366–370.

31. Jackson, M.L., 1973, Soil chemical analysis—advanced course: Dept. Soil Sci., Univ. of Wisconsin, Madison, Wisc., 894 p.

32. Jackson, T.A., and Keller, W.D., 1970, A comparative study of the role of lichens and "inorganic" processes in the chemical weathering of recent Hawaiian lava flows: Amer. Jour. Sci., v. 269, 446–466.

33. Jenny, H., 1941, Factors of soil formation: McGraw-Hill, New York, 281 p.

34. ————, 1950, Origin of soils, p. 41–61 *in* P.D. Trask, ed., Applied sedimentation: John Wiley and Sons, New York, 707 p.

35. Krauskopf, K.B., 1967, Introduction to geochemistry: McGraw-Hill, New York, 721 p.

36. Lehman, D.S., 1963, Some principles of chelation chemistry: Soil Sci. Soc. Amer. Proc., v. 27, p. 167–170.

37. Likens, G.E., Bormann, F.H., Pierce, R.S., Eaton, J.S., and Johnson, N.M., 1977, Biogeochemistry of a forested ecosystem: Springer-Verlag, New York, 146 p.

38. Loughnan, F.C., 1969, Chemical weathering of the silicate minerals: American Elsevier Publ. Co., Inc., New York, 154 p.

39. Meierding, T.C., 1981, Marble tombstone weathering rates: A transect of the United States: Physical Geog. v. 2, p. 1–18.

40. Nettleton, W.D., Flach, K.W., and Nelson, R.E., 1970, Pedogenic weathering of tonalite in southern California: Geoderma, v. 4, p. 387–402.

41. Nixon, R.A., 1979, Differences in incongruent weathering of plagioclase and microcline—cation leaching versus precipitates: Geology, v. 7, p. 221–224.

42. Ollier, C.D., 1969, Weathering: Oliver and Boyd, Edinburgh, 304 p.

43. Parker, A., 1970, An index of weathering for silicate rocks: Geol. Mag., v. 107, p. 501–504.

44. Peel, R.F., 1974, Insolation weathering: Some measurements of diurnal temperature changes in exposed rocks in the Tibesti region, central Sahara: Zeitschrift für Geomorph. Supplementband 21, p. 19–28.

45. Petrović, R., Berner, R.A., and Goldhaber, M.B., 1976, Rate control in dissolution of feldspars—1. Study of residual feldspar grains by X-ray photoelectric spectroscopy: Geochim. et Cosmochim. Acta, v. 40, p. 537–548.

46. Ponomareva, V.V., 1969, Theory of podzolization: Israel Program for Scientific Translations, Jerusalem, 309 p.

47. Reiche, P., 1943, Graphical representation of chemical weathering: Jour. Sed. Petrol., v. 13, p. 58–68.

48. Reynolds, R.C., Jr., and Johnson, N.M., 1972, Chemical weathering in the temperate glacial environment of the northern Cascade Mountains: Geochim. et Cosmochim. Acta, v. 36, p. 537–554.

49. Rice, A., 1976, Insolation warmed over: Geology, v. 4, p. 61–62.

50. Rode, A.A., 1962, Soil science: Israel Program for Scientific Translations, Jerusalem, 517 p.

51. Schalascha, E.B., Appelt, H., and Schatz, A., 1967, Chelation as a weathering mechanism—I. Effect of complexing agents on the solubilization of iron from minerals and granodiorite: Geochim. et Cosmochim. Acta, v. 31, p. 587–596.

52. Schatz, A., 1963, Chelation in nutrition, soil microrganisms and soil chelation. The pedogenic action of lichens and lichen acids: Jour. Agri. and Food Chem., v. 11, p. 112–118.

53. Schwertmann, U., and Taylor, R.M., 1977, Iron oxides, p. 145–180 *in* J.B. Dixon and S.B. Weed, eds., Minerals in soil environments: Soil Sci. Soc. Amer., Madison, Wisconsin.

54. Siever, R., and Woodward, N., 1979, Dissolution kinetics and the weathering of mafic minerals: Geochim. et Cosmochim. Acta, v. 43, p. 717–724.

55. Smith, B.J., 1977, Rock temperature measurements from the northwest Sahara and their implications for rock weathering: Catena, v. 4, p. 41–63.

56. Stevens, R.E., and Carron, M.K., 1948, Simple field test for distinguishing minerals by abrasion pH: Amer. Mineralogist, v. 33, p. 31–49.

57. Strakhov, N.M., 1967, Principles of lithogenesis: Oliver and Boyd, London, 245 p.

58. Thorn, C.E., 1979, Bedrock freeze-thaw weathering regime in an alpine environment, Colorado Front Range: Earth Surf. Proc., v. 4, p. 211–228.

59. Todd, T.W., 1968, Paleoclimatology and the relative stability of feldspar minerals under atmospheric conditions: Jour. Sed. Petrol., v. 38, p. 832–844.

60. Trudgill, S.T., 1976, Rock weathering and climate: quantitative and experi-

mental aspects, p. 59–99 *in* E. Derbyshire, ed., Geomorphology and climate: John Wiley and Co., New York.

61. Wahrhaftig, C., 1965, Stepped topography of the southern Sierra Nevada, California: Geol. Soc. Amer. Bull., v. 76, p. 1165–1190.

62. Wakatsuki, T., Furukawa, H., and Kyuma, K., 1977, Geochemical study of the redistribution of elements in soil—1. Evaluation of degree of weathering of transported soil materials by distribution of major elements among the particle size fractions and soil extract: Geochim. et Cosmochim. Acta, v. 41, p. 891–902.

63. White, S.E., 1976, Is frost action really only hydration shattering? A review: Arc. Alp. Res., v. 8, p. 1–6.

64. Winkler, E.M., 1965, Weathering rates as exemplified by Cleopatra's Needle in New York City: Jour. Geol. Educ., v. 13, p. 50–52.

65. ———, 1975, Stone: properties, durability in man's environment: Springer-Verlag, New York, 230 p.

66. Wolff, R.G., 1967, Weathering of Woodstock granite near Baltimore, Maryland: Amer. Jour. Sci., v. 265, p. 106–117.

67. Wollast, R., 1967, Kinetics of the alteration of K-feldspar in buffered solutions at low temperature: Geochim. et Cosmochim. Acta, v. 31, p. 635–648.

4

The products of weathering

Materials released during weathering either are removed from the system in leaching water or react in the system to form a variety of crystalline and amorphous products. The most commonly observed reaction products are the clay minerals and hydrous oxides of aluminum and iron. These products can occur alone or in combination, and their distribution with depth can be uniform or highly variable. Characterization of these products is important because, of all the properties of a soil, they probably best reflect the long-term effect of the chemical and leaching environment of the soil. Their genesis, however, is varied; it can range from relatively simple ion exchange reactions to the more complex combination of aluminum and silicon from solution or a gel to form a crystalline clay mineral. Now it has been possible to construct phase diagrams to test for mineral-water equilibria. This approach may prove to be very useful for predicting stable products of weathering.

These minerals are generally so finely divided that their identification is difficult; it can be accomplished, however, by a variety of chemical, thermal, X-ray, and electron microscope techniques.[14,72] Only those products most commonly found in soils will be dealt with here.

CLAY MINERALS

The common clay minerals are hydrated silicates of aluminum, iron, and magnesium arranged in various combinations of layers.[14,21,32,39,51] They are called layer silicates or phyllosilicates. Two kinds of sheet structures, the tetrahedral and the octahedral, make up the clay minerals, and variations in combinations of these structures and in their chemical makeup give rise to the multitude of clay minerals. The basic difference between these two sheets is in the geometrical arrangement of Si, Al, Fe, Mg, O, and OH. The arrangements differ with the cation because it is the size of the cation that determines how many O or OH ions surround it.

The silicon tetrahedron is the basic structural unit of the tetrahedral sheet. It consists of one Si^{4+} surrounded by four O ions (Fig. 4-1). A sheet is formed when the basal O ions are shared between adjacent tetrahedra and the remaining unshared O ions all point in the same direction. These sheets have a net negative charge.

In the octahedral sheet, the basic unit consists of the octahedral arrangement of six O or OH ions about a central Al, Mg, or Fe ion (Fig. 4-1). In the octahedral sheet, the O and OH ions are shared between adjacent octahedra. If a trivalent ion is the central cation, two-thirds of the available cation sites are filled and the structure is said to be dioctahedral; in contrast, divalent cations will fill all available sites and exhibit trioctahedral structure.

The variety of clay minerals results in part from different arrangements of the tetrahedral and octahedral sheets. The two adjacent sheets are bound together tightly by the mutual sharing of the O ions of the tetrahedral sheet. Thus in the octahedral sheet of a clay mineral, the six ions surrounding the central cation include both O and OH ions. Three structural groupings of the two basic sheets are recognized. The first consists of a tetrahedral sheet attached to one side of an octahedral sheet to form a 1:1 layer phyllosilicate. The second is the symmetrical arrangement of two tetrahedral sheets about a central octahedral sheet to give a 2:1 layer phyllosilicate. A third arrangement is the presence of an octahedral sheet between adjacent 2:1 layers, and these are known as 2:1:1 layer phyllosilicates. In a clay mineral, each layer combination, hereafter

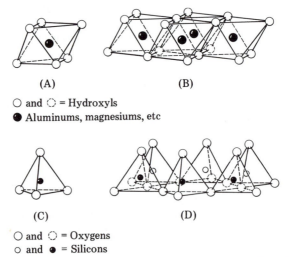

(A)   (B)

○ and ◌ = Hydroxyls
● Aluminums, magnesiums, etc

(C)   (D)

○ and ◌ = Oxygens
o and ● = Silicons

Fig. 4-1   Diagram showing (A) octahedral unit, (B) octahedral sheet, (C) tetrahedral unit, and (D) tetrahedral sheet. (Taken from Grim,[32] Figs. 4-1 and 4-2.)

called a layer, extends for varying distances in the direction of the a and b crystallographic axes.

The 1:1 and 2:1 layers are stacked in the c-axis direction, and bonds between the layers are formed by more than one mechanism. One mechanism involves a bonding of the external O ions of one layer with the external OH ions of the adjacent layer. Bonding probably is due to the mutual sharing of the H ion between the O ions of both sheets to form a hydrogen bond. Bonding can also occur with polar water molecules forming a hydrogen bond between the two layers. The bonding here is as follows: OH ion (one sheet) to H ion (positive pole of water molecule) to OH ion (negative pole of water molecule) to H ion (adjacent sheet). Another mechanism for bonding involves isomorphous substitution within the clay lattice. Although silicon and aluminum are the more common cations, other cations of near similar size can replace them. The common substitutions are Al^{3+} for Si^{4+} in the tetrahedral sheet, and Fe^{3+}, Fe^{2+}, and Mg^{2+} for Al^{3+} in the octahedral sheet. These substitutions, with the exception of Fe^{3+} for Al^{3+}, are marked by an ion of lower valency substituting for one of higher valency, and the net result is a negative charge in the layer. This charge is balanced by some combination of substituting OH^- for O^{2-}, filling vacant cation sites in the octahedral sheets, and adsorbing hydrated cations at the edges of layers and between layers. Cations between layers can bond the layers together, especially if the opposing layers both have external O ions. The kind of interlayer bonding varies from clay mineral to clay mineral and can be correlated with some properties of the clay minerals.

Kaolinite and halloysite are the common 1:1 layer clay minerals.[20] They are structurally similar (Fig. 4-2), the only difference being that halloysite commonly has a layer of water between successive layers; it is then called hydrated halloysite. Hydrogen bonding with a water interlayer (hydrated halloysite), or without a water interlayer (kaolinite, halloysite), binds the adjacent layers together. Isomorphous substitution is negligible.

The smectite group of clay minerals make up a wide variety of 2:1 layer clay minerals (Fig. 4-2). Isomorphous substitution is common, and this gives rise to many minerals that differ in chemical composition. Although Al^{3+} does substitute for Si^{4+}, the more common substitution is Fe^{2+} and Mg^{2+} for Al^{3+} in the octahedral sheet. The resulting net negative charge is partly balanced by interlayer hydrated cations that bond adjacent layers. Because the negative charge is due mainly to substitution in the octahedral layer, and because that layer is some distance from the interlayer-bonding cation, bonding is relatively loose. Therefore, exchange of cations and water layers readily occurs. These minerals expand or contract as water layers are added or removed, and the amount of such change depends, in part, upon the cation present.

Three forms of smectite occur in soils, based on their composition.[12] Montmorillonite is the Mg variety of smectite, with both Al and Mg (Al

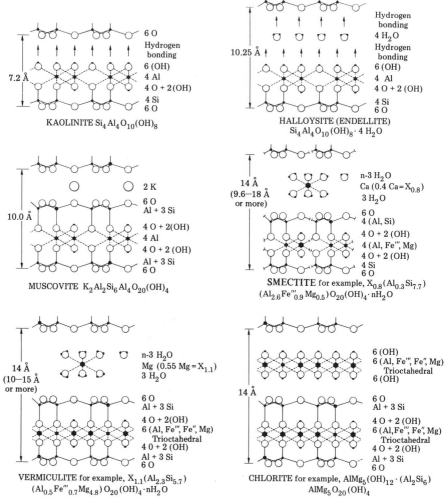

Fig. 4-2 Crystal structure of the common clay minerals. Illite structure is similar to that of muscovite. The *b* crystallographic axis is horizontal and the *c* crystallographic axis is vertical in these diagrams. Taken from Jackson,[39] Figs. 2.3, 2.8, 2.9, 2.10, 2.14, and 2.15 *in* Chemistry of the Soil by F. E. Bear, ed., © 1964 by Litton Educational Publishing, Inc. Reprinted by permission of Van Nostrand Reinhold Company.)

$>$ Mg) in the octahedral sheet. Beidellite is the Al variety, with Al in the octahedral sheet, and Al substituting for some Si in the tetrahedral sheet. Finally, nontronite is the Fe variety, with Fe in the octahedral sheet, and Al substituting for some Si in the tetrahedral sheet.

The illite group of 2:1 layer clay minerals are intermediate in composition between ideal smectite and muscovite[25] (Fig. 4-2). Some illites seem to be micas intermixed with layers of smectite, vermiculite, and chlorite. Although the basic structure is similar to that of montmorillonite, most of the isomorphous substitution is Al^{3+} for Si^{4+} in the tetrahedral sheet. The interlayer bonding cation is K^+, and because it is close to the site of the negative charge, bonds are especially strong, and interlayer expansion does not occur.

Vermiculite is also a 2:1 layer clay mineral[22] (Fig. 4-2). Like smectite, both dioctahedral and trioctahedral types are recognized, and like illite, the major isomorphous substitution is in the tetrahedral layer. This results in a fairly tight bond between the adjacent layers and the inter-layer hydrated cations, and thus interlayer expansion is limited.

Chlorite is the 2:1:1 clay mineral common to soils[5] (Fig. 4-2). It con-sists of alternating 2:1 layers and octahedral layers. The 2:1 layers are commonly trioctrahedral, and substitutions are mainly Al^{3+} for Si^{4+} in the tetrahedral sheet, which results in a negative charge. The octahedral interlayer contains Mg^{2+} and Fe^{2+} in addition to Al^{3+}, which results in a positive charge. Bonding between the two oppositely charged layers is strong, it is enhanced by hydrogen bonding, and expansion does not take place.

Mixed-layer minerals result from the interstratification of 2:1 and 2:1:1 layer clay minerals and aluminum hydroxide and magnesium hydroxide (gibbsite and brucite layers, respectively).[60] Interlayering is possible because many of the clay-mineral species are structurally similar, and because some expand readily in water or upon slight weathering. Mix-ing varies from a regular repetition of the components in the direction of the c-axis to a wholly random arrangement. Such mixed-layer clays are quite common, and their detailed identification is difficult.[14,21,39,55]

Two clay minerals have a fibrous morphology[74] in contrast to the platy morphology of the above clays; the one exception to the latter is halloysite, which exhibits a tubular morphology. Although the fibrous minerals have a 2:1 structure, the lateral extension of the layer is limited in one direc-tion, thus producing chains and a fibrous morphology. The two minerals of this group are palygorskite and sepiolite, with the former having near equal proportions of Al and Mg in the octahedral sheet, whereas the latter has mostly Mg. There is some substitution of Al for Si in the tetrahedral sheet.

Weaver and Pollard[71] provide details on the chemistry of the various clay minerals.

Various quantitative methods using X-ray diffraction data have been employed to determine the content of clay minerals in soils and sedi-ments.[59,61,62] Unfortunately, many commonly used methods do not give comparable results.[53] Attainment of meaningful results within 1 percent seems unlikely, and parts in ten might be a more reasonable goal. An alternate procedure would be to report results in peak heights or peak areas, or in qualitative terms such as abundant and moderately abundant, accompanied by sketches of representative diffractograms.

CRYSTALLINE AND NON-CRYSTALLINE COMPOUNDS OF ALUMINUM AND IRON

Products that form in soils can be crystalline, such as the clay minerals mentioned above, or amorphous, with no regular arrangement of atoms.[70] It has been shown, however, that there is short-range atomic order in some seemingly amorphous materials and the term non-crystalline is suggested for such materials. Obviously, identification of the non-crystalline materials is difficult, in contrast to the relative ease of identifying clay minerals by X-ray diffraction methods.

Allophane and imogolite are the common non-crystalline aluminosilicate materials.[70] Approximate compositions are $Al_2O_3 \cdot 2SiO_2 \cdot nH_2O$ for allophane and $Al_2O_3 \cdot SiO_2 \cdot 2.5H_2O$ for imogolite. Both form mainly from the weathering of volcanic ash in a variety of environments, which are usually relatively humid. Because these materials are less abundant in older (pre-Holocene) sediments and soils, it is thought that they are transitional products and that they transform to crystalline products with time.

Crystalline aluminum minerals are present in some environments. Common ones are gibbsite $(Al(OH)_3)$ and boehmite $(AlOOH)$.

Several iron minerals may form during weathering (Table 4-1); of the common minerals, only magnetite does not form in soil environments.[63] Goethite is the most common iron mineral in most well-drained soils with yellowish brown color. In contrast, lepidocrocite forms the orange mottles and bands in non-calcareous, clay-rich, gleyed soils associated with a high water table, reducing conditions, and Fe^{2+} ions. Hematite is the second most important iron mineral, and seems to be the pigment responsible for the red color of old soils and of soils formed under relatively high surface temperatures. Maghemite is common in soils formed from basic igneous rocks in climates ranging from temperate to tropical. Finally, ferrihydrite is most likely the Fe mineral present in Bs horizons, and is probably that material dissolved in the ammonium oxalate extraction procedure (Ch. 5).

Table 4-1
Characteristics of Iron Minerals Common in Soils

Mineral Name	Formula	Color
Goethite	α-FeOOH	7.5YR–10YR, yellowish-brown
Lepidocrocite	γ-FeOOH	5YR–7.5YR, orange
Hematite	α-Fe$_2$O$_3$	5R–2.5YR, bright red
Maghemite	γ-Fe$_2$O$_3$	Reddish-brown
Ferrihydrite	$Fe_5HO_8 \cdot 4H_2O$ or $Fe_5(O_4H_3)_3$	5YR–7.5YR, reddish-brown

(Taken from Schwertmann and Taylor,[63] Table 5-1.)

CATION EXCHANGE CAPACITY OF INORGANIC AND ORGANIC COLLOIDAL PARTICLES

Colloidal material in soils carries an electrical charge. Although both negative- and positive-charge sites exist, the origin of the negative sites has been studied most extensively.

The negative charge is approximated by the cation exchange capacity (CEC) of the material, and most of the cations attracted to these sites are exchangeable. The CEC varies in amount and origin with the soil material. In humus, the charge originates from the dissociation of H^+ from carboxyl and phenolic groups at the particle surface and within the particle. In clay particles, charge originates at the clay edge or along the surfaces parallel to the *ab* crystallographic plane.[15] Charges along the clay edge originate from unsatisfied (broken) bonds at the edge of the particle, say, between $Si-O$ or $Al-OH$, or from the dissociation of H^+ from OH^-. Charge originating in this manner is common in the 1:1 clays and depends on particle size because the number of exposed edges increases as particle size decreases. Ionic substitution in the clay lattice commonly results in a negative charge, mainly along the interlayer surfaces. This is probably the main origin of charge in the 2:1 clays, although some charge originates at the clay edge.

The CEC varies with the clay mineral, it is expressed in milliequivalents (me)* per unit dry weight of the material, and the highest values are for organic matter (Table 4-2). The variation with clay mineral is due to a combination of ionic substitution and its extent, the degree of hydration, and the number of exchange sites at the edges of particles. Because organic matter can have such a high CEC, the presence of a small amount of organic matter can greatly affect the CEC of the soil. In contrast, non-clay minerals and rock fragments have a negligible effect on the CEC; zeolites, because of their structure, are an exception. Thus, one can obtain a rough estimate of the CEC of the clay fraction, and hence of the possible clay minerals present, by knowing the CEC of the soil and the amount of clay present (in percent), and by sampling deep enough to avoid sizable amounts of organic matter. For example, if a soil has a CEC of 20 me/100 g soil and is 20 percent clay, the CEC for the clay fraction would be approximately 20 me/20 g clay, or 100 me/100 g clay. The clay mineral could be smectite or vermiculite (Table 4-2).

Anion exchange seems to be important to the overall charge of 1:1 clays mainly because it is thought to occur at clay edges. Grim[32] reviews

*One milliequivalent is defined as 1 mg of H^+ or the amount of any other cation that will displace it. If, for example, the CEC is 1 me/100 g oven-dry soil, 1 mg of H^+ is adsorbed. If Ca^{2+} displaces the H^+, the amount of Ca^{2+} has to be equivalent to 1 me of H^+; this amount can be calculated: Ca has an atomic weight of 40, compared to 1 for H, with a positive valency of 2; one Ca^{2+} ion, therefore, is equivalent to two H^+ ions because the latter ion has only one charge. The amount of Ca^{2+} required to displace 1 me of H^+ is 40/2, or 20 mg, the weight of 1 me of Ca^{2+}.

Table 4-2
Representative Cation Exchange Capacities for Various Materials

Material	Approximate cation exchange capacity (me/100 g dry weight)
Organic matter	150–500
Kaolinite	3–15
Halloysite	5–10
Hydrated halloysite	40–50
Illite	10–40
Chlorite	10–40
Smectite	80–150
Vermiculite	100–150+
Palygorskite	5–30
Sepiolite	20–45
Allophane	25–70
Hydrous oxides of aluminum and iron	4
Feldspars	1–2
Quartz	1–2
Basalt	1–3
Zeolites	230–620

(Taken from Carroll[15] and Grim[32]; palygorskite and sepiolite data from Weaver and Pollard.[71])

the subject and cites data suggesting that the average ratio of cation to anion exchange capacity is 0.5 for kaolinite, 2.3 for illite, and 6.7 for montmorillonite. Anion exchange sites are believed to originate from OH^- dissociation and from the linkage of complex ions, like phosphate, to silica sheet edges. Such complex ions would exchange for the silica tetrahedra because both have a similar geometry.

ION MOBILITIES

The behavior of ions, once released by weathering to the soil solution, varies; some ions participate in reactions involving mineral synthesis, some are adsorbed to colloid surfaces, and some are removed with the downward-moving water. Only when specific ions in the soil solution occur in the right proportions can mineral synthesis take place. Here we look into the factors governing the mobility of the more common ions.

Two approaches are used to rank ions on their mobility. One is to calculate the ratio of the percent of an element in stream or spring water to the percent in the parent rock. When this is done,[26,57] the ranking usually obtained is $Ca^{2+} > Na^+ > Mg^{2+} > K^+ > Si^{4+} > Fe^{3+} > Al^{3+}$, and the values for the latter two under the normal soil pHs are so low that their mobilities are almost negligible. Other studies indicate that this ranking can vary with rock type.[16,54] The other approach[49] is to calculate the ionic potential of the ions, which is the ratio of the charge in valency units to the ionic radius in Angstrom units (Å). Ionic potential, therefore, is a measure of the intensity of the positive charge. Ions with a ratio of 3 or less ($K^+ = 0.75$; $Na^+ = 1.0$; $Ca^{2+} = 2.0$; $Fe^{2+} = 2.7$; $Mg^{2+} = 3.0$) can remain in solution as ions, whereas those between 3 and 9.5 precipitate out as hydroxides ($Fe^{3+} = 4.7$; $Al^{3+} = 5.9$). This behavior can be explained as a result of the attraction of the ion in solution for the O ion of the H_2O molecule. If the attraction of the ion for O^{2-} is weak, the ion remains in solution surrounded by water molecules. If, however, the attraction of the ion for the O^{2-} is comparable to that of the H^+ for the O^{2-}, one H^+ from the water molecule is expelled, and the ion is precipitated as an hydroxide. Much of the silicon in natural waters forms non-ionized silicic acid, H_4SiO_4, and so the abundance of silicon is due to other factors. Thus, the data indicate that iron and aluminum can remain close to the site of weathering and partake in synthesis reactions, whereas other ions can be transported away. This precipitation of iron and aluminum hydroxides takes place under oxidizing conditions over the pH range of most normal soils (Fig. 4-3).

When the data for the relative mobilities are compared with those for the ionic potential, the ions are not ranked in the same order. This apparent discrepancy is no doubt due to many factors. One factor is that min-

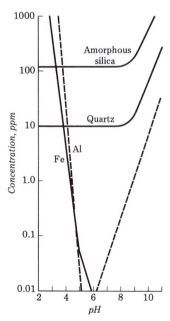

Fig. 4-3 Relationship between pH and solubility of aluminum, iron, amorphous silica, and quartz. Data for amorphous silica and quartz are from Krauskopf[44] (Fig. 6-3), and data for iron and aluminum are from Black[10] (Figs. 1 and 2, © 1967, John Wiley and Sons.)

erals weather at different rates and, because minerals vary in their composition, so too will leaching waters. Another factor would be the fixation of certain ions into the crystal lattices of the clay minerals. A familiar example is the fixation of potassium between layers of illite where it is not readily exchangeable. Another example is the introduction of magnesium into the octahedral layer of montmorillonite. Ion exchange is also an important factor. Certain ions can replace others on clay-mineral surfaces, and although the usual replacement series is $M^{2+} > M^+$, where M refers to any cation, replacement rank may be modified by a number of factors.[15,73] Another important factor is the selective uptake of ions by plants. In short, the difference in ranking by the two methods is explained mostly by processes operative once the ions enter the soil solution.

GENESIS OF CRYSTALLINE PRODUCTS

The genesis of minerals in the soil is complex due to the variety of weathering environments, the possibility of unlike microenvironments in close proximity to one another, and the fact that some minerals are derived by precipitation from solution, whereas others are derived from solid-state alteration of preexisting phyllosilicates. To further complicate matters, some minerals are of eolian origin and may have no relationship with the chemistry of the soil. It is not surprising, therefore, that any soil sample commonly is characterized by several clay-mineral species. This is compounded by the facts that (a) reaction rates are very slow at surface conditions, (b) duplication of natural conditions in the laboratory is difficult, if not impossible, and (c) some minerals may not be part of the current mineral-water equilibrium, but rather may have formed in the past under different conditions or even have been derived from the parent material. However, some general statements can be made regarding the chemical conditions favoring the formation of certain minerals.

Formation of secondary iron-bearing minerals in the soil depends partly on pH-Eh conditions[29] (Eh is a measure of the ability of the environment to bring about either oxidation or reduction). Under oxidizing conditions, the Fe^{2+} released from primary minerals during weathering is oxidized to Fe^{3+}. Because ferric compounds are insoluble in oxygenated waters over the normal pH range for soils (Fig. 4-3), precipitation takes place close to the source of iron. In contrast, if conditions are reducing, Fe^{2+} is the stable aqueous species, and under appropriate conditions, it can migrate far from the site of release by weathering before precipitation takes place. Reducing conditions usually are associated with water saturation (for example, below the water table).

Schwertmann and Taylor[63] have been able to estimate conditions for formation of secondary iron minerals because many of them can be synthesized in the laboratory. Under oxidizing conditions with relatively slow release of Fe by weathering and low soluble, organic matter content, goethite is usually formed. If, however, much soluble organic matter is present, and Fe release is more rapid, ferrihydrite probably will form. Ferrihydrite seems to be a necessary precursor for the formation of hematite. Adsorbed organic matter will retard this transformation, which takes place by dehydration and rearrangement. Goethite does not convert to hematite, nor does hematite convert to goethite, strictly by dehydration or hydration. In the first case, goethite must dissolve, ferrihydrite must form, and the latter then converts to hematite, a transformation enhanced by high temperatures, near-neutral pH, and well-aerated conditions. Hematite can convert to goethite, but only by dissolution and reprecipitation; a change to cool-moist climatic conditions could initiate such a conversion. Under reducing conditions with low partial pressure of CO_2, a $Fe^{2+}Fe^{3+}$ hydroxy compound can form, and subsequent transformation to other minerals is complex. Apparently if dehydration precedes oxidation, maghemite can form, but if oxidation precedes dehydration, lepidocrocite forms. Transformation of these to other iron minerals requires dissolution.

Clay minerals can form in several ways.[27,56] If, on one hand, the primary minerals undergoing weathering are not phyllosilicates, the formation of the clay mineral involves weathering with the release of cations and silica and alumina, followed by their recombination into phyllosilicate clay minerals. If, on the other hand, the weathering primary mineral is a phyllosilicate from the original igneous rock or from a soil or sediment, the alteration may take place essentially in the solid state. These are called transformations. Both conditions will be explored here.

The type of clay mineral that forms in a soil solution mostly depends on the content of silica, the kind and concentration of cations present, the soil pH, and the amount of leaching.[6,21,41,48,51,52] Both iron and aluminum released by weathering precipitate as oxides and hydrous oxides over the normal soil pH range and so generally remain within the soil for possible reaction to form clay minerals. As mentioned earlier, many of the other common cations in the soil have relative mobilities much greater than iron and aluminum, and they can be leached from the soil environment and thus will not react. Because silicon and aluminum form the basic framework for many clay minerals, their presence in the correct proportions is essential. Davis[19] has shown that silicon is common in most natural waters. Thus, the amount of silicon available for reaction depends less on its rate of release during weathering and more on the rate of leaching from the soil environment. The same can be said for many other cations, but their presence probably depends more on rock type.

The Si/Al molar ratio of both primary and secondary minerals indicates the amount of silica leaching required to form most secondary products. For example, various analyses of primary minerals give a wide range of ratios (muscovite and anorthite = 2; biotite, orthoclase and albite = 5 to 6; hornblende = 9; augite = 29; hypersthene and olivine > 200). In contrast, the secondary clays have a relatively narrow range in Si/Al ratio (kaolinite and dioctahedral chlorite = 2; 2:1 clay minerals and triocta-hedral chlorite = 3 to 5; and palygorskite = 9). If the soil solution is characterized by a very low silica content, gibbsite and possibly boehmite would form, and the conditions required for such formation are quite complex.[37] Leaching conditions that would keep silicon content low would also result in low cation concentrations. With increasing amounts of silicon, aluminosilicates form. Relatively high leaching, a low cation content, and a pH less than 7 favors the formation of kaolinite and halloysite. Halloysite forms primarily from volcanic materials and, in time, seems to convert to kaolinite. The conditions necessary to form the 2:1 clay minerals are quite varied. Montmorillonite forms in near neutral to alkaline pH and relatively high concentrations of Ca^{2+}, Mg^{2+}, and Na^+. In contrast, biedellite seems to form under conditions of acid pH and high leaching, such as in the E horizons of some Spodosols.[30,45,50,58] Illite forms under these same conditions when the concentration of K^+ is high. If the solution is high in Na_2CO_3 and $NaHCO_3$, extremely alkaline conditions prevail and zeolites form.[4,33,34] Palygorskite[65] and sepiolite are now considered possible products of pedogenesis and are favored by an alkaline pH (8–9), good drainage, abundant $(Ca,Mg)CO_3$ to supply the Mg^{2+}, and a low Al^{3+} content.[3] They are most common in fairly old carbonate-enriched horizons (Bk and K). Vermiculite formation is favored by a high concentration of Mg^{2+} and a pH less than 7. Chlorite with a gibbsite-layer interlayer would probably form under these same or more acid conditions; if, however, conditions are more alkaline and Mg^{2+} is abundant, the interlayer probably would be brucite-like. This latter species also could be inherited from the parent material.

There is much conjecture regarding the manner in which clay minerals form from a solution,[6,21,39,40] because the reactions are very difficult to reproduce under laboratory conditions that even approach those in the field (for laboratory studies see Siffert,[64] Grim[32] and Millot[51]). One way in which crystallization might proceed is by the precipitation of colloidal SiO_2 and $Al(OH)_3$. One problem here is the orientation of the combining constituents as tetrahedral and octahedral sheets. One way in which this could take place is by adsorption on both non-clay and clay-mineral surfaces. Orientation into sheets may be facilitated with substrates with an atomic lattice structure similar to that of the forming clay minerals. Because clay-mineral surfaces carry a negative charge, one might visualize colloidal $Al(OH)_3$ first being attracted to the mineral surface followed by colloidal SiO_2 to build up a particular clay mineral.

Various processes have been proposed for the alteration of micas derived from rock to clay minerals or for alteration from one clay mineral to another.[25,39] These changes almost always involve the interlayer areas because that is where ions can be exchanged, hydroxy ions introduced, or a silica sheet removed. Reactions usually occur preferentially along a layer or are initiated at the crystal edge and proceed inward.

Alteration among the 2:1 and 2:1:1 groups of clay minerals probably takes place quite readily because of their similar structures and inter-layer-bonding mechanisms. Ion exchange probably is the simplest reaction to envisage (Fig. 4-4, A). By this process, ions from the soil solution replace interlayer ions, and there may be some replacement within the crystal lattice. The resulting product is mostly a function of the host mineral and the replacing ion. For example, replacement of interlayer K^+ with

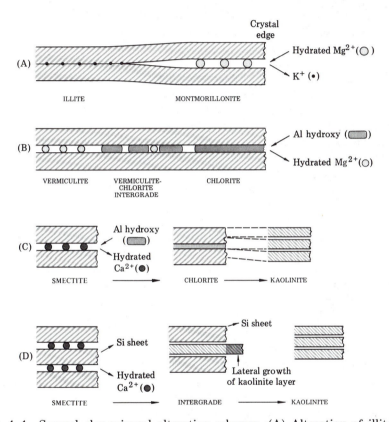

Fig. 4-4 Several clay-mineral alteration schemes. (A) Alteration of illite to montmorillonite by ion exchange along the edges of the layers. (B) Alteration of vermiculite to chlorite by replacement of interlayer ions by aluminum hydroxy groups to form a gibbsite layer. (Modified from Jackson,[39] Fig. 2.12 *in* Chemistry of the Soil by F. E. Bear, ed., © 1964 by Litton Educational Publishing, Inc. Reprinted by permission of Van Nostrand Reinhold Company.) (C) Alteration of smectite to chlorite to kaolinite. (Modified from Brindley and Gillery,[13] Fig. 1; see also Glenn and others.[31]) The silica tetrahedra in the middle kaolinite layer depicted must invert so that the interlinked bases of the tetrahedra all face upward. (D) Alteration of smectite to kaolinite by stripping silicon sheets from smectite. (Modified from Altschuler and others.[2] © 1963 by the American Association for the Advancement of Science.) Kaolinite growth can take place laterally from a newly formed kaolinite layer, and additional layers in turn can become oriented on these layers. Hydroxyls occupy the oxygen sites in the octahedral layers left vacant by silicon sheet removal.

Mg^{2+} could alter muscovite or illite to montmorillonite or biotite to vermiculite. If, however, hydroxy ion groups (Al) are present, they could replace the interlayer ions, because they too carry a positive charge, and form chlorite (Fig. 4-4, B). In a similar manner, if the soil solution is rich in Mg^{2+}, brucite layers could form and become interlayered with 2:1 clays to form chlorite. Thus, trioctahedral montmorillonite and vermiculite could be altered to chlorite by these mechanisms. Such an alteration might proceed more rapidly for montmorillonite than for vermiculite, because the former is characterized by an expandable lattice. This expansion would allow for more rapid interchange of interlayer ion and hydroxy ion, or hydroxide sheet.

The alteration from 2:1 to 1:1 clay minerals involves a greater amount of structural reorganization. Two mechanisms proposed for this transformation are (a) gibbsite interlayering to form a chlorite-montmorillonite intergrade, followed by kaolinite formation, and (b) the removal of silica tetrahedral sheets from montmorillonite layers (Fig. 4-4, C and D).

It is also known that gibbsite and kaolinite can form from one another. The formation of gibbsite from kaolinite involves the removal of a silicon sheet, perhaps in the manner shown in Fig. 4-4, D. The reverse reaction, the formation of kaolinite from gibbsite, is thought to involve partial dehydration of the gibbsite structure, the entry of silicon-enriched solutions between gibbsite layers, with the oxygen of the silicon tetrahedra occupying the lattice positions vacated by hydroxide ions during dehydration.[67]

The reactions between the 2:1 clay minerals, between 2:1 and 2:1:1 clay minerals, and between kaolinite and gibbsite seem to be reversible. The reaction 2:1 to 1:1 clay, however, may not be readily reversible under surface conditions. The reverse reaction would involve addition of a silicon sheet between each kaolinite layer, and this might be difficult because the hydrogen bond between the kaolinite layers is fairly strong, the layers are held closely together, and there are no interlayer cations. Perhaps the reverse reaction proceeds by solution of the original kaolinite followed by precipitation of the constituents as montmorillonite.

Bachman and Machette[3] review the origin of the fibrous clay minerals and point out that opinions vary from precipitation from solution to transformation. Their studies suggest that montmorillonite and mixed-layer illite-montmorillonite transform to palygorskite. This transformation is indicated by the usual interpretation of X-ray diffraction data; that is, the intensities of the peaks of the former minerals decrease as that of palygorskite increases systematically in a soil profile. A further transformation of palygorskite → sepiolite takes place only in fairly old soils (middle Pleistocene or older).

To summarize some of the above, clay minerals can form both by transformation and by precipitation from solution. However, because mica is so common in rocks, transformation may be the main process responsible for most soil clay minerals. Mica, or other clay minerals, can also be introduced to the soil by eolian processes, and this will further confound our ability to decipher the origin of specific clay minerals. Even if the original mineral is no longer present, that does not rule out transformation as a main process; it does make it more difficult to prove, however. Finally, in a recent study Colman[18] challenges the assumption that well-crystallized clay minerals form in abundance in soils from primary minerals and rock particles in some soils about 10^5 years old or older. He bases this on the fact that weathering rinds of basaltic and andesitic clasts contain few crystalline clay minerals, whereas the clay fraction of the Bt horizon that the clasts occur in has abundant crystalline clay minerals. His study should force workers to look more carefully for the origin of soil clay minerals and to not always accept weathering of non-clay minerals as the primary source of such minerals.

Phase diagrams have been constructed to help quantify the relationship between the ionic concentration (or activity) of the solution and the mineral or minerals present.[21,23,28,29,35,36,43,46] Data on free energies of formation and on the chemistry of the species are used to construct such diagrams; Fig. 4-5 is one example. The diagrams can be considered to be only approximations because the crystallinity of the minerals varies, because some data on the activity–activity ratios are not available for low temperatures and must be extrapolated from higher temperature data, and because the precision of free energy data used to calculate the equilibrium constants varies. At any rate, the diagrams are useful in that they show the approximate stability fields for clay minerals precipitated from solutions of varying chemical composition. They also indicate the stability

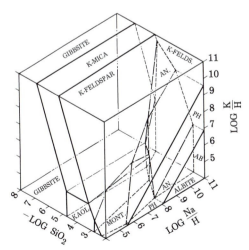

Fig. 4-5 Phase diagram for the system $K_2O - Na_2O - Al_2O_3 - SiO_2 - H_2O$ at 25°C and 1 atmos. KAOL, kaolinite; MONT, montmorillonite; PH, phillipsite; AN, analcite; and AB, albite. (Taken from Hess,[36] Fig. 1.)

relationships of one or more minerals; minerals adjacent to one another in the diagram can be in equilibrium (for example, kaolinite–montmorillonite), but those not adjacent to one another are not in equilibrium (for example, gibbsite–montmorillonite). Furthermore, the use of such diagrams helps to predict possible successive alteration products with gradual variation in ion–activity ratios, and activity of H_4SiO_4. Kittrick[42] discusses some possible applications and relates them to field conditions. In any study involving equilibrium diagrams, it is important to compare the mineralogical data with the water data from the same horizon; river water collected several miles away may not always be representative.

A usual extension of this mineral–equilibria work is to plot the water data on the same diagram to determine if the minerals are in equilibrium with the solution. If the water-chemistry data plot in a particular mineral field and the X-ray data confirm the presence of that particular mineral, the mineral probably is in equilibrium with the solution. This is the standard interpretation. However, equilibrium conditions may not exist for examples in which the water and X-ray mineralogical data do not coincide. As an example, plots have been made for perennial and ephemeral springs in granitic terrain of the Sierra Nevada (Fig. 4-6). K-mica and montmorillonite are present in some samples, but the water data all plot in the kaolinite field. It is suggested that this means that the water composition may later shift to activity values at which all three minerals are in equilibrium, or perhaps the minerals present will in time convert to more stable phases.[26] Garrels and Mackenzie[29] plot data from many rock types and environments, and most fall within the kaolinite field. It is suggested that kaolinite is the stable end product in the chemical weathering of silicate minerals. This is not always borne out by field studies, however; perhaps more time is needed to reach equilibrium, or some of the basic data on free energies are not too accurate (for example, compare Fig. 4-6 with that on p. 17 in Helgeson and others[35]). In contrast, Bohn and others[11] plot the extreme ranges of soil solutions on such a diagram (their Fig. 4.6) and find that they include parts of the stability fields for gibbsite, kaolinite, and montmorillonite. Finally, in a broad study covering several major environments and rock types, Tardy[68] found good correspondence between clay mineralogy and water chemistry in only some cases.

The applicability of these water–mineral equilibrium data is being tested for soils. In any such application, it will be extremely important to know which, if any, minerals were inherited from the parent material, which are of eolian origin, and the direction in which any mineral alteration schemes might be proceeding. Cleaves and others,[17] for example, report that because montmorillonite and chlorite are present in soils they studied and stream–water data plot in the stability fields for both minerals, the latter could have been synthesized in the present pedologic environment. Analogous data can be used to help prove a non-pedogenic origin for some clay minerals. For example, mica is present in humid soils

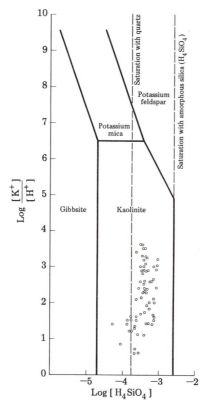

Fig. 4-6 Phase diagram for the system $K_2O - Al_2O_3 - SiO_2 - H_2O$ at 25 °C and 1 atmos. Open circles are analytical data for water from springs or seeps in a granitic terrain in the Sierra Nevada. (Taken from Feth and others.[26] Fig. 13.)

formed from mica-free basalt in Hawaii, but because the soil–solution data do not plot in the mica field of the stability diagram, a contemporary pedogenic origin is not supported.[66] Although arguments were made relative to equilibrium conditions, or whether or not the thermodynamic data were correct, later study showed the micas to be of eolian origin and to be much older by K-Ar dating than the island.[24]

Within any one soil or soil horizon there is usually a variety of clay minerals.[1,7,8,9,38,39,48] Various hypotheses have been advanced to explain this diversity. One explanation is that the clay minerals in the assemblage are not in equilibrium with present conditions and are being altered to other more stable clay minerals. Some of the clay minerals, for example, could have been inherited from the parent material, or be of eolian origin. Another possible explanation is that some of the clay minerals could have formed in the past under different environmental conditions, and reaction rates being so slow at surface conditions, these clay minerals are metastable in the present environment. If one considers the diverse microenvironmental conditions within a soil, perhaps diversity of clay minerals should be expected. For example, ionic concentration might vary from place to place due to the variation in charge of the particles. Furthermore, the primary minerals weather at slightly different rates, releasing a variety of substances into the soil. Thus, there will be times when the ionic species do not have a uniform distribution. Furthermore, there could be slight variations in pH that could be important to any ongoing reactions involving mineral synthesis. Such variation could be caused by proximity to a root, as they commonly have adsorbed H^+, or root CO_2 respiration and the local concentration of H_2CO_3. Data on abrasion pH of various primary minerals (Table 3-2), as well as their composition, also might lead one to suspect that similar local variations in the soil exist during mineral weathering and synthesis. Barshad[6] suggests that plant-nutrient cycling might control cation content and pH of the solution in some soils to such an extent that montmorillonite might form during periods of inactive plant growth and kaolinite might form during periods of active plant growth and nutrient uptake.

The above examples do not exhaust the list of possible chemical conditions at the microscopic level. They are mentioned to point out that a soil may have a variety of clay minerals due to a variety of local conditions. If this is the case, one may have to look carefully at the use of clay–mineral stability diagrams to determine whether a clay mineral or a clay–mineral assemblage is stable or not. The soil–solution chemical data needed to test for equilibrium on any diagram represent only the average conditions within the soil. On a microscopic level, the average may be difficult to find, controlled as it is by the rate of ion release by weathering, the chemical conditions at the site, and the rate of leaching.[69]

CLAY–MINERAL DISTRIBUTION WITH DEPTH IN SOIL PROFILES

Clay minerals vary in amount with depth in some soil profiles, but they remain relatively constant throughout in others. Such relationships seem to be closely associated with the leaching conditions within the soil. Three leaching conditions can be examined. With extensive leaching, many cations and silica may be carried to great depth. This can result in relatively uniform chemical conditions with depth in the soil and therefore a rather uniform clay–mineral distribution. The same uniformity involving different clay minerals would be expected in arid regions, because the little leaching that takes place may not change the chemical conditions enough to favor differences in clay mineralogy with depth. Between these two extremes, however, differences in leaching with depth might produce enough variation in chemistry to favor the formation of different clay minerals. Put another way, the ions and colloids released by weathering in the higher levels of the profile become reactants in mineral synthesis at depth.

The most common vertical variation is one in which the amount of silica and bases in the weathering product increases with depth. Thus, gibbsite at the surface grading downward to kaolinite or kaolinite at the surface grading downward to montmorillonite are common trends.[47] Although it is possible that one or the other mineral may no longer be in equilibrium with the present environmental conditions, this need not be the case. Both minerals in the vertical sequence can be in equilibrium with the environmental conditions, and because the latter changes with depth, this change is reflected in the clay mineralogy. If conditions change, for example, due to erosional lowering of the surface with time, the montmorillonite at depth eventually may be in that part of the profile most conducive to kaolinite formation, and it may alter to kaolinite (Fig. 4-7). The variation of clay mineral with depth does not by itself constitute proof that one mineral is changing over to another. In this case, the low-

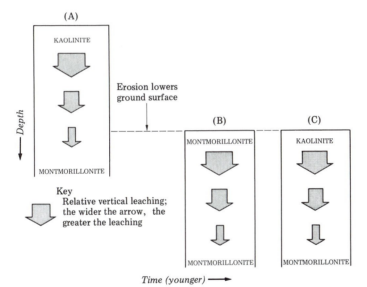

Time (younger) ⟶

Fig. 4-7 Vertical distribution of clay minerals as a function of leaching conditions within the soil. (A) Leaching conditions favor the formation of kaolinite and montmorillonite; each forms at different levels in the soil profile. (B) Rapid lowering of the surface by erosion results in the montmorillonite residing in a soil environment of high leaching. (C) With time, the montmorillonite in the surface becomes desilicated and alters to kaolinite. Rates of ground-surface lowering, mineral alteration, and amount of leaching are important to stable end product clay-mineral formation. With slow ground-surface lowering, alteration may proceed at the same rate, and the clay-mineral distribution may always appear as in (A). If the rate of erosion far exceeds the rate at which the clay mineral can change, a distribution like (B) occurs temporarily.

ering of the ground surface so changed the conditions within the soil that the original clays were no longer stable; they therefore equilibrated with the environment.

REFERENCES

1. Allen, B.L., 1977, Mineralogy and soil taxonomy, p. 771–796 *in* J.B. Dixon and S.B. Weed, eds., Minerals in soil environments: Soil Science Society of America, Madison, Wisc.
2. Altschuler, Z.S., Dwornik, E.J., and Kramer, H., 1963, Transformation of montmorillonite to kaolinite during weathering: Science, v. 141, p. 148–152.
3. Bachman, G.O., and Machette, M.N., 1977, Calcic soils and calcretes in the southwestern United States: U.S. Geological Survey Open-File Report 77–794, 163 p.
4. Baldar, N.A., and Whittig, L.D., 1968, Occurrence and synthesis of soil zeolites: Soil Sci. Soc. Amer. Proc., v. 32, p. 235–238.
5. Barnhisel, R.I., 1977, Chlorites and hydroxy interlayered vermiculite and smectite, p. 331–356 *in* J.B. Dixon and S.B. Weed, eds., Minerals in soil environments: Soil Science Society of America, Madison, Wisc.
6. Barshad, I., 1964, Chemistry of soil development, p. 1–70 *in* F.E. Bear, ed., Chemistry of the soil: Reinhold Publ. Corp., New York, 515 p.
7. ———, 1966, The effect of a variation in precipitation on the nature of clay mineral formation in soils from acid and basic igneous rocks: Proc. Internat. Clay Conf., Israel, 1966, v. 1, p. 167–173.
8. Birkeland, P.W., 1969, Quaternary paleoclimatic implications of soil clay mineral distribution in a Sierra Nevada–Great Basin transect: Jour. Geol., v. 77, p. 289–302.
9. Birkeland, P.W., and Janda, R.J., 1971, Clay mineralogy of soils developed from Quaternary deposits of the eastern Sierra Nevada: Geol. Soc. Amer. Bull., v. 82, p. 2495–2514.
10. Black, A.B., 1967, Applications: Electrokinetic characteristics of hydrous oxides of aluminum and iron, p. 247–300 *in* S.D. Faust and J.V. Hunter, eds. Principles and applications of water chemistry: John Wiley and Sons, New York, 643 p.
11. Bohn, H.L., McNeal, B.L., and O'Connor, G.A., 1979, Soil chemistry: John Wiley and Sons, Inc., New York, 329 p.
12. Borchardt, G.A., 1977, Montmorillonite and other smectite minerals, p. 293–330 *in* J.B. Dixon and S.B. Weed, eds., Minerals in soil environments: Soil Science Society of America, Madison, Wisc.
13. Brindley, G.W., and Gillery, F.H., 1954, A mixed-layer kaolin-chlorite structure: Proc., 2nd National Conf., Clays and clay minerals (Columbia, Mo., Oct. 15–17, 1953), Publ. 327 of the committee on clay minerals of the National Acad. Sci., National Res. Coun., Washington, D.C., p. 349–353.
14. Brindley, G.W., and Brown, G., eds., 1980, Crystal structures of clay minerals and their X-ray identification: Mineralogical Soc. (London) Monograph No. 5, 495 p.

15. Carroll, D., 1959, Ion exchange in clays and other minerals: Geol. Soc. Amer. Bull., v. 70, p. 749–780.

16. ———, 1970, Rock weathering: Plenum Press, New York, 203 p.

17. Cleaves, E.T., Fisher, D:W., and Bricker, O.P., 1974, Chemical weathering of serpentinite in the eastern Piedmont of Maryland: Geol. Soc. Amer. Bull., v. 85, p. 437–444.

18. Colman, S.M., 1982, Clay mineralogy of weathering rinds and possible implications concerning the sources of clay minerals in soils: Geology, v. 10, p. 370–375.

19. Davis, S.N., 1964, Silica in streams and ground water: Amer. Jour. Sci., v. 262, p. 870–891.

20. Dixon, J.B., 1977, Kaolinite and serpentine group minerals, p. 357–403 *in* J.B. Dixon and S. B. Weed, eds., Minerals in soil environments: Soil Science Society of America, Madison, Wisc.

21. Dixon, J.B., and Weed, S.B., eds., 1977, Minerals in soil environments: Soil Science Society of America, Madison, Wisc., 948 p.

22. Douglas, L.A., 1977, Vermiculites, p. 259–292 *in* J.B. Dixon and S.B. Weed, eds., Minerals in soil environments: Soil Science Society of America, Madison, Wisc.

23. Drever, J.I., 1982, The geochemistry of natural waters: Prentice-Hall, Inc., Englewood Cliffs, N. J., 388 p.

24. Dymond, J., Biscaye, P.E., and Rex, R.W., 1974, Eolian origin of mica in Hawaiian soils: Geol. Soc. Amer. Bull., v. 85, p. 37–40.

25. Fanning, D.S., and Keramidas, V.Z., 1977, Micas, p. 195–258 *in* J.B. Dixon and S.B. Weed, eds., Minerals in soil environments: Soil Science Society of America, Madison, Wisc.

26. Feth, J.H., Roberson, C.E., and Polzer, W.L., 1964, Sources of mineral constituents in water from granitic rocks, Sierra Nevada, California and Nevada: U.S. Geol. Surv. Water-Supply Pap. 1535-I, 70 p.

27. Fieldes, M., and Swindale, L.D., 1954, Chemical weathering of silicates in soil formation: New Zealand Jour. Sci. and Tech., v. 36, p. 140–154.

28. Garrels, R.M., and Christ, C.L., 1965, Solutions, minerals, and equilibria: Harper & Row, New York, 450 p.

29. Garrells, R.M., and Mackenzie, F.T., 1971, Evolution of sedimentary rocks: W.W. Norton and Co., Inc., New York, 397 p.

30. Gjems, O., 1960, Some notes on clay minerals in podzol profiles in Fennoscandia: Clay Minerals Bull., v. 4, p. 208–211.

31. Glenn, R.C., Jackson, M.L., Hole, F.D., and Lee, G.B., 1960, Chemical weathering of layer silicate clays in loess-derived Tama silt loam of southwestern Wisconsin: Clays and Clay Minerals, v. 8, p. 63–83.

32. Grim, R.E., 1968, Clay mineralogy: McGraw-Hill, New York, 596 p.

33. Hay, R.L., 1963, Zeolitic weathering in Olduvai Gorge, Tanganyika: Geol. Soc. Amer. Bull., v. 74, p. 1281–1286.

34. ———, 1964, Phillipsite of saline lakes and soils: Amer. Mineralogist, v. 49, p. 1366–1387.

35. Helgeson, H.C., Brown, T.H., and Leeper, R.H., 1969, Handbook of theoretical activity diagrams depicting chemical equilibria in geologic systems involving an aqueous phase at one atm and 0°–300°C: Freeman, Cooper and Co., San Francisco, Calif., 253 p.

36. Hess, P.C., 1966, Phase equilibria of some minerals in the $K_2O-Na_2O-Al_2O_3-SiO_2-H_2O$ system at 25°C and 1 atmosphere: Amer. Jour. Sci., v. 264, p. 289–309.

37. Hsu, P.H., 1977, Aluminum hydroxides and oxyhydroxides, p. 99–143 *in* J.B. Dixon and S.B. Weed, eds., Minerals in soil environments: Soil Science Society of America, Madison, Wisconsin.

38. Jackson, M.L., Tyler, S.A., Willis, A.L., Bourbeau, G.A., and Pennington, R.P., 1948, Weathering sequence of clay-size minerals in soils and sediments—I. Fundamental generalizations: Jour. Phys. Colloid. Chem., v. 52, p. 1237–1260.

39. ———, 1964, Chemical composition of soils, p. 71–141 *in* F.E. Bear, ed., Chemistry of the soil: Reinhold Publ. Corp., New York, 515 p.

40. ———, 1965, Clay transformations in soil genesis during the Quaternary: Soil Sci., v. 99, p. 15–22.

41. Keller, W.D., 1964, Processes of origin and alteration of clay minerals, p. 3–76 *in* C.I. Rich, and G.W. Kunze, eds., Soil clay mineralogy: Univ. No. Carolina Press, Chapel Hill.

42. Kittrick, J.A., 1969, Soil minerals in the $Al_2O_3-SiO_2-H_2O$ system and a theory of their formation: Clays and Clay Minerals, v. 17, p. 157–167.

43. ———, 1977, Mineral equilibria and the soil system, p. 1–25 *in* J.B. Dixon and S.B. Weed, eds., Minerals in soil environments: Soil Science Society of America, Madison, Wisconsin.

44. Krauskopf, K.B., 1967, Introduction to geochemistry: McGraw-Hill, New York, 721 p.

45. Lee, R., ed., 1980, Soil groups of New Zealand, Part 5, podzols and gley podzols: New Zealand Society of Soil Science, Lower Hutt, 452 p.

46. Lindsay, W.L., 1979, Chemical equilibria in soils: John Wiley and Sons, New York, 449 p.

47. Loughnan, F.C., 1969, Chemical weathering of the silicate minerals: American Elsevier Publishing Co., Inc., New York, 154 p.

48. Marshall, C.E., 1977, The physical chemistry and mineralogy of soils—vol. II: Soils in place: John Wiley and Sons, New York, 313 p.

49. Mason, B.H., 1966, Principles of geochemistry: John Wiley and Sons, New York, 329 p.

50. McKeague, J.A., Ross, G.J., and Gamble, D.S., 1978, Properties, criteria of classification, and concepts of genesis of podzolic soils in Canada, p. 27–60 *in* W.C. Mahaney, ed., Quaternary soils: Geo Abstracts Ltd., University of East Anglia, Norwich, England.

51. Millott, G., 1970, Geology of clays: Springer-Verlag, New York, 429 p.

52. Pedro, G., Jamagne, M., and Begon, J.C., 1969, Mineral interactions and transformations in relation to pedogenesis during the Quaternary: Soil Sci., v. 107, p. 462–469.

53. Pierce, J.W., and Siegel, F.R., 1969, Quantification in clay mineral studies of sediments and sedimentary rocks: Jour. Sed. Petrol., v. 39, p. 187–193.

54. Reiche, P., 1950, A survey of weathering processes and products: Univ. New Mex. Publ. in Geol., no. 3, 95 p.

55. Reynolds, R.C., and Hower, J., 1970, The nature of interlayering in mixed-layer illite-montmorillonites: Clays and Clay Minerals, v. 18, p. 25–36.

56. Rich, C.I., and Thomas, G.W., 1960, The clay fraction of soils: Advances in Agron., v. 12, p. 1–39.

57. Rode, A.A., 1962, Soil Science: Israel Program for Scientific Translations, Jerusalem, 517 p.

58. Ross, G.J., 1980, The mineralogy of Spodosols, p. 127–143 *in* B.K.G. Theng, ed., Soils with variable charge: New Zealand Society of Soil Science, Lower Hutt.

59. Ruhe, R.V., and Olson, C.G., 1979, Estimate of clay–mineral content: Additions of proportions of soil clay to constant standard: Clays and Clay Minerals, v. 27, p. 322–326.

60. Sawnhey, B.L., 1977, Interstratification in layer silicates, p. 405–434 *in* J.B. Dixon and S.B. Weed, eds., Minerals in soil environments: Soil Science Society of America, Madison, Wisconsin.

61. Schultz, L.G., 1964, Quantitative interpretation of mineralogical composition from X-ray and chemical data for the Pierre Shale: U.S. Geol. Surv. Prof. Pap. 391-C, 31 p.

62. ———, 1978, Mixed-layer clay in the Pierre Shale and equivalent rocks, northern Great Plains region: U.S. Geol. Surv. Prof. Pap. 1064-A, 28 p.

63. Schwertmann, U., and Taylor, R.M., 1977, Iron oxides, p. 145–180 *in* J.B. Dixon and S.B. Weed, eds., Minerals in soil environments: Soil Science Society of America, Madison, Wisc.

64. Siffert, B., 1967, Some reactions of silica in solution: Formation of clay: Israel Program for Scientific Translations, Jerusalem, 100 p.

65. Singer, A., and Norrish, K., 1974, Pedogenic palygorskite occurrences in Australia: Amer. Mineral., v. 59, p. 508–517.

66. Swindale, L.D., and Uehara, G., 1966, Ionic relationships in the pedogenesis of Hawaiian soils: Soil Sci. Soc. Amer. Proc., v. 30, p. 726–730.

67. Tamura, T., and Jackson, M.L., 1953, Structural and energy relationships in the formation of iron and aluminum oxides, hydroxides, and silicates: Science, v. 117, p. 381–383.

68. Tardy, Y., 1971, Characterization of the principal weathering types by the geochemistry of waters from some European and African crystalline massifs: Chem. Geol., v. 7, p. 253–271.

69. Tardy, Y., Bocquier, G., Paquet, H., and Millot, G., 1973, Formation of clay from granite and its distribution in relation to climate and topography: Geoderma, v. 10, p. 271–284.

70. Wada, K., 1977, Allophane and imogolite, p. 603–638 *in* J.B. Dixon and S.B. Weed, eds., Minerals in soil environments: Soil Science Society of America, Madison, Wisc.

71. Weaver, C.E., and Pollard, L.D., 1973, The chemistry of clay minerals: Elsevier Scientific Publishing Co., New York.

72. Whittig, L., 1965, X-ray diffraction techniques for mineral identification and mineralogical composition, p. 671–698 *in* C.A. Black, ed., Methods of soil analysis (part I): Amer. Soc. Agron., Madison, Series in Agron., no. 9, 770 p.

73. Wiklander, L., 1964, Cation and anion exchange phenomena, p. 163–205 *in* F.E. Bear, ed., Chemistry of the soil: Reinhold Publ. Corp., New York, 515 p.

74. Zelazny, L.W., and Calhoun, F.G., 1977, Palygorskite (attapulgite), sepiolite, talc, pyrophyllite, and zeolites, p. 435–470 *in* J.B. Dixon and S.B. Weed, eds., Minerals in soil environments: Soil Science Society of America, Madison, Wisc.

5

Processes responsible for the development of soil profiles

Since many processes act together to form any one soil profile, it is difficult to discuss soil formation as a function of a specific process. The formation of a soil profile is viewed by Simonson[137] as the combined effect of additions to the ground surface, transformations within the soil, vertical transfers (up or down) within the soil, and removals from the soil (Fig. 5-1). For any one soil, the relative importance of these processes varies, and the result is the variety of profiles seen in any landscape. The main additions to most soils are organic matter from the surface vegetation and their contained elements, ions and solid particles introduced with rainfall, and particles carried by the wind. Transformations include the multitude of organic compounds that form during organic matter decomposition, the weathering of primary minerals, and the formation of secondary minerals and other products. Transfers generally involve the movement of ions and substances with the moving soil water. Soluble substances move with the percolating water unless changing chemical conditions or dehydration cause them to precipitate out of solution. In places where capillary rise of water is important, ions can be transferred upward and be precipitated high in the soil profile. In addition, biological activity is an important agent in the upward movement of materials. Ions can move upward through plants and be returned to the surface with litterfall. Soil-dwelling fauna can actively move solid particles in any direction. Finally, when water moves through the profile, it removes substances still in solution; these substances then become part of the dissolved constituents of the groundwater or surface waters. Because transformations have been discussed elsewhere in this book, the emphasis here will be on other processes that form the various horizons diagnostic of soil profiles.

It should be pointed out that it is very difficult to determine soil pro-

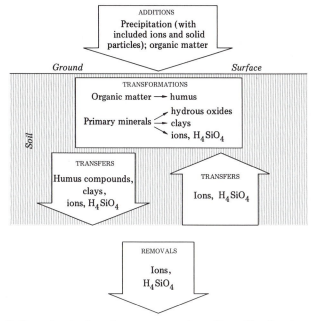

Fig. 5-1 A flow chart of major processes in soil-profile development (after Simonson[137]). Surface erosion might be added here, because erosion rates must be less than soil-formation rates for profiles to form.

cesses because few actual measurements can or have been made. Consider, for example, clay translocation from the A to the B horizon. This process probably is so slow that measurements taken over a year or two may not be representative. In addition, laboratory simulations may or may not be good approximations of what happens in nature. In lieu of these approaches, we commonly infer the processes from combined field and laboratory data. Process interpretation is confounded because, although process is commonly tied to particular climates, the Quaternary record is clear that climates have varied during the formation of many soils; hence, what one might infer to be a contemporary process actually could have been active only at some time in the past. One should not always expect variation in process or its rate with climatic change, for it might be that neither of these changed markedly in the past, even though climatic change seems fairly well documented.

In this chapter, both laboratory chemical methods and trends with depth will be mentioned as they are used to help identify various processes. The reader should become familiar with the expected trends, for any deviations from such trends for seemingly uniform profiles might indicate subtle parent laterial layering, eolian additions at the surface, buried soils, climatic change, or more complicated processes.

FORMATION OF THE A HORIZON

The organic matter content of a soil is a function of gains in organic matter to the soil surface and upper layers of the soil and losses that accompany decomposition. In the ideal case, a newly formed surface has no organic matter. Once vegetation is established, an A horizon can begin to form. Early in the formation of a soil, the gains exceed the losses, organic matter gradually accumulates, and the A or A and O horizons thicken. With time a steady-state condition is reached, and the gains equal the losses. Thereafter, even though the system is a dynamic one with continuing gains and losses, the amount of organic matter in the soil and its distribution with depth remain essentially constant. The length of time needed to reach the steady-state condition and the amount of organic matter in the soil at that time vary with the soil-forming factors. The maximum time to reach the steady state probably is about 3000 years, but in many environments it is reached much sooner.

Because of the dynamic nature of the A horizon its age is difficult to define. However, in some stratigraphic studies, workers have tried to differentiate the A horizon from underlying horizons by age. Thus, one might see a description of a soil in which a modern A horizon overlies a Sangamon B horizon. Such differentiation is possible if the modern A horizon has formed from a younger parent material that was deposited on the Sangamon B. If this is not the case, such age differentiation is difficult. For example, the primary mineral fraction of both the A and B horizons are the same age in a soil formed from one parent material in a stable landscape. Clays and other weathering products are younger than the primary minerals, and the organic constituents are youngest of all. The organic constituents, however, vary from vegetative matter just deposited, and in the initial stages of decomposition, to humus that may have resided in the soil several thousand years. In this sense, then, all A horizons at the land surface are modern or have a considerable modern component. If the interpretation is made that the A horizon is modern because the older A horizon was removed at some time in the past by erosion, evidence for the erosion that removed the A horizon may be preserved in the stratigraphic record as unconformities.

TRANSLOCATION OF IRON AND ALUMINUM (PODZOLIZATION)

Many soils show evidence of podzolization, a process that involves a pronounced downward translocation of iron, aluminum, and organic matter to form an eluvial E horizon overlying an illuvial spodic or B horizon (Figs. 1-2 and 2-3). In many podzolized soils, the processes behind such translocation are not readily apparent because the pH, although acid, can be in the range in which Fe^{3+} and Al^{3+} are essentially insoluble (Fig. 4-3). Furthermore, many of these soils are characterized by an oxidizing environment, which means that iron cannot move as the more soluble ionic species Fe^{2+}. A fair amount of work has gone into the study of the processes by which iron and aluminum move in Spodosols or Spodosol-like soils.[83,97,99,117,144] Petersen[115] has reviewed much of the literature on podzols (Spodosols), and presents data on some in Denmark, as well as a theory on their formation.

Before proceeding with this discussion, the various Fe and Al extracts of pedologic significance should be mentioned.[98] In a very general way, three chemical extractions of soils can be performed to aid in identifying the forms of Fe in a soil. Dithionite-citrate treatment is thought to remove the total free Fe that is not included in the silicate minerals; these are the crystalline oxides (goethite and hematite), amorphous hydrous oxides, and organic-bound iron. Oxalate treatment removes the amorphous hydrous oxides (much of which probably is ferrihydrite[133]) and some of the organic-bound Fe, and pyrophosphate extracts the organic-bound Fe. Because the Fe forms are not always consistent with the extracts, it might be best to list Fe amounts by extract procedure, rather than by implied form. In contrast to the above, the extracts are less specific to forms of Al.

The Fe and Al extracts can be performed on soils in any environment to quantify Fe and Al relationships, but they are most commonly used in defining Spodosols of Soil Taxonomy and Podzols of other classifications. For example, with increasing degree of podzolization, amounts of all extracts increase and display eluvial–illuvial relations down the profile (Table 5-1). In addition, some workers use the ratio of the oxalate-extractable Fe to the dithionite–citrate-extractable Fe, called the Fe activity ratio,[13] to indicate the relative amount of total free Fe that is amorphous; this ratio also increases with degree of podzolization. As regards soil classification using these extracts, the Spodosols of Soil Taxonomy are defined on the basis of both pyrophosphate and dithionite–citrate extracts of both Fe and Al, whereas in Canada, the Podzol is defined on the pyrophosphate extracts of both elements; the specific definitions, however, are so detailed that readers should consult the references.[97,138] Organic C content is important to both classifications because it or organic matter is high in the B horizons of Spodosols and Podzols (Table 5-1).

Because of the association between spodic horizons and organic matter, Daly[45] has proposed a new and simple method for differentiating podzolized soils from other soils by measuring the optical density of the oxalate extracts of the soils. It is assumed that the optical density is due mainly to extracted fulvic acids and that appreciable amounts of the latter indicate podzolization as a major process. He suggests calculating the optical density for the B horizon and the A horizon; if the ratio is greater than 1, then podzolization has occurred.

Many workers suggest that iron and aluminum probably move in the soil as soluble metallo-organic chelating complexes (Fig. 3-7). Fulvic acids

Table 5-1
Iron and Aluminum Extracts of Some Canadian Soils

Approximate classification	Horizon	Organic matter(%)	Organic carbon(%)	Dithionite Fe (%)	Dithionite Al (%)	Oxalate Fe (%)	Oxalate Al (%)	Pyrophosphate Fe (%)	Pyrophosphate Al (%)	Fe activity ratio
Mollisol*	A	14.99	—	1.28	0.34	0.38	0.34	—	—	0.30
	Bw1	2.83	—	1.51	0.34	0.39	0.36	—	—	0.24
	Bw2	1.61	—	1.14	0.26	0.32	0.29	—	—	0.28
	C	0.19	—	0.64	0.11	0.14	0.07	—	—	0.22
Spodosol**	E	1.88	—	0.18	0.16	0.05	0.15	—	—	0.28
	Bs1	7.26	—	3.26	1.40	2.60	1.38	—	—	0.80
	Bs2	2.13	—	1.22	0.79	0.56	1.01	—	—	0.46
	C	0.22	—	0.56	0.38	0.18	0.18	—	—	0.32
Spodosol***	E	—	1.4	0.1	—	—	—	0.02	0.01	—
	Bhs	—	12.0	3.3	—	—	—	2.6	1.6	—
	Bs	—	2.8	1.5	—	—	—	0.6	1.1	—
	Cx	—	0.2	0.6	—	—	—	0.03	0.2	0.31
Spodosol****	O	—	23.04	0.70	0.10	0.22	0.08	0.05	0.06	0.31
	E	—	2.15	0.34	0.10	0.22	0.10	0.06	0.10	0.65
	Bs	—	3.98	2.35	1.50	1.34	2.50	0.27	0.07	0.57
	2C1	—	0.27	2.30	0.23	0.70	0.25	0.16	0.10	0.30
	2C2	—	1.10	2.30	0.10	0.54	0.04	0.06	0.10	0.24

*Harkaway soil.[96]
**Holmesville soil.[99]
***Laurentide soil.[99]
****Profile 2.[82]

are thought to be the common chelating agents for a number of soils.[83,131,159] The stability of the chelate that forms is a function of size and valency of the metal ion, with increasing stability associated with smaller sizes and higher valencies. The process of translocation envisaged begins with the production of fulvic acids in the O or A horizon. A stable fulvic acid chelate is formed with Al^{3+} and Fe^{3+} and because these chelates are water-soluble, they move downward with the percolating soil water. Precipitation of these complexes can be brought about by a number of conditions at some depth in the soil profile. In some cases quite small changes in ionic content can bring about precipitation. For example, Wright and Schnitzer[159] report experimentally determined values of 0.13 ppm Ca^{2+} and 4.5 ppm Mg^{2+} to flocculate a complex of iron and organic matter and 10 and 45 ppm of the same cations, respectively, to flocculate a complex of aluminum and organic matter. Because of these relationships, it is possible that complexes of aluminum and organic matter could move deeper in the profile before flocculation, than can complexes of iron and organic matter; this relationship is seen in some Spodosols. Yet another way to precipitate Fe or Al or both is by increasing the ratio of chelated ion to fulvic acid, for at relatively high ratios precipitation occurs.[88] Petersen[115] presents a theory that the complexes that move have variable ratios of organic matter to metal. These complexes move in the descending water, but those with high organic matter to metal ratios precipitate first, whereas those with lower ratios can move to greater depths before encountering conditions for precipitation. Thus, a spodic horizon can be relatively enriched in organic matter near the upper boundary, a common attribute of Spodosols (Table 5-1). The position of the spodic horizon, therefore, marks the position of flocculation of the chelating compounds and associated metal ions. Furthermore, this process explains the loss of appreciable iron, aluminum, and organic matter from the E horizons. Another way to precipitate aluminum and iron from the chelating complex at depth is through microbial action. Schuylenborgh,[132] for example, suggests that the organic matter of the chelating complex is decomposed by microorganisms at depth, leaving the metal ions free to precipitate as oxides or hydroxides. A process for deposition that does not involve chemical interaction is deposition during partial dessication, as might be expected to happen during dry seasons.[56]

Conditions for chelate formation and movement of iron and aluminum are best for the Spodosols, but movement can also take place in other soils, as shown by some with E horizons (Table 2-1). In general, the degree of podzolization declines in a transect going from boreal forest to grassland,[83] that is, in a transect from Spodosols to Alfisols to Mollisols. In such a transect, the humic acids:fulvic acids ratio increases from about 0.5 to 1.5–2.0. Furthermore, the humic acids in the Spodosols are similar to the fulvic acids in that they are dispersed and mobile. Calcium–humic acid complexes that are both relatively stable and immobile form in Mollisols and thus cannot translocate iron and aluminum. As will be demonstrated later, evidence for iron, aluminum, and organic matter translocation can be used to suggest past vegetation and possible climatic effects on both relict and buried soils.

Parent material controls the formation of Spodosols, to some extent, because they are most common in sandy materials. Petersen[115] suggests that Ca^{2+}, Al^{3+}, and Fe^{3+} are released slowly from the parent minerals. The Ca^{2+} can be quickly leached from the soil and so does not accumulate to the point where it might cause the metallo–organic complexes to precipitate. In addition, the slow release of Al^{3+} and Fe^{3+} keeps the metal:chelate ratio low enough so that the complex is soluble and moves to greater depths. In contrast, clayey parent material might release the three ions at such a rapid rate that the metallo–organic complexes do not remain mobile.

Despite the voluminous work on podzolization in an attempt to find the connection between organic matter complexes, Fe, and Al, recent work suggests that such complexes may not be always necessary. Recent workers[5,39,53] have found that Al and Si can form a positively charged, soluble hydroxy–aluminum silicate complex, called proto-imogolite, that is stable at low pHs (<5). This proto-imogolite could represent the form in which Al is eluviated in the Bs horizons of Spodosols. Mixed Al_2O_3–Fe_2O_3 sols appear to be stable, with and without silica, and can also be eluviated.[54] These inorganic sols precipitate to form the Bs horizons. Eventually, the Al-silicates convert to imogolite, and the iron to ferrihydrite. This could be the mobile form of some Al and Fe in some Spodosols, and some Fe might move due to the low pH. Soluble organic matter complexes can form and move down and, because they carry a negative charge, are precipitated on the imogolite and ferrihydrite; the same organic matter complexes might carry small amounts of Al and Fe. Thus, the Al and Fe move as inorganic complexes or as ions, precipitate as materials of low crystallinity, and most of the organic matter is precipitated later.

The best way to test the many hypotheses for podzolization, and for many other pedogenic processes, is to analyze samples of water moving through the soil under field conditions. Ugolini and co-workers[49,149] have been doing this in a long-range study of podzolization and the environments in which the process can be detected. One study is on a Spodosollike soil in western Washington (Table 5-2). The soil data meet most, but not all, of the requirements for a Spodosol. Organic data on the soil indicate a buildup of fulvic acids in the B horizons, as well as a higher proportion of pyrophosphate-extractable C; these apparently move downward from the O horizon and are responsible for the movement of Fe and Al from the E horizon. Soil water analyses are used to show that these processes are still going on. Since Si solubility is not affected by organic matter complexes, Si is nearly uniform in amount with depth and is being

Table 5-2
Data for a Spodosol-like Soil in Western Washington

Horizon	Depth (cm)	Soil analysis*						Soil water analysis (mg/ml)**		
		pH	Organic C	Cp (%) <2 mm	$Fe_d + Al_d$	$Fe_p + Al_p$	Humic: Fulvic acids	Fe	Al	Si
01	5–3	3.9	56.0	15.4	—	—	0.36	—	—	—
02	3–0	3.6	53.6	15.7	—	—	0.33	0.04	0.75	3.37
E	0–11	4.3	2.4	0.8	0.3	0.2	0.25	0.03	0.68	4.36
2Bhs	11–31	4.7	6.9	5.4	2.3	2.6	0.10	<0.01	0.35	3.71
3B	31–53	4.8	9.7	6.4	3.9	3.9	0.12	—	—	—
4B	53–100	4.6	18.3	8.4	7.4	4.5	0.13	0.02	0.66	3.61

*p is the pyrophosphate extract, d the dithionite-citrate extract.
**Precipitation values are <0.01 Fe, 0.03 Al, and 0.09 Si, and amounts for the precipitation intercepted by the vegetation are 0.02 Fe, 0.06 Al, and 0.09 Si. Data for soil horizons generally are for the base of the horizon.
(Data from Singer and others[138], Tables 1, 2, and 3.)

leached from the system. In contrast, Fe and Al probably are affected by the complexes, and although not much Fe moves below the 2 Bhs horizon, Al moves deeper, probably to be precipitated in still lower materials.

TRANSLOCATION OF CLAY-SIZE PARTICLES

The distribution of clay-size particles in many moderately to strongly developed soils is marked by relatively low contents in the A and C horizons with the maximum amount in the "B horizon," generally in the upper part of the B. Several processes may account for this distribution. One such process is that the constituents of clay are derived by weathering higher in the profile and that they move downward in solution with the percolating water and precipitate as clay minerals in the B horizon. A second process is that the clays have formed in place from mineral weathering in the B horizon. A third is that the clays have moved as particles in suspension in the downward-percolating water to accumulate in the B horizon because of flocculation or constrictions in the pores through which the water moves, or because the base of the B horizon marks the lower limit of most water movement.[97] No doubt, in most soils, clays in the B horizon form by all three processes, but the relative importance of each may vary from soil to soil. It should be possible to differentiate clay particles translocated from those formed in place, but I know of no criteria that can be used to identify clay precipitated from downward-moving solutions.

Ratios of clay fractions, such as fine clay:total clay, can be used to indicate clay translocation.[106,140,150] The hypothesis is that the finer clay sizes are more mobile and so will move from the A to the B horizon or downward within the B horizon. An excellent example of this is a soil in New York in which the Bt1 horizon shows signs of being degraded to an E horizon. Here the percent clay in the ped interior of both the Bt1 and Bt2 horizons are comparable, but the data on total clay and on the coarse clay:fine clay ratio demonstrate removal of the fine clay fraction from the ped surface in the Bt1 horizon and its deposition as nearly pure films on ped surfaces of the Bt2 horizon (Table 5-3). In most cases, however, the

Table 5-3
Clay Data on a Degraded Bt1 Horizon and the Underlying Bt2 Horizon for a Soil Formed from Till, New York

	Bt1 horizon (41 cm)		Bt2 horizon (68 cm)	
	Clay (%)	Coarse: Fine clay	Clay (%)	Coarse: Fine clay
Ped surface	9.3	2.1	75.2	0.4
Ped interior	20.5	1.1	23.2	1.2

(Taken from Bullock and others,[30] Table 1.)

interpretation may not be so clear, since some ratios can be explained by parent material layering and others because the fine clay could be newly formed clay minerals.

Clay films lining ped surfaces or voids in both field and thin-section studies generally are taken as evidence for clay particle translocation (Fig. 1-11). This translocation can be verified by comparing detailed chemical analyses of the horizon with those of the film; the two usually differ because the film has been emplaced by downward movement more recently, and therefore the film has properties that more closely resemble those of the A horizon than of the B.[32] Thin-section analysis indicates that films of oriented clay along voids and ped surfaces that have sharp boundaries with the soil mass probably result from translocation.[21] Care must be taken, however, to eliminate the possibility of stress as the cause of oriented clay particles. Brewer[21] lists several criteria useful in identifying the origin of clay particle orientation. Thin-section evidence for origin of clay formation in place would be suggested by textural relations suggesting that the clays are pseudomorphous after the parent mineral grains.[151] Subsequent mixing might destroy the evidence, however.

Many kinds of data have been used to assess the relationship between the amount of clay formed in place from that introduced by translocation.[108] One common analysis involves the calculation of the volume of clay films in the soil, by horizon, to determine if the horizon of maximum clay content coincides with the horizon exhibiting the greatest amount of clay films. In many cases the match is not good because the films are commonly best expressed below the zone of maximum clay content.[21,33,104] In other cases, however, the match between the argillic B horizon and maximum expression of clay films is good.[63,106] Furthermore, in some soils the amount of illuviated clay, as evidenced by percent of clay films in thin section, fall short of the total clay in the B horizon (Fig. 5-2), thus suggesting that the B-horizon clay may not be derived entirely by illuvial processes. In another study, there is a poor correlation between clay translocation demonstrated by clay fraction ratios and that shown by thin-section study.[150]

One problem with using clay films as the only basis for demonstrating clay illuviation is that they are sometimes destroyed as soon as they form, or later. Nettleton and others[105] report clay-film destruction due to shrinkage and swelling in soils formed in the desert and Mediterranean climates of the southwestern United States. They found, for example, that soils with a low shrink-swell potential and a lower than 40 percent clay content can retain clay films and that soils with a higher shrink-swell potential and a higher than 40 percent clay content cannot retain films. Thus, many soils may have translocated clay, but the thin-section evidence for it is destroyed as the films become incorporated into the B-horizon matrix. Gile and co-workers[63,66] note that clay films in the desert region are most stable on sand and pebble surfaces, but they can be

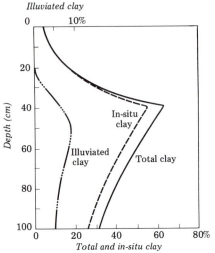

Fig. 5-2 Relationship among total clay, illuviated clay (recognized by oriented clay films), and, by difference, clay formed in place. (Taken from Brewer,[22] Fig. 3.) Because only some clay distribution is accounted for by illuviation, the clay in place represents that formed by weathering and/or that originally in the parent material.

destroyed during the accumulation of $CaCO_3$ in the soil. Clay-film destruction may also occur through mixing of the soil by roots and fauna.

Clay bands seem to provide good evidence for clay translocation.[50,60] Such bands usually consist of many thin layers with relatively high amounts of clay (~10 percent) that have an abrupt contact with interband layers of similar thickness, but much less clay (~1 percent). The banded material is in the B-horizon position and is common in sandy parent material, such as eolian sand. Although geological layering and *in situ* weathering are possible origins for the banded materials, the field and laboratory relations generally support a pedologic translocation origin. Several processes that could be responsible for the deposition of bands include (a) depth of water penetration for particular storm events or seasons, (b) subtle changes in grain size to retard downward movement of water, and (c) flocculation of clay at particular levels, say by $CaCO_3$.

Finally, the scanning electron microscope (SEM) might be used to help identify translocation soil clays, as Walker and others[151] have demonstrated for young sediments (Fig. 5-3). Chemically precipitated clay minerals form a crystalline texture on the host grain, and the morphology of the crystalline clay varies with mineral species. In contrast, mechanically translocated clays are recognized by platelets aligned parallel to the grain surfaces on which they are deposited and by platelet deposits that have the overall shape of menisci between grains, with an orientation more or less parallel to the former meniscus surface. For sediments, and

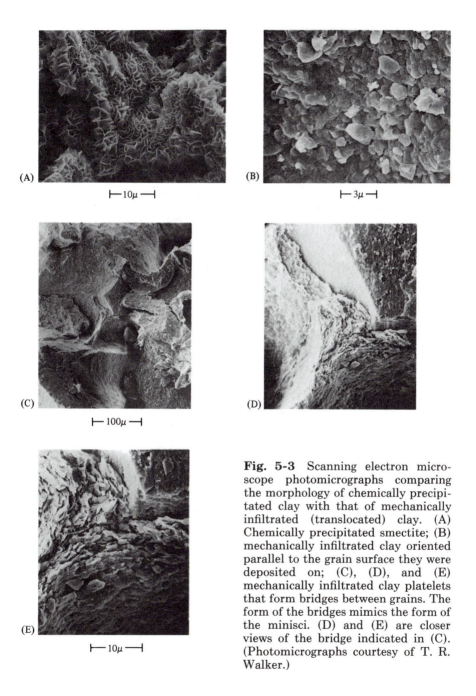

(A)

├─10μ─┤

(B)

├─3μ─┤

(C)

├─100μ─┤

(D)

(E)

├─10μ─┤

Fig. 5-3 Scanning electron micro-scope photomicrographs comparing the morphology of chemically precipitated clay with that of mechanically infiltrated (translocated) clay. (A) Chemically precipitated smectite; (B) mechanically infiltrated clay oriented parallel to the grain surface they were deposited on; (C), (D), and (E) mechanically infiltrated clay platelets that form bridges between grains. The form of the bridges mimics the form of the minisci. (D) and (E) are closer views of the bridge indicated in (C). (Photomicrographs courtesy of T. R. Walker.)

perhaps for soils, they point out that SEM photomicrographs point more clearly to a translocation origin than does thin-section evidence.

Quantitative mineral-analysis techniques have been developed by various workers to determine weight gains and losses of both clay and non-clay (or primary) minerals within the profile; these techniques can be used to assess clay formation in place from a clay distribution due to translocation,[9,21,94] but they can only be used for soils in which the parent material is uniform throughout the soil profile (Ch. 7). Unfortunately, it turns out that few soils have a uniform parent material, and this limits the usefulness of these techniques. All changes within the soil are based on comparisons with the unweathered parent material, which might be either rock or unconsolidated deposits at depth. A resistant immobile primary mineral, such as zircon or quartz in the sand fraction, is used as an index mineral. If one assumes that the amount of weathering of the index mineral is negligible, the ratios of all other primary minerals to the index mineral can be calculated. These ratios will decrease from the parent material upward into the soil as the less-resistant primary minerals weather, and they should be lowest where the weathering has been most intense. There should be some relationship between the loss in primary minerals and the formation of clay-size particles. Once the relationship is established, one can estimate the amount of clay formed per horizon and the amount of clay now present per horizon; the difference is the amount of clay that has been gained or lost by translocation (Fig. 5-4).

Brewer[21] has raised some objections to the above approach and called for the integration of both mineralogical and thin-section data to determine the formation and migration of clay-size material. Such a study was made on an Ultisol in Australia.[20] Knowing the approximate amounts of the rock-forming minerals that have weathered, as well as the clay mineralogy, one can calculate the maximum amount of clay that could form due to weathering in place (Fig. 5-5); this amount of clay will vary with the chemical composition of the clay fraction. Calculations were made for both kaolinite and for "complex" clay minerals (a variety of 2:1 minerals considered end products of weathering and found to be present on X-ray analysis). A comparison of the weight of kaolinite or of the "complex" clay mineral that could form from weathering in place with the actual clay present showed that much of the material released by weathering was removed from the soil and that only a small portion remained to form the clays. It is, therefore, not necessary to hypothesize clay illuviation to explain present clay distribution. Thin-section study revealed a lack of oriented clay and, thus, no evidence of clay illuviation. In this study, therefore, the weight of the evidence fails to support much clay illuviation as an explanation of the present clay distribution with depth. There is the possibility, however, that the evidence for clay movement has been destroyed, because the B horizons contain 30 to 43 percent clay.

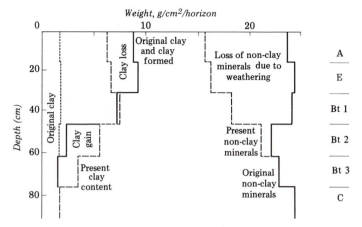

Fig. 5-4 Comparison between non-clay (or primary) mineral loss due to weathering and clay formation from constituents released during weathering, by the index-mineral method. (Data from Barshad,[9] Table 1.13 *in* Chemistry of the Soil by F. E. Bear, ed., © 1964 by Litton Educational Publishing, Inc. Reprinted by permission of Van Nostrand Reinhold Company.) Non-clay mineral losses and clay-mineral formation are estimated to be restricted to the upeer 61 cm; this thickness would depend upon the parent material choice. Differences between the distribution of the original clay plus the clay formed and the present distribution give approximate values for overall clay migration. Although Brewer[21] has some doubts about the uniformity of the parent material for this profile and the mineral content in the parent material, these are the kind of results one can obtain from analysis of the clay and non-clay fractions on a volume basis.

The examples given here demonstrate how difficult it is to find evidence that translocation of clay-size particles in soil profiles has occurred. Oriented clays in thin section generally are accepted as evidence for such movement. However, several studies show that subsequent pedogenesis can destroy this evidence. Clay and non-clay mineral relationships also can be used, but it is extremely difficult to calculate the gains and losses accurately. I would suspect that, as shown by the study of Brewer,[20] the losses of material upon weathering in most soils will far outweigh the gains due to clay formation. To prove clay translocation by these methods, one would have to demonstrate that the clay content in a particular horizon exceeds the maximum amount that could form only by weathering in place. However, any mineral weathering should be quantified by thin-section studies, and some of the weathered mineral forms shown in Figure 8-6 might be expected. As a final example combining many of the approaches described here, Chittleborough and Oades[40,41,42] present detailed morphological, micromorphological, chemical, and mineralogical data for a soil in Australia to determine the origin of marked clay increase with depth: the clay content of the A horizon is 10 to 20 percent, whereas

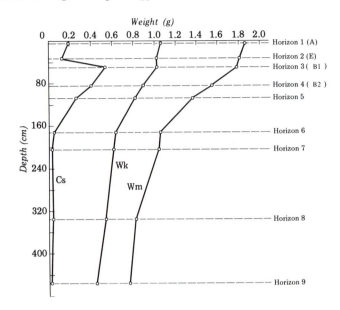

Fig. 5-5 Variation with depth in the weight of clay and fine slit (<5 μ) present in the soil, in the maximum weights of <5 μ material that could form from the amounts of mineral weathering and the specific clay minerals that have formed. (Taken from Brewer,[20] Fig. 4.) Here, Cs is the weight of <5 μ particles present in the soil derived from 1 cm^3 of parent rock and Wk and Wm are the weights of kaolinite and "complex" clay minerals, respectively, that could have formed from the observed weathering of 1 cm^3 of parent rock.

that in the B horizon is 80 percent. They conclude that translocation of clay inherited from parent material is the most likely reason for the extreme contrast in particle size with depth.

Clay migration requires that the clay be dispersed so that it can remain in suspension and be transported by water moving slowly through pores or cracks in the soil. Dispersion is favored by several factors, among them a low electrolyte content in the soil solution and the absence of positively charged colloids, such as iron and aluminum hydroxides.[9] Flocculation is induced by high electrolyte content in the soil solution and the presence of positively charged colloids; clays under these conditions cannot migrate.

The dispersion or flocculation of clay-size particles also depends on the thickness of the ion layer that satisfies the negative charge of the particle, and this thickness can vary with the ion present.[109] Ions attracted to particle surfaces are distributed so that their concentration is highest close to the surface, and concentration diminishes away from the surface (Fig. 5-6, A). When two particles, each with positive ions attracted to their surfaces, move toward one another, the initial reaction is one of repulsion (Fig. 5-6, B). Thus, the clays are dispersed. If, however, the clay particles

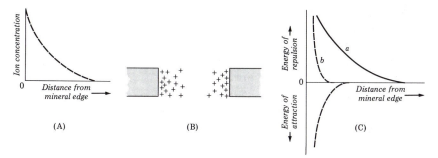

Fig. 5-6 Conditions for clay dispersion and flocculation. (Modified from Olphen,[109] Figs. 9 and 12 © 1963, John Wiley and Sons.) (A) Concentration of ions adjacent to a charged surface. Concentration is highest near the surface and decreases away from the surface. The total positive ion charge satisfies the negative clay particle charge. (B) Distribution of positive ions attracted to clay surfaces. If the two clays are brought close together, and the ion layers are thick enough, flocculation cannot take place and the system remains dispersed. (C) Energy relations for repulsive and attractive forces between two clay particles with positive surface ions. In plot a, the ion layer is thick; flocculation does not occur because the surfaces are too far apart for short-range van der Waals forces of attraction to operate. In plot b, the ion layer is thin; because of either higher electrolyte content or higher valency cations, van der Waals forces operate once the clays come in close contact and flocculation occurs.

can move closer together, van der Waals force takes over, and attraction and flocculation occur. The particles also can come close together if the thickness of the ionic layer is reduced (Fig. 5-6, C). This reduction can be brought about in two ways. One is to increase the electrolyte concentration of the soil solution and the other is to replace the ion layer with ions of higher valency.

These flocculation-dispersion effects can be seen in soils.[97] Calcium clays have a thin ion layer, they commonly are flocculated, and thus, commonly, do not migrate. Soils high in $CaCO_3$, for example, show little evidence of clay migration; migration can take place only after the carbonate has been leached from the soil and some of the Ca^{2+} in the exchange complex replaced. In contrast, clays with appreciable amounts of Na^+ in the exchange complex remain dispersed because the ionic layer is thick. The pedogenic translocation and subsequent accumulation of Na-clays results in a natric horizon, which requires the same clay-increase relations as an argillic horizon. The dispersive effects of Na^+, however, is such that in some desert basin deposits of the western United States, natric horizons can form in about one-half the time it takes argillic horizons to form.[2,116] Furthermore, coastal regions are characterized by high inputs of atmospheric Na^+,[77,161] and for reasons given above, the rate of formation of Bt horizons is accelerated.[135] At higher amounts of Na^+, flocculation results because the high electrolyte content in the soil solution compresses the adsorbed ion layer.

Some experiments have been performed to determine the factors responsible for clay migration.[23,74] Besides verifying, to some extent, the cation relationships discussed above, the experiments indicate an upper limit of 20 to 40 percent clay, depending on the clay mineral, above which migration in the pore spaces virtually ceases. At these clay contents, the pore spaces may be small enough to limit movement, or the swelling of clays upon hydration may close off some pores. Clay migration at greater clay contents would have to take place along soil-structure discontinuities, and as mentioned earlier, soils with such high clay contents tend to lose the clay-film evidence for migration quite rapidly.

Conditions for clay migration can vary with depth in the profile, or even seasonally. Quite commonly, the electrolyte content will increase with depth because the upper parts of the profile have more water moving through them. Thus, clays could be dispersed near the surface, but be flocculated at depth. In soils with a high clay content, high shrink–swell potential, and distinct dry and wet periods, clay movement might take place only at the onset of the wet season, when an open prismatic soil structure with wide cracks reaches the surface and thus provides avenues for rapid clay translocation.

Still another way in which clays can be translocated in a soil is by suspension originally in rainwater. Holliday[78] mentions this as a mechanism to form argillic horizons within about 350 years at Lubbock, Texas. In that area it is not uncommon for a dust storm to precede a thunderstorm, and the subsequent rain commonly comes through the dust, producing a mud rain. Such occurrences are not mentioned in the literature dealing with translocated clay, but they might be quite important. One appealing aspect of this process is that it does not require that clays already present in the soil be brought into suspension before translocation by slowly moving soil water. A similar mechanism might also be responsible for the movement of clays in calcareous materials. It was argued above that such movement would not be expected, but it has been demonstrated in an artificial field experiment.[70]

TRANSLOCATION OF SILT

Some coarse-grained soils display an increase in silt with depth. An example of this is a soil formed in till at least 100,000 years old in the eastern Canadian Arctic, in which the silt-enriched horizon underlies the Bw horizon (Table 5-4). Commonly, the silt-enriched horizons also have silt caps (Appendix 1) on the clasts.[36] This particle-size distribution could reflect parent material layering or be due to some post-depositional process. Because many of these soils are in cold climates, one must consider frost sorting as a possible mechanism to redistribute silt.[15,145,154] That is not the only mechanism, however, for soils at high latitudes and altitudes or for other areas. Some sites are on well-drained moraine crests, where

Table 5-4
Data for a Soil with a Silt-enriched Layer (Cox1), Baffin Island, Eastern Canadian Arctic

Horizon	Depth (cm)	Color	Particle-size distribution in <2 mm fraction (%)		
			Sand	Silt	Clay
A	0–1	10YR 4/2.5	82.3	14.6	3.1
Bw	1–12	10YR 4/4.5	91.6	5.4	3.0
Cox1	12–28	2.5Y 5.5/4	66.4	29.3	4.3
Cox2	28–106	5Y 7/4	90.1	7.3	2.6
Cox3	106–116+	5Y 7/3	89.4	7.8	2.8

(Data from Birkeland,[11] Table 5.)

arguments can be made that frost stirring is minimal. Locke[89] has done the most recent work on these accumulations in arctic soils and favors translocation by soil water. Work in the laboratory demonstrates the movement of silt, provided the voids are of the proper size, such as those in coarse-grained soils.[160] The silt accumulation could be (a) a redistribution of materials of an originally uniform parent material or of silt particles produced by mechanical weathering, (b) derived from a surface layer of loess, or (c) materials infiltrating the soil as rapidly as they are delivered to the surface. Oxygen-isotope composition of quartz grains might help identify any materials in the soil suspected of being of eolian origin from a distant source. Wilding and others[157] review the literature on this and point out that different geological environments in which quartz forms are characterized by different compositions.

FRAGIPAN AND ITS ORIGIN

Fragipan, as reviewed by Grossman and Carlisle[72] and Soil Survey Staff,[140] is a subsoil horizon that is brittle and rigid when moist, yet yields suddenly in a brittle fashion when pressure is applied to a sample. The style of breakage is likened to that of a dry graham cracker. Such behavior would make one suspect a cementing material, but if one exists it must be weak, for the material slakes in water. Bulk density is usually greater than 1.6 g/cm^3. Thickness of fragipan ranges from 15 to 200 cm, with the top at about 25 to 150 cm depth. Fragipan is most common in loamy materials, in climates characterized by water moving through the soil at some time of the year, warm or cold temperatures, and tree vegetation being most common. Very coarse prismatic structure is common, as are bleaching and mottling along the prism walls.

Fragipan is found in four soil orders: Alfisols, Inceptisols, Spodosols, and Ultisols. In wetter soils, it is found beneath the eluvial horizon. In many other soils, however, there is an intervening spodic, cambic, or argillic horizon, and the fragipan is in the B or C horizon. Commonly, an E horizon is at the top of the fragipan, formed perhaps by lateral flowage of water because water movement is impeded in the fragipan. The horizon sequence gives rise to bisequal soils, that is soils with two vertical and separate eluvial–illuvial sequences. The lower horizons in a bisequal soil are denoted by a prime. Thus, a bisequal soil with fragipan can have a A/E/Bt/E'/Btx horizon sequence, with x designating fragipan (Table 1-1).

The origin of fragipan is obscure.[72] There seems to be no chemical or mineralogical association, other than the fact that fragipans are low in organic matter and usually noncalcareous. The binding material might be silicate clay minerals, hydrated oxides of Al and Fe,[73] or silica. The latter seems to have many proponents.[126,143] Some workers cite climatic change as an aid in the formation of some fragipans. However, many fragipans seem to have formed during the Holocene,[126] so climatic-change arguments are restricted to these changes documented for the Holocene.[127] Other fragipans have been shown to predate the Holocene.[31]

ORIGIN OF OXIC HORIZONS

The origin of oxic horizons has been reviewed by several workers.[35,51,71,92,94,112,139,140] The oxic horizon is characterized by the extreme chemical alteration of the original parent material. In places, the alteration can extend for tens of meters. The profiles are red, clay content is high (many have greater than 50 percent clay), with there being little evidence for clay illuviation, and horizonation is weak. Subspherical nodules up to 1 to 2 cm across are reported in some of these horizons. Despite the high clay content and the high rainfall in areas where oxic horizons are forming, movement of soil moisture is rapid because of a well-developed microstructure. Oxic horizons can form in place, or the materials can be transported from previously weathered landscapes. It is difficult to differentiate between these two modes of formation, considering the thickness of the materials and the likelihood of adequate exposures, but the presence of stone lines is commonly cited as being helpful in identifying erosional surfaces at the bases of such deposits.[124]

Chemical analyses commonly show the near total depletion of cations and silica and the enrichment of iron, aluminum, and titanium, alone or in combination, relative to the amounts in the primary aluminosilicate minerals (Fig. 5-7). Some horizons are characterized by extreme aluminum enrichment, others by extreme iron enrichment.[71] These differences could be attributed to parent material differences or to biochemical conditions during weathering.[51] The silica that remains in the soil generally is seen as the original quartz grains, secondary quartz,[157] or is combined in secondary clay minerals. Oxides and oxyhydroxides of iron and aluminum, alone or in combination, are common; in many places they consti-

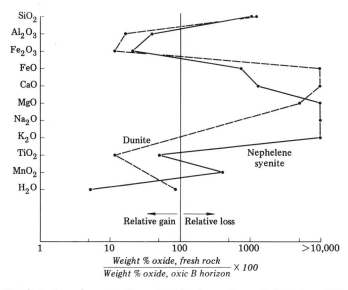

Fig. 5-7 Gain-loss diagram for oxic B horizons formed from two different parent materials. Water gain is due to its incorporation into the crystal lattices of the sesquioxides and clay minerals that form; FeO loss could be real or due in part to oxidation to Fe_2O_3. (Data from Maignien,[92] Tables 3 and 4, *Review of Research on Laterities, Natural Resources Research IV*, 1966, reproduced by permission of UNESCO.)

tute most of the material present. The common minerals of these latter constituents are goethite, hematite, and gibbsite, and kaolinite is the common aluminosilicate clay mineral.[51,76] Weathering of primary minerals often has been so intense that few are left, and there is very little cation reserve from further weathering of the primary minerals. What cations there are occur at exchangeable sites of the colloidal fraction or in the plant tissue. Plant nutrition, therefore, depends upon the cycling of nutrients between the upper part of the soil and the plant itself. If this cycling is disturbed, the cations can be leached and the soil rendered essentially devoid of nutrients.[58]

Many of the processes that produce oxic horizons operate in other soil horizons. Most cations and silica are soluble under the usual soil conditions and so can be leached from the soil. In many areas where these horizons occur, the leaching conditions are so intensive and the landscapes so old, dating perhaps to the Tertiary, that near total removal of cations and silica is not unexpected. However, under usual biochemical conditions, the sesquioxides released from the parent rock are relatively insoluble and can accumulate. In some soils, however, the sesquioxide content, especially that of iron, is greater than would be expected as a residual from the weathering of the parent material. Under these circumstances, iron may be derived from overlying horizons, perhaps now partly eroded away, or be transported in with laterally moving groundwater from higher parts of the landscape (Fig. 9-9). Combination with organic compounds, perhaps as chelates, could solubilize iron, as could low acidity and reducing conditions. A high or fluctuating water table seems to be important mainly in producing conditions for more intensive alteration and the production of colorful mottling, some of which is due to the segregation of iron. Where clay minerals are not too plentiful, leaching conditions may have been too intense or the biochemical environment inappropriate for their formation. Once clays form, they do not seem to migrate very far, even though most oxic horizons are relatively porous and can translocate large quantities of rainfall. Perhaps the mutual attraction of the negatively charged clay and the positively charged susquioxides to give stable aggregates prevents much clay translocation, since this seems to be the reason why dispersion of laboratory samples is so difficult.

Some oxic horizons are hard due to exposure to drying conditions.[3] The term "laterite" is restricted to these materials. In contrast, plinthite, as used in Soil Taxonomy, is a soft sesquioxide-rich material capable of hardening irreversibly on drying, and Daniels and others[48] give clues for its recognition in the field. Sufficient iron and dehydration seem to be necessary for hardening to occur, concomitant with increased crystallinity and continuity of the crystalline phase of already existing iron compounds. This results in the formation of goethite and hematite in a rigid network of crystals that cement the material together. Dehydration necessary for the hardening may come about by a natural change in vegetation from forest to savanna, by the clearing of forests for agricultural and other purposes, or by exposure by erosion.

ORIGIN OF CaCO$_3$-RICH HORIZONS, AND THOSE WITH MORE SOLUBLE SALTS

Soils in semiarid and arid regions commonly have carbonate-rich horizons at some depth below the surface, or if the climate is dry enough or the surface erosion intensive enough, these horizons may extend to the surface. Although several origins have been presented for some of these horizons,[12,19,28,71,118,119] our concern here is with CaCO$_3$-rich horizons of pedogenic origin.[8] Some of these horizons are the caliche and calcrete of present and past geologic literature. Pedologists call these accumulations Bk and K horizons, and because they have recognized and defined stages in the buildup of CaCO$_3$-bearing horizons[61,62] (see Appendix 1), this more precise terminology seems preferable to the general term caliche or calcrete. Both calcium and magnesium carbonates are present in these soils, with the former dominant.

The origin of carbonate horizons involves carbonate–bicarbonate equilibria, as discussed by Jenny[80] and Krauskopf,[84] and shown by the following reactions

$$CO_2 + H_2O$$
$$g \qquad 1$$
$$\Updownarrow$$
$$CaCO_3 + H_2CO_3 \rightleftharpoons Ca^{2+} + 2HCO_3^-$$
$$c \qquad aq \qquad aq \qquad aq$$

An increase in CO_2 content in the soil air or a decrease in pH will drive the reaction to the right; carbonate will dissolve and move as Ca^{2+} and HCO_3^- with the soil water. Dissolution is also favored by increasing the amount of water moving through the soil, as long as the water is not already saturated with respect to $CaCO_3$. Precipitation of carbonate occurs under conditions that drive the reaction to the left, that is, a lowering of CO_2 pressure, a rise in pH, an increase in ion concentration to the point where saturation is reached and precipitation takes place, or evapotranspiration of the soil moisture.

All the above conditions are found in soils in which $CaCO_3$ has accumulated. Carbon dioxide partial pressures in soil air are 10 to more than 100 times that in the atmosphere[12,27,29]; this decreases the pH which, in turn, increases $CaCO_3$ solubility (Fig. 5-8, A). The partial pressure of CO_2 is high as a result of CO_2 produced by root and microorganism respiration and organic matter decomposition. Thus, one would expect the highest CO_2 partial pressure to be associated with the A horizon, with values diminishing down to the base of the zone of roots. The amount of water leaching through the soil also is greater near the surface than at depth, so as the water moves vertically through the soil, the Ca^{2+} and HCO_3^- content might increase to the point of saturation after which further dissolution of $CaCO_3$ is not possible. Combining the effects of high CO_2 partial pressure and downward-percolating water, we might visualize the formation of a $CaCO_3$-rich horizon as follows. In the upper parts of the soil, Ca^{2+} may already be present or may be derived by weathering of calcium-bearing minerals. Due to plant growth and biological activity, CO_2 partial pressure is high and forms HCO_3^- upon contact with water. Water leaching through the profile can carry the Ca^{2+} and HCO_3^- downward in the profile. Precipitation as a $CaCO_3$-rich horizon would take place by a combination of decreasing CO_2 partial pressure below the zone of rooting and major biological activity and the progressive increase in concentration with depth in Ca^{2+} and HCO_3^- in the soil solution as (a) the water percolates downward and (b) water is lost by evapotranspiration. The position of the $CaCO_3$-bearing horizon is, therefore, related to depth of leaching, which, in turn, is related to the climate.[71] Recent work suggests that the carbon- and oxygen-isotope composition of pedogenic carbonate can be used to differentiate between precipitation by CO_2 loss and that by

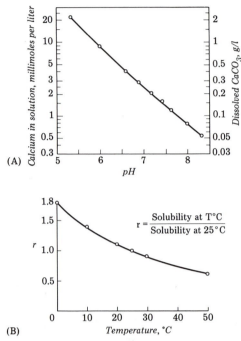

Fig. 5-8 Some factors affecting $CaCO_3$ solubility. (Taken from Arkley,[7] Figs. 6 and 7, © 1963, The Williams & Wilkins Co., Baltimore.) (A) Solubility of $CaCO_3$ with varying pH in an equilibrium solution of $CaCO_3 - CO_2 - H_2O$ at 25 °C. (B) Relationship between temperature and $CaCO_3$ solubility.

evapotranspiration.[129,130] A more sophisticated model for the formation of carbonate horizons has been presented by McFadden.[95]

Temperature also affects $CaCO_3$ equilibria. Because CO_2 is less soluble in warm water than in cold water, $CaCO_3$ solubility decreases with rising temperature (Fig. 5-8, B). This temperature effect may not be too great in one profile, but it is important in comparing the depths to the tops of Bk or K horizons between regions of contrasting temperature.

Several stages in the buildup of carbonate horizons are recognized.[61,62] In gravelly material, the first stage is the appearance of carbonate coatings on the undersides of gravel particles; in non-gravelly material, the first stage is the occurrence of thin filaments (Fig. A-1, Appendix 1). The undersides of gravel probably are favored sites initially because down-ward-moving water would tend to collect there. With time, in both parent materials, the horizon is increasingly impregnated by carbonate deposition on solids until the voids become plugged and water percolation through the horizon is greatly restricted. At this point, water tends to collect periodically over the plugged horizon; the resulting solution and reprecipitation produces the laminated part of the upper K horizon. At this point in the development the horizon builds upward and so is younger in that direction. During the buildup of a K horizon, the volume of pedogenic carbonate eventually exceeds the original volume of the pores. In most cases, this is attributed to the carbonate, upon crystallization, forcing the silicate grains and gravel apart,[8,156] but there also are reported instances of carbonate replacing silicate minerals.[75] The buildup of $CaCO_3$ in the diagnostic stages (Appendix 1) is more rapid in gravel than in nongravelly material because gravel has less pore space.[55,66] Finally, progressively higher stages are usually attained in the older soils of a chronosequence; in any one soil profile, however, the highest stage attained is usually near the top of the carbonate-enriched horizon, and this changes to progressively lesser stages with depth.

Several other morphological features are present in indurated K horizons that usually have morphologies of stages IV, V, and VI. One is ooids, which are sand-size round particles in which a nucleus is surrounded by one or more concentric layers of fine-grained calcite grains (micrite). The origin of ooids is not clear, but those layers studied by Hay and Reeder[75] seem to result from carbonate replacing clay coats on sand grains. Another common feature is pisoliths, which are subangular to spherical bodies, 0.5 to more than 10 cm across, surrounded by many thin layers of micrite. The nucleus of pisoliths can be anything: rock fragments to pieces of broken and rotated indurated K horizon material, some of which display internal laminar layers and pisoliths from previous pedogenesis. Both of these features usually are set into massive carbonate cement (Fig. 5-9).

The chemistry and mineralogy of K horizons has interested workers because some of the best developed ones approach the composition of limestone (Table 5-5); in these $CaCO_3$ dominates the analyses, and in the best developed K horizons approaches 90 percent. Some of the MgO is in the form of dolomite,[79] part of which could be primary, but in one instance, dolomite has been shown to be pedogenic.[75] The rest of the MgO can be in original silicate grains as exchangeable ions of the clay minerals or as pedogenic clay minerals. In this last group are the pedogenic Mg-enriched fibrous clay minerals. Bachman and Machette[8] point out that in K horizons of stages IV to VI with high carbonate contents, the Ca:Mg molar ratio of the carbonate fraction increases relative to that of the parent material and that these increases coincide with the clay–mineral changes to the more Mg-rich clays:

Increasing total pedogenic carbonate $\longrightarrow$

Montmorillonite
Illite-Montmorillonite $\rightarrow$ Palygorskite $\rightarrow$ Sepiolite

The ratio for carbonate in the parent material can be 10 to 20 and increase to 100 to 160 in the best developed K horizons. In short, the Mg is moving from the carbonate to the clay mineral fraction. The rest of the oxides probably reside in silicate grains or in engulfed B-horizon material, but some of the SiO_2 could be pedogenic opal, chalcedony, or silica gel.

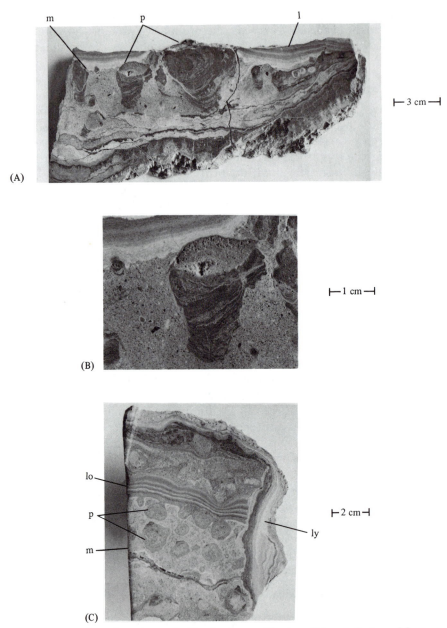

Fig. 5-9 Polished sections of a Km horizon with stage VI morphology, Mormon Mesa, Nevada. A. Sample here is almost entirely $CaCO_3$, with pisoliths (p) set in a matrix of massive carbonate (m) and the top of this horizon marked by laminar carbonate (l). B. Closer view of pisolith in (A). C. Sample with complex carbonate relations in which old pisoliths (p) are set in a massive groundmass (m), and still younger are two sets of laminar carbonate deposition (lo is older and ly is younger).

Table 5-5
Comparison of Chemical Analyses for Calcretes, K Horizons, and Limestones

Oxide	Average of 300 calcretes	K horizons		Average of 345 marine limestones
		Pleistocene age, Tanzania	Pliocene age, New Mexico	
CaO	42.6	44.4	45.4	42.6
SiO_2	12.3	8.5	13.2	5.2
MgO	3.1	1.1	0.4	7.9
Al_2O_3	2.1	2.5	1.4	0.8
Fe_2O_3	2.0	2.7	0.5	0.5
K_2O	—	1.0	0.2	0.3
Na_2O	—	0.6	0.1	0.1
CO_2	—	36.0	36.0	41.6
Total $CaCO_3$	79.3	81.8*	81.8*	94.5*

*Based on CO_2 content and includes $MgCO_3$.
(Taken from Goudie,[71] Hay and Reeder,[75] Aristarian,[6] and Clark[44]).

The $CaCO_3$ of the carbonate horizons may come from several sources.[8,46,119] Thus, Ca^{2+} released by weathering could combine with HCO_3^- deeper in the profile, and if this is the origin, there should be a close relationship between the CaO content of the parent material, the amount of weathering in the upper part of the soil to release Ca^{2+}, and the amount of $CaCO_3$ in the Bk or K horizon. For most K horizons, derivation of the carbonate by parent material weathering is highly unlikely. Gardner[59] has calculated that 37 to 90 m of material would have to be weathered to release the amount of Ca present in the K horizon of Mormon Mesa, Nevada. Not only does the lack of mineral weathering negate such an origin, but geomorphologically, it is highly unlikely. For example, tens of meters of weathered material have to be removed from the mesa and the carbonate concentrated in a horizon of several meters, while a well-preserved terrace formed with a surface that has a consistent relationship with the position of the K horizon. In many places, an external source of Ca^{2+} seems most likely. Detailed study in the Las Cruces region of New Mexico indicates that the atmosphere is an important source for Ca^{2+} and $CaCO_3$.[62,66,125] Dust-trap data[66] at Las Cruces suggest that the dust contains less than 5 percent carbonate and that the latter is added to the soil at a yearly rate of about 0.2 to 0.4×10^{-4} g/cm^2. However, the amount of Ca^{2+} in yearly precipitation[81] is such to produce perhaps three times that amount of carbonate, giving total annual carbonate production of about 2×10^{-4} g/cm^2. This can only be an estimate because, although the Ca^{2+} in precipitation probably is carried into the soil, the carbonate in dust could be carried to other sites before precipitation events dissolve it and move it into the soil. A similar atmospheric origin for most pedogenic carbonate is favored for semiarid to arid regions in the western United States[8] and other parts of the world.[46,163] A locally important source for the Ca^{2+} in carbonate is atmospheric gypsum, for on dissolution, Ca^{2+} is released.[86]

In places, a pedogenic carbonate horizon will form in carbonate parent material. In many of these cases, it is difficult to determine what fraction of the carbonate has gone through dissolution and reprecipitation and can thus be termed pedogenic. Carbon-isotope composition can help solve this problem, since it has been shown that the isotopic composition of the parent material carbon can be much different from that of the CO_2 in the soil atmosphere. If the carbonate equilibria equation is recalled, a pedogenic composition will probably be between the two isotopic values, and from this it is possible to calculate the percent of the total carbonate that is pedogenic.[91,129,130]

Accumulation of $CaCO_3$ by capillary rise from a perched high water table may explain some $CaCO_3$ occurrences,[93] but such occurrences are not thought to be widespread. This is because capillary rise in coarse alluvium is nil, and impermeable beds necessary to perch the water table are uncommon. Also in many areas, streams have downcut following deposition of alluvium, which has lowered the water table far below the level that could possibly be reached by the capillary rise of water with its dissolved carbonate salts.

Finally, it is important to be able to differentiate between pedogenic carbonate and groundwater carbonate when working with buried soils because it is obvious that the latter is of little pedologic or geomorphical importance. Pedogenic carbonate should have features like those of the six morphological stages with the correct depth distribution. In addition, in most cases there should be overlying soil horizons, and their development ideally should bear some relationship to the carbonate stage attained (see descriptions in Gile and Grossman[65]). For soils with high amounts of carbonate (K horizons), pedogenic carbonate is mainly micrite and has dry colors of 7.5YR 7/4 to 10YR 8/3, and the original grains have been forced apart.[8] In contrast, groundwater carbonate is coarser grained, fills only the original pore spaces, and because of overburden pressure has not forced the original grains apart.

In areas of aridity (precipitation < 300 mm/year), salts more soluble than the carbonates may accumulate at depth in well-drained soils above the influence of the water table.[47,107,110,155] They even occur in polar regions.[14,16,43,145] Common salts are gypsum ($CaSO_4 \cdot 2H_2O$) and halite (NaCl) (Table 5-6). Such accumulations may qualify as gypsic or salic horizons, respectively, if the salt concentrations are high enough, and the former is termed petrogypsic if cemented.[140] Some of the latter can contain over 90 percent gypsum, and at such high contents the original silicate grains are forced aside during gypsum crystallization. These salts

Table 5-6
Analysis of a Gypsiorthid Formed in Gravel in the Central Sinai with about 50-mm Annual Precipitation

Horizon	Depth (cm)	Electrical Conductivity (mmho/cm)	Cations* meq/100 g soil				Anions*			Gypsum (meq/100 g soil)	pH
			Ca	Mg	K	Na	HCO$_3$	SO$_4$	Cl		
A1	0–2	9.4	0.8	0.38	0.051	2.5	0.093	1.31	2.3	0.3	7.8
A2	2–9	34.9	8.0	4.47	0.028	5.8	0.079	6.44	11.8	8.1	7.4
Bz1	9–19	46.0	14.2	3.65	0.024	7.6	0.07	8.05	17.4	7.7	7.4
Bz2	19–35	86.3	23.3	7.23	0.028	43.1	0.09	23.22	50.4	139.0	7.3
Byz	35–47	182.6	18.8	5.34	0.018	142.4	0.097	32.00	134.5	103.0	7.2
Cy1	74–75	24.6	6.9	3.40	0.013	11.9	0.13	10.46	11.6	404.0	7.8
Cy2	75–100	48.1	7.6	3.88	0.012	21.3	0.10	9.60	23.1	293.0	7.8
Cy3	100–110	44.0	7.8	3.52	0.012	19.1	0.10	8.12	22.2	331.0	7.7
C	110–120	51.8	7.8	4.31	0.011	18.0	0.08	8.08	22.0	207.0	7.7

*Water extract from saturated paste.
(Data from Dan and Yaalon,[47], Table 4.)

accumulate at depths reached by mean annual soil water, or by extreme rainfall events, and if both salt accumulations occur in one profile, the depth of maximum amount of halite probably is deeper than that for gypsum because the former is more soluble.

The source of the salts is a problem. Some are derived from salts in the rock or the sediment the soils form in. Another source for sulfur would be release during the weathering of pyrite. If the parent materials are sufficiently low in salts, however, pedogenic salts probably have an atmospheric origin, either as solid matter or dissolved in precipitation.

It is not uncommon for the horizon of carbonate or gypsum accumulation to occur at the surface. This might be due to extreme aridity, to derivation from a former shallow water table, or to erosion of all overlying horizons. Knowing the present field relations, the geological history of the area, and the mean annual precipitation might favor one origin over another. For example, the precipitation in many parts of the western United States is such that Km horizons should be at some depth; when they are not, erosion of overlying horizons usually is suspected.

HORIZONS ASSOCIATED WITH HIGH WATER TABLE OR PERCHED WATER TABLE

Some soil horizons are associated with high water contents within the soil due to high regional or local water tables or to horizons within the soil that impede the downward movement of water. Quite commonly, soils under the influence of a water table occur in the lower parts of the landscape, juxtaposed with soils at a higher topographic level exhibiting more normal drainage conditions. In some places, however, they can be quite extensive.

Poor water drainage and the accompanying low oxygen content, in the presence of organic matter, leads to reducing conditions in a soil.[18,121,123] The Eh-pH diagrams show that iron and manganese will be in the reduced state and as such can migrate with the soil solutions. If reduction is the dominant process, the soil has the characteristic gray, greenish gray, and bluish colors that are due to some combination of iron loss and the compounds that form in such an environment. Reduced soils with the above colors are commonly termed gleyed. A high regional water table will produce extensive gleyed soils. In some profiles, however, layers with sufficient textural contrast can produce a local perched water table, and thus gleyed horizons, in an otherwise well-drained profile. The contrasting layers can be parent material layering, or they can be pedogenically formed; an example is the gleyed basal Btg horizon above a duripan in Southern California.[147]

Brinkman[25] has coined the term ferrolysis for a complex process that results in acid soils in which E horizons with poor structure form and grow downward at the expense of the underlying B horizon. A relatively impermeable barrier within the soil, such as a Bt horizon or fragipan, causes alternating oxidation and reduction in the overlying material, so the iron alternates between the Fe^{2+} and Fe^{3+} state. In the reducing mode, Fe^{2+} displaces the other exchangeable ions on the clay minerals, and the latter, as well as some Fe^{2+}, can be leached from the soil by laterally moving waters. In the oxidizing mode, however, ferric hydroxide forms as a precipitate and the soil becomes more acid. With progressively more acid conditions, some clay minerals are destroyed and others, because of the relatively high solubility and availability of aluminum at low pH, can be converted to Al-chlorite. The result is an acid soil with a prominent E horizon that has an abrupt and wavy contact with the underlying mottled Bt horizon. Although the impermeable barrier that sets up the process can have many origins, Pedro and others[114] describe how the process can act on an originally uniform loamy parent material, and the Bt horizon that forms with time eventually sets up the ferrolysis conditions that result in the destruction of the upper parts of the B. The sequence is

Eutrocrept → Hapludalf → Fraglossudalf → Fragiaqualf

They also describe the similarities and differences between this evolutionary sequence and the one involving podzolization.

Other soils may show a similar process. In Alaska[148] and New Zealand,[122] iron-cemented B horizons, formed by podzolization processes, have so restricted soil drainage in the overlying material that iron is depleted to the point that an E horizon forms. In the Alaska example, continuing reducing conditions eventually solubilize the cementing materials, the B horizon disappears, and a bog results.

A fluctuating water table can produce very colorful horizons in which the drab colors that characterize reduced conditions are mixed with the bluish black of manganese compounds and the bright yellows and reds of oxidized iron compounds. Under this water-table regime conditions fluctuate between reducing and oxidizing, and the iron and manganese alternate between being mobile as Fe^{2+} and Mn^{2+} or precipitated in the Fe^{3+}, Mn^{3+} and Mn^{4+} form. The result can be a net loss of iron and manganese,[52] but also with local iron and manganese enrichment in the brightly colored mottles. The position of mottles helps indicate the position of the water table.[136] An extreme example of mottling is produced in soils with these water regimes in humid tropical regions.[90,101] Some of these conditions can lead to the formation of hard iron and manganese concretions; in other places, plinthite is formed.

In areas of relatively high evapotranspiration, under the influence of a high local or regional water table, salts can accumulate to appreciable concentrations.[4,24] Common cations are Ca^{2+}, Na^+, Mg^{2+}, and K^+ and common anions are Cl^-, SO_4^{2-}, S^{2-}, and HCO_3^-. The position of the salt concentration can be either at the surface or at some depth, depending upon the depth to the water table and the height of capillary rise. The latter will differ with the texture of the soil, being greater for finer-grained materials.

In geomorphological studies, it is important to differentiate salt accumulations of this origin from those mentioned earlier in well-drained soils. In the latter case, the salts commonly have an external origin, their position is very sensitive to mean annual precipitation, and the more soluble ones occur at greater depth. With a capillary rise origin, however, the position of the salt accumulation would not bear much relationship to mean annual precipitation, the more soluble salts might be closest to the surface,[111] and the rate of salt accumulation could be more rapid.

PHOSPHORUS TRANSFORMATIONS

In New Zealand, there has been a considerable amount of work on the forms of phosphorus that are significant in soil genesis.[153] The total P in a soil is in four fractions, as follows:

Pca: P of the original minerals, such as apatite

Poc: Occluded P, or that fraction that occurs within coatings or concretions of oxides and hydrous oxides of iron and aluminum

Pnoc: Non-occluded P, or those P ions adsorbed at the surfaces of the above oxides and hydrous oxides and $CaCO_3$

Po: That fraction bound with the organic matter

An earlier fractionation scheme that is probably just as useful for pedologic studies recognizes three fractions, with acid-extractable P (Pa) being approximately equal to Pca + Pnoc, Pf to Poc, and the remainder reported as Po.[152] In areas of relatively high leaching, the original Pca declines in total amount and converts to the other forms (Fig. 5-10). As will be shown later, climate and the amount of leaching dictate the rate of transformation and the amounts in each form. Generally in a leached soil profile, Pca and Pa are progressively depleted toward the surface, whereas Po increases toward the surface in a fashion similar to organic matter trends.

PROCESSES THAT DESTROY SOIL HORIZONS

Most of this chapter has been spent on processes that produce soil horizonation; however, a few processes subsequently can destroy them. It has been mentioned that ferrolysis can destroy parts of a previously formed Bt or Bs horizon.[114] Another way to destroy a Bt horizon in sandy parent material is to initially form a A/Bt soil profile, perhaps by increments of airfall dust, but eventually the clay content becomes so high that the soil becomes a Vertisol, mixes with shrinking and swelling, and so loses its marked textural horizonation.[34,103] Tree windthrow is a process

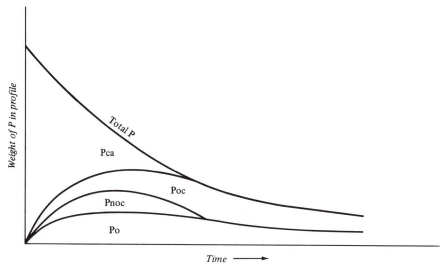

Fig. 5-10 Relative change in the amounts of forms of phosphorus (P) in soils with time. (Taken from Walker and Syers,[153] Fig. 1.)

that will periodically mix horizons and thereby slow down the rate of marked horizon formation.[102] Climatic change could so change the vegetation that the soil is susceptible to removal by wind erosion. Freeze-thaw cycles can have several effects. One is to disrupt horizonation.[145] The other is that if needle ice can form, especially if the surface layer is loess, the surface is much more susceptible to wind erosion.[141,142] In areas of K horizon development, the top of the K builds up and engulfs the lower part of the B horizon.[65] Animals can also mix and destroy some horizons. The above is but a short list of destructive processes, but it is important to keep these processes in mind when trends in soil chronosequences that are considered predictable are not found in the field. An example of this is the absence of many trends with age for soils near Laramie, Wyoming, in spite of some deposits being over 0.5 million years old.[120] The reasons for this are not clear, but strong winds and paleopermafrost conditions[100] might be contributing factors.

Some of these processes are what Torrent and Nettleton[146] call feedback or self-terminating processes. Examples mentioned are the eventual destruction of duripans and petrocalcic horizons. These features eventually impede the vertical movement of soil waters, but in some places, water can escape downward along cracks. In time, dissolution causes the cracks to expand to form windows through the cemented pans, and a steady-state number of windows per unit area can develop. Although such windows may develop in response to climatic change, I can envisage situations in which the development of windows is part of the natural progression of soil formation.

RADIOCARBON DATING OF SOIL HORIZONS

Because soils contain both organic and inorganic carbon, radiocarbon dates can be obtained in an attempt to date the soil or some of its features (see papers in Yaalon[162]). Such dates should not be accepted at face value, however, because the systems are very complex, and both new and old carbon can be introduced or exchanged in the soil. Graphs have been prepared to estimate the error due to contamination.[37]

A radiocarbon date of organic carbon from the A horizon of a surface soil includes a mixture of organic matter varying from that fraction being added daily to that synthesized and resynthesized over several thousand years. The dates that reflect this dynamic system are mean residence times (MRT). The MRT for western Canada range from about 200 to 2000 years,[113] and values of 200 to 400 years are reported for the central United States.[26] Dates for various fractions of the organic matter yield different values, a reflection of the relative stability of the fractions.[113,134] The MRT have little meaning with respect to the age of a surface soil, except that they do give minimum ages. They are important, however, in dating buried A horizons to obtain a limiting date on the overlying deposits, because the MRT prior to burial will be a built-in error.[10] The error will vary with the soil and is difficult to estimate. In any sampling of buried soils, the uppermost part of the A horizon should be sampled because the MRT usually decreases upward (Fig. 5-11).

B-horizon radiocarbon dates on organic carbon also are contaminated with younger carbon translocated from above. In many soils, the B-horizon dates are greater than the A-horizon dates (Fig. 5-11). However, with

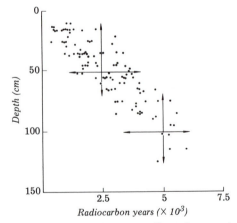

Fig. 5-11 Radiocarbon dates on humus collected at various depths from 24 Mollisols. (Taken from Scharpenseel,[134] Fig. 2; Published by Israel Universities Press, Jerusalem.)

Spodosols there is essentially no relationship of radiocarbon age with depth because the organic compounds are so mobile.[134]

Recent work discloses other problems with radiocarbon dates on organic material, which are important to soil research as well as to the interpretation of any radiocarbon date. Goh and co-workers in New Zealand[67,68,69] report that the commonly used techniques for removing contaminants from materials prior to dating may not always be sufficient. They have treated samples with different strong chemicals prior to dating and obtained dates that are older than those obtained with more traditional pretreatments. In one case, a peat deposit that was originally dated at 31,000 years[128] eventually ended up being > 49,700 years old with nontraditional pretreatment. The idea is that the oldest date is probably closest to the true age of the material. Charcoal is considered an ideal material for dating, but older ages are obtained when it is so pretreated.[68] It is hard to predict which environmental settings might give the most contaminants that have to be so pretreated. New Zealand, for example, is humid and such environments have percolating waters with soluble organic matter. Alexander and Price[1] report a number of reversals in ages of buried organic horizons in Alaskan solifluction deposits; perhaps some of these reversals are due to contaminants that could be removed by the Goh pretreatment. Dry environments might be less susceptible to contamination.

Radiocarbon dates on inorganic carbon of $CaCO_3$-enriched horizons have been obtained in an effort to date soils.[17,57,158] The problem here is that $CaCO_3$ is readily soluble, and solution and re-precipitation can take place; every time this happens, new carbon is added to the system because of the CO_2 in the soil air. Moreover, the carbon in airfall carbonate may be of any age. Gile and others[64] have studied these problems in considerable detail; dates on one of the soils they have studied are given in Fig. 5-12. Although the $CaCO_3$ is mainly of airfall origin and enters the soil by solution and re-precipitation, paired dates on initial pedogenic $CaCO_3$ in young soils and on charcoal in the deposits from which the soils have formed indicate an initial age at time of deposition of less than 3000 years. The dates on the Btk horizon are consistent, because carbonate first coats pebble bottoms and later begins to accumulate in the fine-grained matrix. The soft upper laminae at the top of the K horizon probably undergoes occasional solution and re-precipitation as water is held at that impermeable interface, and thus it is younger than the B horizon pebble coatings. Hard laminae in the upper K horizon is older, probably because percolating waters seldom have a chance to dissolve the more indurated carbonate. One important reversal in the dates is apparent here: organic carbon sealed in the hard laminae is younger than that of the surrounding carbonate sealant. This reversal brought forth the warning by Ruhe[125] to be very cautious in the use of inorganic carbonate dates. One explanation of this discrepancy might be that roots in cracks in the K horizon introduced

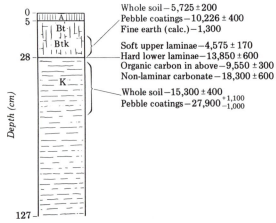

Fig. 5-12 Radiocarbon dates on a soil near Las Cruces, New Mexico. This is the Terino soil; additional laboratory data are presented in Fig. 1-2. Radiocarbon age data (from Gile and others,[64] p. A14), are on inorganic carbon of the $CaCO_3$ fraction, with the exception of one date on organic carbon.

the material later; however, this problem requires further study. At greater depth in the K horizon, dates on pebble coatings are consistent with other dates, but that on the whole soil is much younger, again perhaps the result of solution and re-precipitation. The original stratigraphic work suggested that these dates are not all worthless because some workers thought that deposition ceased and soil formation began about 30,000 years ago. A more recent interpretation of the radiocarbon dates on pedogenic carbonate from Las Cruces is that they are bare minimums, especially those over 10,000 years old.[66] In contrast, in the more arid parts of Australia, there is a fairly good match between dates on pedogenic carbonate and on the parent material deposits.[158]

Because contamination is so likely, one should be cautious about using radiocarbon dates on pedogenic carbonate to indicate the age of the soil, unless other supporting data are available. The detailed interpretation of a series of dates on pedogenic carbonate in two profiles in New Zealand[87] could represent more enthusiasm with the method than is warranted by our current understanding of what is really being dated.

Finally, a recent study in California demonstrates the use of the disequilibrium relationships between Th^{230}, U^{234}, and U^{238} to date pedogenic carbonate.[85] Samples were carefully taken of the innermost carbonate coating pebbles, and the dates are considered to be for the time the carbonate precipitated from solution shortly after the deposits were laid down. The dates obtained are encouraging, they seem to match the estimated geomorphological ages for the deposits. The potential useful age range of the method is a few thousand to about 350,000 years.

REFERENCES

1. Alexander, C.S., and Price, L.W., 1980, Radiocarbon dating of the rate of movement of two solifluctuation lobes in the Ruby Range, Yukon Territory: Quaternary Res., v. 13, p. 365–379.
2. Alexander, E.B., and Nettleton, W.D., 1977, Post-Mazama Natrargids in Dixie Valley, Nevada: Soil Sci. Soc. Amer. Jour., v. 41, p. 1210–1212.
3. Alexander, L.T., and Cady, J.G., 1962, Genesis and hardening of laterite in soils: U.S. Dept. Agri. Tech. Bull. 1282, 90 p.
4. Alperovitch, N., and Dan, J., 1972, Sodium-affected soils in the Jordan Valley: Geoderma, v. 8, p. 37–57.
5. Anderson, H.A., Berrow, M.L., Farmer, V.C., Hepburn, A., Russell, J.D., and Walker, A.D., 1982, A reassessment of podzol formation processes: Jour. Soil Sci., v. 33, p. 125–136.
6. Aristarain, L.F., 1970, Chemical analyses of caliche profiles from the High Plains, New Mexico: Jour. Geol., v. 78, p. 201–212.
7. Arkley, R.J., 1963, Calculation of carbonate and water movement in soil from climatic data: Soil Sci., v. 96, p. 239–248.
8. Bachman, G.O., and Machette, M.N., 1977, Calcic soils and calcretes in the southwestern United States: U.S. Geol. Surv. Open-File Report 77-794, 163 p.
9. Barshad, I., 1964, Chemistry of soil development, p. 1–70 *in* F.E. Bear, ed., Chemistry of the soil: Reinhold Publ. Corp., New York, 515 p.
10. Benedict, J.B., 1966, Radiocarbon dates from a stone-banked terrace in the Colorado Rocky Mountains: Geogr. Annaler, v. 48A, p. 24–31.
11. Birkeland, P.W., 1978, Soil development as an indication of relative age of Quaternary deposits, Baffin Island, N.W.T., Canada: Arc. Alp. Res., v. 10, p. 733–747.
12. Black, C.A., 1957, Soil–plant relationships: John Wiley and Sons, New York, 332 p.
13. Blume, H.P., and Schwertmann, U., 1969, Genetic evaluation of profile distribution of aluminum, iron, and manganese oxides: Soil Sci. Soc. Amer. Proc., v. 33, p. 438–444.
14. Bockheim, J.G., 1979, Relative age and origin of soils in eastern Wright Valley, Antarctica: Soil Sci., v. 128, p. 142–152.
15. ———, 1979, Properties and relative age of soils of southwestern Cumberland Peninsula, Baffin Island, N.W.T., Canada: Arc. Alp. Res., v. 11, p. 289–306.
16. ———, 1980, Properties and classification of some desert soils in coarse-grained glacial drift in the Arctic and Antarctic: Geoderma, v. 24, p. 45–69.
17. Bowler, J.M., and Polach, H.A., 1971, Radiocarbon analyses of soil carbonates: An evaluation from paleosols in southeastern Australia, p. 97–108 *in* D.H. Yaalon, ed., Paleopedology: Israel Univ. Press, Jerusalem, 350 p.
18. Breeman, N. van, and Brinkman, R., 1976, Chemical equilibria and soil formation, p. 141–170 *in* G.B. Bolt and M.G.M. Bruggenwert, eds., Soil chemistry: Elsevier Sci. Publ. Co., Amsterdam, 281 p.
19. Bretz, J.H., and Horberg, L., 1949, Caliche in southeastern New Mexico; Jour. Geol., v. 57, p. 491–511.

20. Brewer, R., 1955, Mineralogical examination of a yellow podzolic soil formed on granodiorite: Commonwealth Scientific and Industrial Res. Organ. (Australia), Soil Publ. no. 5, 28 p.

21. ———, 1964, Fabric and mineral analysis of soils: John Wiley and Sons, New York, 470 p.

22. ———, 1968, Clay illuviation as a factor in particle-size differentiation in soil profiles: 1968 Internat. Cong. Soil Sci. Trans., v. 4, p. 489–499.

23. Brewer, R., and Haldane, A.D., 1957, Preliminary experiments in the development of clay orientation in soils: Soil Sci., v. 84, p. 301–309.

24. Bridges, E.M., and Burnham, C.P., 1980. Soils of the State of Bahrain: Jour. Soil Sci., v. 31, p. 689–707.

25. Brinkman, R., 1970, Ferrolysis, a hydromorphic soil forming process: Geoderma, v. 3, p. 199–206.

26. Broecker, W.S., Kulp, J.L., and Tucek, C.S., 1956, Lamont natural radiocarbon measurements III: Science, v. 124, p. 154–165.

27. Brook, G.A., Folkoff, M.E., and Box, E.O., 1983, A world model of soil carbon dioxide: Earth Surface Proc. and Landforms, v. 8, p. 79–88.

28. Brown, C.H., 1956, The origin of caliche on the northeastern Llano Estacado, Texas: Jour. Geol., v. 64, p. 1–15.

29. Buckman, H.O., and Brady, N.C., 1969, The nature and properties of soils: The Macmillan Co., Toronto, 653 p.

30. Bullock, P., Milford, M.H., and Cline, M.G., 1974, Degradation of argillic horizons in Udalf soils of New York State: Soil Sci. Soc. Amer. Proc., v. 38, p. 621–628.

31. Buntley, G.J., Daniels, R.B., Gamble, E.E., and Brown, W.T., 1977, Fragipan horizons in soils of the Memphis-Loring-Grenada sequence in west Tennessee: Soil Sci. Soc. Amer. Jour., v. 41, 400–407.

32. Buol, S.W., and Hole, F.D., 1959, Some characteristics of clay skins on peds in the B horizon of a Gray-Brown Podzolic soil: Soil Sci. Soc. Amer. Proc., v. 23, p. 239–241.

33. ———, 1961, Clay skin genesis in Wisconsin soils: Soil Sci. Soc. Amer. Proc., v. 25, p. 377–379.

34. Buol, S.W., Hole, F.D., and McCracken, R.J., 1973, Soil genesis and classification: The Iowa State Univ. Press, Ames, 360 p.

35. Buringh, P., 1968, Introduction to the study of soils in tropical and subtropical regions: Centre for Agricultural Publishing and Documentation, Wageningen, 118 p.

36. Burns, S.F., 1980, Alpine soil distribution and development, Indian Peaks, Colorado Front Range: unpubl. Ph.D. thesis, Univ. of Colorado, Boulder, 360 p.

37. Campbell, C.A., Paul, E.A., Rennie, D.A. and McCallum, K.J., 1967a, Factors affecting the accuracy of the carbon-dating method in soil humus studies: Soil Sci., v. 104, 81–85.

38. ———, 1967b. Applicability of the carbon-dating method of analysis to soil humus studies: Soil Sci., v. 104, p. 217–224.

39. Childs, C.W., Parfitt, R.L., and Lee, R., 1983, Movement of aluminum as an inorganic complex in some podzolized soils, New Zealand: Geoderma, v. 29, p. 139–155.

40. Chittleborough, D.J., and Oades, J.M., 1979, The development of a red-brown earth. I. A reinterpretation of published data: Austr. Jour. Soil Res., v. 17, p. 371–381.

41. ———, 1980, The development of a red-brown earth. II. Uniformity of the parent material: Australian Jour. Soil Res., v. 18, p. 375–382.

42. ———, 1980, The development of a red-brown earth. III. The degree of weathering and translocation of clay: Austr. Jour. Soil Res., v. 18, p. 383–393.

43. Claridge, G.G.C., and Campbell, I.B., 1968, Some features of Antarctic soils and their relation to other desert soils: 9th Internat. Cong. Soil Sci. Trans., v. 4, p. 541–549.

44. Clark, F.W., 1924, The data of geochemistry: U.S. Geol. Surv. Bull. 770, 841 p.

45. Daly, B.K., 1982, Identification of podzols and podzolized soils in New Zealand by relative absorbance of oxalate extracts of A and B horizons: Geoderma, v. 28, p. 29–38.

46. Dan, J., Gerson, R., Koyumdjisky, H., and Yaalon, D.H., 1981, Aridic soils of Israel: Special Publication no. 190, Agricultural Research Organization, The Volcani Center, Israel, 353 p.

47. Dan, J., and Yaalon, D.H., 1982, Automorphic saline soils in Israel: Catena, Supplement 1, p. 103–115.

48. Daniels, R.B., Perkins, H.F., Hajek, B.F., and Gamble, E.E., 1978, Morphology of discontinuous phase plinthite and criteria for its field identification in the southern United States: Soil Sci. Soc. Amer. Jour., v. 42, p. 944–949.

49. Dawson, J.J., Ugolini, F.C., Hrutfiord, B.F., and Zachara, J., 1978, Role of soluble organics in the soil processes of a podzol, central Cascades, Washington: Soil Sci., v. 126, p. 290–296.

50. Dijkerman, J.C., Cline, M.G., and Olson, G.W., 1967, Properties and genesis of textural subsoil lamellae: Soil Sci., v. 104, p. 7–16.

51. Eswaran, H., and Tavernier, R., 1980, Classification and genesis of Oxisols, p. 427–442 *in* B.K.G. Theng, ed., Soils with variable charge: New Zealand Soc. Soil Sci., Lower Hutt.

52. Fanning, D.S., Hall, R.L., and Foss, J.E., 1972, Soil morphology, water tables, and iron relationships in soils of the Sassafras drainage catena in Maryland, p. 71–79, *in* E. Schlichting and U. Schwertmann, eds., Pseudogley and gley: Trans. Commissions V and VI of Internat. Soc. Soil Sci., Verlag Chemie, GmbH, Weinheim/Bergstr.

53. Farmer, V.C., Russell, J.D., and Berrow, M.L., 1980, Imogolite and proto-imogolite allophane in spodic horizons—evidence for a mobile aluminum silicate complex in podzol formation: Jour. Soil Sci., v. 31, p. 673–684.

54. Farmer, V.C., and Fraser, A.R., 1982, Chemical and colloidal stability of sols in the $Al_2O_3-Fe_2O_3-SiO_2-H_2O$ system: Their role in podzolization: Jour. Soil Sci., v. 33, p. 737–742.

55. Flach, K.W., Nettleton, W.D., Gile, L.H., and Cady, J.G., 1969, Pedocementation: Induration by silica, carbonates, and sesquioxides in the Quaternary: Soil Sci., v. 107, p. 442–453.

56. Flach, K.W., Holzhey, C.S., De Coninck, F., and Bartlett, R.J., 1980, Gen-

esis and classification of Andepts and Spodosols, p. 411–426 *in* B.K.G. Theng, ed., Soils with variable charge: New Zealand Soc. Soil Sci., Lower Hutt.

57. Frye, J.C., Glass, H.D., Leonard, A.B., and Coleman, D.D., 1974, Caliche and clay mineral zonation of the Ogallala Formation, central-eastern New Mexico: New Mexico Bureau of Mines and Mineral Resources, Circular 144, 16 p.

58. Fyfe, W.S., and Kronberg, B.I., 1980, Nutrient conservation, the key to agricultural strategy: Mazingira 13, v. 4, p. 64–69.

59. Gardner, L.R., 1972. Origin of the Mormon Mesa caliche: Geol. Soc. Amer. Bull., v. 83, p. 143–156.

60. Gile, L.H., 1979, Holocene soils in eolian sediments of Bailey County, Texas: Soil Sci. Soc. Amer. Proc., v. 43, p. 994–1003.

61. Gile, L.H., Peterson, F.F., and Grossman, R.B., 1965, The K horizon: A master soil horizon of carbonate accumulation: Soil Sci., v. 99, p. 74–82.

62. ———, 1966, Morphological and genetic sequences of carbonate accumulation in desert soils: Soil Sci., v. 101, p. 347–360.

63. Gile, L.H., and Grossman, R.B., 1968, Morphology of the argillic horizon in desert soils of southern New Mexico: Soil Sci. v. 106, p. 6–15.

64. Gile, L.H., Hawley, J.W., and Grossman, R.B., 1970, Distribution and genesis of soils and geomorphic surfaces in a desert region of southern New Mexico: Soil Sci. Soc. Amer. Guidebook, soil-geomorphology field conferences, Aug. 21–22, 29–30, 1970, 156 p.

65. Gile, L.H., and Grossman, R.B., 1979, The desert soil monograph: Soil Conservation Service, U.S. Department of Agriculture, 984 p.

66. Gile, L.H., Hawley, J.W., and Grossman, R.B., 1981, Soils and geomorphology in the Basin and Range area of southern New Mexico—Guidebook to the Desert Project: New Mexico Bureau of Mines and Mineral Resources, Mem. 39, 222 p.

67. Goh, K.M., and Pullar, W.A., 1977, Radiocarbon dating techniques for tephras in central North Island, New Zealand: Geoderma, v. 18, p. 265–278.

68. Goh, K.M., Tonkin, P.J., and Rafter, T.A., 1978, Implications of improved radiocarbon dates of Timaru peats on Quaternary loess stratigraphy: New Zealand Jour. Geol. and Geophys., v. 21, p. 463–466.

69. Goh, K.M., Molloy, B.P.J., and Rafter, T.A., 1977, Radiocarbon dating of Quaternary loess deposits, Banks Peninsula, Canterbury, New Zealand: Quaternary Res., v. 7, p. 177–196.

70. Goss, D.W., Smith, S.J., and Stewart, B.A., 1973, Movement of added clay through calcareous materials: Geoderma, v. 9, p. 97–103.

71. Goudie, A., 1973, Duricrusts in tropical and subtropical landscapes: Oxford University Press, London, 174 p.

72. Grossman, R.B., and Carlisle, F.J. 1969, Fragipan soils of the eastern United States: Adv. Agron., v. 21, p. 237–275.

73. Hallmark, C.T., and Smeck, N.E., 1979, The effect of extractable aluminum, iron, and silicon on strength and bonding of fragipans of northeastern Ohio: Soil Sci. Soc. Amer. Jour., v. 43, p. 145–150.

74. Hallsworth, E.G., 1963, An examination of some factors affecting the movement of clay in an artificial soil: Jour. Soil Sci. v. 14, p. 360–371.

75. Hay, R.L., and Reeder, R.J., 1978, Calcretes of Olduvai Gorge and the Ndolanya Beds of northern Tanzania: Sedimentology, v. 25, p. 649–673.

76. Herbillon, A.J., 1980, Mineralogy of Oxisols and oxic materials, p. 109–116 *in* B.K.G. Theng, ed., Soils with variable charge: New Zealand Soc. Soil Sci., Lower Hutt.

77. Hingston, F.J., and Gailitis, V., 1976, The geographic variation of salt precipitated over Western Australia: Aust. Jour. Soil Res., v. 14, p. 319–335.

78. Holliday, V.T., 1982, Morphological and chemical trends in Holocene soils at the Lubbock Lake Archeological Site, Texas: Ph.D. thesis, University of Colorado, Boulder, 285 p.

79. Hutton, J.T., and Dixon, J.C., 1981, The chemistry and mineralogy of some South Australian calcretes and associated soft carbonates and their dolomitisation: Jour. Geol. Soc. Aust., v. 28, p. 71–79.

80. Jenny, H., 1941, Calcium in the soil: III. Pedologic relations: Soil Sci. Soc. Amer. Proc., v. 6, p. 27–35.

81. Junge, C.E., and Werby, R.T., 1958, The concentration of chloride, sodium, potassium, calcium, and sulphate in rain water over the United States: Jour. Meteorology, v. 15, p. 417–425.

82. King, R.H., and Brewster, G.R., 1976, Characteristics and genesis of some subalpine Podzols (Spodosols), Banff National Park, Alberta: Arc. Alp. Res., v. 8, p. 91–104.

83. Kononova, M.M., 1961, Soil organic matter: Pergamon Press, New York, 450 p.

84. Krauskopf, K.B., 1967, Introduction to geochemistry: McGraw-Hill, New York, 721 p.

85. Ku, T., Bull, W.B., Freeman, S.T., and Knauss, K.G., 1979, Th^{230}–U^{234} dating of pedogenic carbonates in gravelly desert soils of Vidal Valley, southeastern California: Geol. Soc. Amer. Bull., v. 90, p. 1063–1073.

86. Lattman, L.H., and Lauffenburger, S.K., 1974, Proposed role of gypsum in the formation of caliche: Zeitschrift fur Geomorphologie N. F., Suppl. Bd. 20, p. 140–149.

87. Leamy, M.L., 1974, The use of pedogenic carbonate to determine the absolute age of soils and to assess rates of soil formation: Trans. 10th Internat. Congr. Soil Sci., v. 6 (2), p. 331–339.

88. Lee, R., 1980, ed., Soil groups of New Zealand, Part 5: Podzols and gley podzols: New Zealand Society of Soil Science, Lower Hutt, 452 p.

89. Locke, W.W., III, 1980, The Quaternary geology of the Cape Dyer area, southeasternmost Baffin Island, Canada: Ph.D. thesis, Univ. of Colorado, Boulder, 331 p.

90. Loughnan, F.C., 1969, Chemical weathering of the silicate minerals: American Elsevier Publ. Co., New York, 154 p.

91. Magaritz, M., and Amiel, A.J., 1980, Calcium carbonate in a calcareous soil from the Jordan Valley, Israel: Its origin as revealed by the stable carbon isotope method: Soil Sci. Soc. Amer. Jour., v. 44, p. 1059–1062.

92. Maignien, R., 1966, Review of research on laterites: UNESCO, Natural resources research, no. 4, 148 p.

93. Malde, H.E., 1955, Surficial geology of the Louisville Quadrangle, Colorado: U.S. Geol. Surv. Bull. 996-E, p. 217–259.

94. Marshall, C.E., 1977, The physical chemistry and mineralogy of soils. Vol. II: Soils in place: John Wiley and Sons, Inc., New York, 313 p.

95. McFadden, L.D., 1982, The impacts of temporal and spatial climatic changes on alluvial soils genesis in Southern California: Ph.D. thesis, University of Arizona, Tucson, 430 p.

96. McKeague, J.A., and Day, J.H., 1966, Dithionite and oxalate-extractable Fe and Al as aids in differentiating various classes of soils: Can. Jour. Soil Sci., v. 46, p. 13–22.

97. McKeague, J.A., and St. Arnaud, R.J., 1969, Pedotranslocation: Eluviation-illuviation in soils during the Quaternary: Soil Sci., v. 107, p. 428–434.

98. McKeague, J.A., Brydon, J.E., and Miles, N.M., 1971, Differentiation of forms of extractable iron and aluminum in soils: Soil Sci. Soc. Amer. Proc., v. 35, p. 33–38.

99. McKeague, J.A., Ross, G.J., and Gamble, D.S., 1978, Properties, criteria of classification, and concepts of genesis of podzolic soils in Canada, p. 27–60 *in* W.C. Mahaney, ed., Quaternary Soils: Geo Abstracts, Univ. of East Anglia, Norwich, England.

100. Mears, B., Jr., 1981, Periglacial wedges and the late Pleistocene environment of Wyoming's intermontane basins: Quaternary Res., v. 15, p. 171–198.

101. Mohr, E.C.J., van Baren, F.A., and van Schuylenborgh, J., 1973, Tropical soils: Mouton-Ichtiar Baru-van Hoeve, The Hague, Netherlands, 481 p.

102. Mueller, O.P., and Cline, M.G., 1959, Effects of mechanical soil barriers and soil wetness on rooting of trees and soil-mixing by blow-down in central New York: Soil Sci., v. 88, p. 107–111.

103. Muhs, D.R., 1980, Quaternary stratigraphy and soil development, San Clemente Island, California: Ph.D. thesis, Univ. of Colorado, Boulder, 221 p.

104. Nettleton, W.D., Flach, K.W., and Borst, G., 1968, A toposequence of soils in tonalite grus in the southern California Peninsular Range: U.S. Dept. Agri., Soil Surv. Investigation Rept. no. 21, 41 p.

105. Nettleton, W.D., Flach, K.W., and Brasher, B.R., 1969, Argillic horizons without clay skins: Soil Sci. Soc. Amer. Proc., v. 33, p. 121–125.

106. Nettleton, W.D., Witty, J.E., Nelson, R.E., and Hawley, J.E., 1975, Genesis of argillic horizons in soils of desert areas of the southwestern United States: Soil Sci. Soc. Amer. Proc., v. 39, p. 919–926.

107. Nettleton, W.D., Nelson, R.E., Brasher, D.R., and Derr, P.H., 1982, Gypsiferous soils in the western United States, p. 147–168 *in* J.A. Kittrick, D.S. Fanning, and L.R. Hossner, eds., Acid sulphate weathering: Soil Science Society of America, Madison, Wisconsin.

108. Oertal, A.C., 1968, Some observations incompatible with clay illuviation: 1968 Internat. Cong. Soil Sci. Trans., v. 4, p. 481–488.

109. Olphen, H. van, 1963, An introduction to clay colloid chemistry: John Wiley and Sons, New York, 301 p.

110. Page, W.D., 1972, The geological setting of the archeological site at Oued El Akarit and the paleoclimatic significance of gypsum soils, southern Tunisia: Ph.D. thesis, Univ. of Colorado, Boulder, 111 p.

111. Parakshin, Y.P., 1982, Some regularities in the formation and development

of Solonetz soils, Kokchetav Upland, Kazakh SSR: Pochvovedeniye, No. 11, p. 8–16.

112. Paramananthan, S., and Eswaran, H., 1980, Morphological properties of Oxisols, p. 35–43 *in* B.K.G. Theng, ed., Soils with variable charge: New Zealand Soc. Soil Sci., Lower Hutt.

113. Paul, E.A., 1969, Characterization and turnover rate of soil humic constituents, p. 63–76 *in* S. Pawluk, ed., Pedology and Quaternary research: Univ. of Alberta Printing Dept., Edmonton, Alberta, 218 p.

114. Pedro, G., Jamagne, M., and Begon, J.C., 1978, Two routes in genesis of strongly differentiated acid soils under humid, cool-temperate conditions: Geoderma, v. 20, p. 173–189.

115. Petersen, L., 1976, Podzols and podzolization: DSR Forlag, Copenhagen, 293 p.

116. Peterson, F.F., 1980, Holocene desert soil formation under sodium salt influence in a playa-margin environment: Quaternary Res., v. 13, p. 172–186.

117. Ponomareva, V.V., 1969, Theory of podzolization: Israel Program for Scientific Translations, Jerusalem, 309 p.

118. Reeves, C.C., Jr., 1970, Origin, classification, and geologic history of caliche on the southern High Plains, Texas and eastern New Mexico: Jour. Geol., v. 78, p. 352–362.

119. ———, 1976, Caliche-origin, classification, morphology and uses: Estacado Books, Lubbock, Texas, 233 p.

120. Reider, R.G., Kuniansky, N.J., Stiller, D.M., and Uhl, P.J., 1974, Preliminary investigation of comparative soil development on Pleistocene and Holocene geomorphic surfaces of the Laramie Basin, Wyoming: p. 27–33 *in* M. Wilson, ed., Applied geology and archeology: The Holocene of Wyoming: Geol. Surv. of Wyoming, Report of Investigations 10.

121. Rijkse, W.C., ed., 1978, Soil groups of New Zealand, Part 3: Gley soils: New Zealand Society of Soil Science, Lower Hutt, 127 p.

122. Ross, C.W., Mew, G., and Searle, P.L., 1977, Soil sequences on two terrace sequences in the North Westland area, New Zealand: New Zealand Jour. Sci., v. 20, p. 231–244.

123. Rowell, D.L., 1981, Oxidation and reduction, p. 401–461 *in* D. J. Greenland and M.H.B. Hayes, eds., The chemistry of soil processes: John Wiley and Sons, Inc., New York.

124. Ruhe, R.V., 1959, Stone lines in soils: Soil Sci., v. 87, p. 223–231.

125. ———, 1967, Geomorphic surfaces and surficial deposits in southern New Mexico; New Mex. Bur. Mines and Mineral Resources, Memoir 18, 66 p.

126. ———, 1983, Aspects of Holocene pedology in the United States, *in* H.E. Wright, Jr., ed., Late-Quaternary environments of the United States, Vol. 2. The Holocene: Univ. Minnesota Press, Minneapolis, Minn.

127. ———, Brunson, K.L., and Hall, L.E., 1975, Fragisols and Holocene environments in midwestern U.S.A.: Ann. Acad. Brasil. Cienc. 47 (Suplemento), p. 119–126.

128. Runge, E.C.A., Goh, K.M., and Rafter, T.A., 1973, Radiocarbon chronology and problems in its interpretation for Quaternary loess deposits—South Canterbury, New Zealand: Soil Sci. Soc. Amer. Proc., v. 37, p. 742–746.

129. Salomons, W., and Mook, W.G., 1976, Isotope geochemistry of carbonate dissolution and reprecipitation in soils: Soil Sci., v. 122, p. 15–24.

130. Salomons, W., Goudie, A., and Mook, W.G., 1978, Isotopic composition of calcrete deposits from Europe, Africa and India: Earth Surface Processes, v. 3, p. 43–57.

131. Schnitzer, M., 1969, Reactions between fulvic acid, a soil humic compound and inorganic soil constituents: Soil Sci. Soc. Amer. Proc., v. 33, p. 75–81.

132. Schuylenborgh, J. van, 1965, The formation of sesquioxides in soils, p. 113–125 *in* E.G. Hallsworth and D.V. Crawford, eds., Experimental pedology: Butterworth and Co., London, 414 p.

133. Schwertman, U., and Fischer, W.R., 1973, Natural "amorphous" ferric hydroxide: Geoderma, v. 10, p. 237–247.

134. Sharpenseel, H.W., 1971, Radiocarbon dating of soils—problems, troubles, hopes, p. 77–88 *in* D.H. Yaalon, ed., Paleopedology: Israel Univ. Press, Jerusalem, 350 p.

135. Shlemon, R.J., and Hamilton, P., 1978, Late Quaternary rates of sedimentation and soil formation, Camp Pendleton–San Onofre State Beach coastal area, Southern California, U.S.A.: Tenth Internat. Cong. on Sedimentology (Jerusalem) (Abstracts), p. 603–604.

136. Simonson, G.H., and Boersma, L., 1972, Soil morphology and water table relations: I. Correlations between annual water table fluctuations and profile features: Soil Sci. Soc. Amer. Proc., v. 36, p. 649–653.

137. Simonson, R.W., 1978, A multiple-process model of soil genesis, p. 1–25 *in* W. C. Mahaney, ed., Quaternary soils: Geo Abstracts, University of East Anglia, Norwich, England.

138. Singer, M., Ugolini, F.C., and Zachara, J., 1978, In situ study of podzolization on tephra and bedrock: Soil Sci. Soc. Amer. Jour., v. 42, p. 105–111.

139. Sivarajasingham, S., Alexander, L.T., Cady, J.G., and Cline, M.G., 1962, Laterite: Advances in Agronomy, v. 14, p. 1–60.

140. Soil Survey Staff, 1975, Soil Taxonomy: U.S. Dept. Agri. Handbook 436, 754 p.

141. Soons, J.M., 1968, Erosion by needle ice in the Southern Alps, New Zealand, p. 217–227 *in* H.E. Wright, Jr. and W.H. Osburn, eds., Arctic and alpine environments: Indiana Univ. Press, Bloomington.

142. Soons, J.M., and Greenland, D.E., 1970, Observations on the growth of needle ice: Water Resources Res., v. 6, p. 579–593.

143. Steinhardt, G.C., and Franzmeier, D.P., 1979. Chemical and mineralogical properties of the fragipans of the Cincinnati catena: Soil Sci. Soc. Amer. Jour., v. 43, p. 1008–1013.

144. Stobbe, P.C., and Wright, J.R., 1959, Modern concepts of the genesis of podzols: Soil Sci. Soc. Amer. Proc., v. 23, p. 161–164.

145. Tedrow, J.C.F., 1977, Soils of the polar landscapes: Rutgers Univ. Press, New Brunswick, N.J., 638 p.

146. Torrent, J., and Nettleton, W.D., 1978, Feedback processes in soil genesis: Geoderma, v. 20, p. 281–287.

147. Torrent, J., Nettleton, W.D., and Borst, G., 1980, Genesis of a Typic Durixeralf of southern California: Soil Sci. Soc. Amer. Jour., v. 44, p. 575–582.

148. Ugolini, F.C., and Mann, D.H., 1979, Biopedological origin of peatlands in southeast Alaska: Nature, v. 281, p. 366–368.

149. Ugolini, F.C., Zachara, J.M., and Reanier, R.E., 1982, Dynamics of soil-forming processes in the Arctic, p. 103–115 *in* Proc. 4th Canadian Permafrost Conf., National Research Council of Canada, Ottawa.

150. Walker, P.H., and Hutka, J., 1979, Size characteristics of soils and sediments with special reference to clay fractions: Aust. Jour. Soil Res., v. 17, p. 383–404.

151. Walker, T.R., Waugh, B., and Crone, A.J., 1978, Diagenesis in first-cycle desert alluvium of Cenozoic age, southwestern United States and northwestern Mexico: Geol. Soc. Amer. Bull., v. 89, p. 19–32.

152. Walker, T.W., 1964, The significance of phosphorus in pedogenesis, p. 295–315 *in* E.G. Hallsworth and D.V. Crawford, eds., Experimental pedology: Buttersworths, London.

153. Walker, T.W., and Syers, J.K., 1976, The fate of phosphorus during pedogenesis: Geoderma, v. 15, p. 1–19.

154. Washburn, A.L., 1979, Geocryology: A survey of periglacial processes and environments: Halsted Press, J.W. Wiley and Sons, Inc., New York, 406 p.

155. Watson, A., 1979, Gypsum crusts in deserts: Jour. Arid Environments, v. 2, p. 3–20.

156. Watts, N.L., 1978, Displacive calcite: Evidence from recent and ancient calcretes: Geology, v. 6, p. 699–703.

157. Wilding, L.P., Smeck, N.E., and Drees, L.R., 1977, Silica in soils: Quartz, cristobalite, tridymite, and opal, p. 471–552 *in* J.B. Dixon and S.B. Weed, eds., Minerals in soil environments: Soil Science Society of America, Madison, Wisc.

158. Williams, G.E., and Polach, H.A., 1971, Radiocarbon dating of arid-zone calcareous paleosols: Geol. Soc. Amer. Bull., v. 82, p. 3069–3086.

159. Wright, J.R., and Schnitzer, M., 1963, Metallo-organic interactions associated with podzolization: Soil Sci. Soc. Amer. Proc., v. 27, p. 171–176.

160. Wright, W.R., and Foss, J.E., 1968, Movement of silt-sized particles in sand columns: Soil Sci. Soc. Amer. Proc., v. 32, p. 446–448.

161. Yaalon, D.H., 1963, On the origin and accumulation of salts in groundwater and in soils of Israel: Bull. Res. Council of Israel, v. 11G, p. 105–131.

162. ———, ed., 1971, Paleopedology: Israel Univ. Press, Jerusalem, 350 p.

163. Yaalon, D.H., and Ganor, E., 1973, The influence of dust on soils during the Quaternary: Soil Sci., v. 116, p. 146–155.

6

Factors of soil formation

Since the early work of Dokuchaev in Russia and Hilgard in the United States, pedologists have been trying to describe the main factors that define the soil system and to determine mathematically the relationship between soil properties and these factors. Jenny[9,11,12,13] has made many important contributions to this facet of pedology; he has also traced its historical development. Major[15] applied the same principles to plant ecology, and Crocker[5] critically analyzed Jenny's factors, especially the biotic factor, and reviewed some of the Russian thinking on the factorial approach. A recent paper by Yaalon[18] brings us up to date on the subject.

FACTORS AND THE FUNDAMENTAL EQUATION

Five factors are usually used to define the state of the soil system. They are climate, organisms, topography, parent material, and time. Other factors may be important locally. The factors theoretically are independent variables, in that field sites can be found in which the factors vary independently. Although the factors can be dependent variables in some field sites, their real value in a rigorous quantitative factorial treatment is as independent variables.

Factors and processes should be clearly distinguished. In previous chapters, we have been concentrating on the processes that are operative in soils. These processes form the soil. The factors, in contrast, define the state of the soil system. If one knew precisely the combination of factors that describe a soil system, the soil properties could be predicted. A change in a factor would change the soil. However, we are not yet to the point in our knowledge of the factors where they can be defined this precisely, and we may never be. Despite this, some valid qualitative and sometimes quantitative predictions can be made.

With the recognition of the factors, equations were formulated to establish the dependence of certain soil properties on the factors. The most widely quoted is the fundamental equation of Jenny[9,13]:

$$S \text{ or } s = f(cl, o, r, p, t, \ldots,)$$

where S denotes the soil, s any soil property, cl the climatic factor, o the biotic factor, r the topographic factor, p the parent material, t the time factor, and the dots after t represent unspecified factors, such as airfall salts, that might be important locally. In this equation, S and s are the dependent variables; the factors are the independent variables. More elaborate forms of this equation have been proposed, but none has been solved, for the reason that, if all factors are allowed to vary, it would not be possible to sort out the effect of each factor on the soil property studied.

Jenny overcame the dilemma of solving the equation by solving an equation for one factor at a time. To do this, one factor is allowed to vary while the others are held constant. He established the following functions:

$$s = f(\underline{cl}, o, r, p, t, \ldots,) \text{ climofunction}$$
$$s = f(\underline{o}, cl, r, p, t, \ldots,) \text{ biofunction}$$
$$s = f(\underline{r}, cl, o, p, t, \ldots,) \text{ topofunction}$$
$$s = f(\underline{p}, cl, o, r, t, \ldots,) \text{ lithofunction}$$
$$s = f(\underline{t}, cl, o, r, p, \ldots,) \text{ chronofunction}$$

To solve each function, the first factor listed (underlined) is allowed to vary while the others remain constant: one therefore determines the dependency of one (or more) soil property on a single factor by appropriate statistical methods. This can be extended to include more factors, and eventually it might be possible to rank them on their relative importance to that soil property or properties.

There are two ways in which a factor can be considered constant: (a) if the range in the state factor is quite small and (b) if variation in the state factor is large, yet has a negligible effect on the soil property. Figure 6-1 illustrates this latter case. The functional relationship between a soil property and factor a is to be determined. In the same sampling area, factor b also varies, so its relationship to the soil property must also be established. The plot indicates a close dependence of the property on both factors between 0 and x on the horizontal axis, and a simple functional relationship cannot be established. Between x and y, however, ds/dF_b approaches zero, and thus variation in factor b has little effect on the soil property. It can be considered a constant. Variation in s, therefore, between x and y is ascribed to variation in factor a in this simplified model. Beyond y, both factors can be considered constants because the slopes of the functions are close to zero. To establish the dependence of s

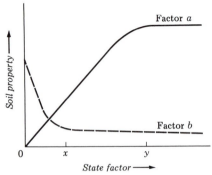

Fig. 6-1 Hypothetical variation in one soil property with variation in two state factors. The vertical scale is not necessarily the same for both plots. On the horizontal axis, x and y are arbitrary values for the state factors.

on factor a between 0 and x, sites would have to be chosen in which variation in factor b is small.

The relative importance of the factors varies from soil to soil. In the early days of soil science, however, rock type was generally considered to be most important. Following the early Russian work, especially that of Dokuchaev, the effect of climate was considered to be most important, and many soil-classification schemes were based on climate and processes thought to be related to climate. Although climate may be the most important factor in the gross, worldwide distribution of soils, the other four factors are equally important in describing soil variation in a landscape.[8,10]

One problem in solving the functions set down by Jenny is to find appropriate field sites at which the factors operate as independent variables. Crocker[5] attributes a part of Jenny's success in solving the functional equations to the definitions given the state factors, definitions that assure that they could be independent of each other. The definition of the climatic factor will illustrate the problem. One can speak of the regional climate or the climate at the ground surface immediately above the soil. The two climates are measurably different. The regional climate is measured above the canopy of vegetation, and so it depends less on any ecosystem property. The climate at the ground surface, however, depends on the vegetation at the site or on the orientation of slope with respect to position of the sun. In this example, the regional climate is the independent variable, whereas the soil climate is a dependent variable. Thus in solving for the functions, the regional climate must be used. The definitions of Jenny are as follows.

Climate *(cl):* The regional climate; because precipitation and temperature are not interdependent, they can be treated as separate functions.

Organism or biotic factor (o): Because vegetation is most important, this factor is the summation of plant disseminules reaching the soil site or the potential vegetation; it is approximated by a list of species growing in the surrounding region that could gain access to the site under appropriate conditions.

Topography (r): Included are the shape and slope of the landscape related to the soil, the direction the slope faces, and the effects of a high water table, the latter being commonly related to the topography.

Parent material or initial state of the system (p): Included are materials, both weathered and unweathered, from which the soil formed. Parent material could also be a soil in the case where one wishes to study the effect of climatic change on a pre-existing soil.

Time (t): Elapsed time since deposition of material, the exposure of the material at the surface or formation of the slope to which the soil relates; if a study is being made of the effect of climatic change on a preexisting soil, the time since the change.

These definitions raise several problems, and these have been discussed at length by Jenny[9,11,13] and Crocker.[5] One problem is that the biotic and parent-material factors cannot be quantified; for the most part, they can only be described qualitatively. An example of what can be done, however, is to determine the function for each of several rock or vegetation types and then compare one function with the other. Some workers have criticized Jenny for including time as one of the factors, because, by itself, time does nothing to a soil. Its importance, however, lies in the fact that most soil-forming processes are so slow that their effect on the soil is markedly time dependent. The one definition that probably has been most controversial, however, is that of the biotic factor, because it is defined and used in two different ways. When deriving functions for the other factors, the biotic factor is taken as the potential vegetation or species pool. The actual vegetation at the site is a function of the same set of factors as is the soil[11,13,15]; both vegetation and soil develop concurrently and can and do influence each other. Therefore vegetation cannot be taken as an independent variable. Biofunctions can be derived, however, for areas in which vegetation is the variable and all other factors are constant and thus do not influence the vegetation. Instances of this are quite rare, but a commonly studied example is the prairie–forest transition. Some workers might prefer a bioclimatic factor to include both present climate and vegetation.

There have been several recent developments in the evaluation of the factorial approach to soil studies. Runge[17] has proposed a more simplified equation that states that a soil is a function of organic matter production, the amount of water available for leaching, and time. Chesworth[3] argues that time is the only independent factor. Both give valid arguments for their cases, and both have been critiqued by Yaalon.[18] However, I feel that from the point of view of a field-oriented geologist–pedologist working with the wide variety of soils at the earth's surface, Jenny's approach is the most practicable and understandable. As one example, Chesworth argues that because the effects of parent material are nullified with time, parent material cannot be considered independent. The argument is that, with long intervals of time, soils with different parent materials converge toward common chemical properties. This is true, but probably only for some very old soils. Surely for most Quaternary soils parent material has a significant effect on the soil properties. Finally, pedologic studies are becoming more quantitative. Some published functions are reviewed by Yaalon,[18] and some will be presented in later chapters.

ARGUMENTS FOR AND AGAINST THE USE OF FACTORS

Several criticisms have been made of the use of factors in the study of soils, some of which are discussed by Crocker[5] and Bunting.[2] One is that the general equation has never been solved. By this I mean that we cannot apply quantitative data on the factors to predict the resulting soil or soil properties adequately. This is true now, and it may never be otherwise. Moreover, many soils have formed under conditions in which several factors have varied, and functions for these soils may never be derived. Individual functions for other soils have been derived, however, and they are very useful in pedology.

Another problem is that many soils are polygenetic; that is, they have formed under more than one set of factors, and thus present-day factors should not be used to define the state of these particular soils. Monogenetic soils (soils that have formed under one set of factors) are certainly preferred; however, these soils are not abundant and generally are restricted to those formed in postglacial time. There are two ways to surmount this problem. One would be to study only those soil properties that form rapidly and reach a steady state in a rather short period of time, that is less than the duration of the postglacial time span. This would limit the kinds of properties that could be studied by the use of factors. Many argillic horizons, for example, take longer than that to form. The other solution is to allow for climatic change and to assume that, at all sampled sites in a particular region, the differences between the climates of the past and the present climate were relatively constant; that is, all sites experienced a similar variation in climate.[7] This seems reasonable as long as soils of similar age are being studied, because all these soils may have gone through similar climatic cycles. And, as pointed out by Jenny,[9] in solving for climofunctions the main interest is in gradients rather than in absolute values of precipitation and temperature. Past climatic gradients may have been similar to present ones in some, but certainly not all regions. It might be difficult to solve some chronofunctions, however, because older soils probably have undergone more changes in climate than have younger soils. Perhaps these problems are not insurmountable when one considers the problems of locating and collecting truly representative samples in the field. Furthermore, as more data become available on past climates and their duration, perhaps these can be taken into account in quantitative studies.

Another problem cited in the use of factors is that one learns a lot about the factors, but little about the soil.[2] However, Barshad[1] feels that this approach is important in studying soil-forming processes, and the rates at which they proceed, and that it is valuable in predicting the soil properties that might be encountered at a particular site. As an example, we can consider the origin of a clay-enriched horizon near the surface. It could be an argillic horizon or a depositional layer. A knowledge of the variation in clay chronofunctions with other factors, and the factors at the locality studied, would help set upper and lower limits on the probable amount of pedogenic clay. If further study indicates that more clay is pedogenic than at first was thought probable, other factors, such as time, would have to be reassessed. Another example might be the determination of the origin of an E horizon in a soil. It could result from podzolization; if it did, it usually would be indicated by a combination of factors. Or, the E could be a carry-over from a past vegetation or climatic factor. It might be found, however, that the E horizon is not related to podzolization, but rather is due to either a layered parent material or a strongly differentiated soil with a perched water table that produced the E horizon by removal of Fe^{2+} in laterally draining soil water (ferrolysis). Although more examples could be cited, the point is made. The use of factors, along with geological data and data on soil processes, provides one with the tools necessary to evaluate the origin of the soil in the field partly because it can keep the mind open to alternate hypotheses.

Although the derived functions are few, and although one may question how quantitative the use of factors is, the approach is basic to geomorphological research. Derived functions can be used in a qualitative sense, and they are commonly used to indicate the trends that one can expect throughout a region. Such qualitative expressions of the variables give rise to what Jenny[11] would call sequences rather than functions. One can study chronosequences or climosequences, and these are not without value in research. If the trends are not those that were predicted, perhaps the factors for that site will have to be redefined.

STEADY STATE

Soils often are described as being mature or in equilibrium with their environment. Lavkulich[14] argues that equilibrium is not a good term in the sense that reactions at dynamic equilibrium go in both directions with

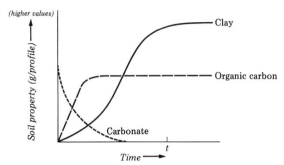

Fig. 6-2 Hypothetical variation in several soil properties with time. The parent material contained carbonate, which has been leached from the soil. Organic carbon reaches the steady-state condition before clay, and the soil profile might be considered at the steady state at time t.

no apparent change in the system. This condition is probably uncommon in the soil because many reactions, such as primary mineral → clay, are not reversible in the soil environment. He believes steady state is a better term to describe soil conditions. At the steady state, energy is being applied to the system, and reactions are going on, but properties are either not changing or their rate of change is too slow to be measured. This is similar to the concept of a soil-forming factor being considered constant even though it varies. That is, for a change in the factor, the soil property does not change.

The time factor is a familiar example of soil steady state. If one plots a soil property as a function of time, the curve is usually steep during the initial stages of soil development (Fig. 6-2). After some time the soil property shows little change, even though reactions are still going on. We can say that the soil property has reached a steady state with the environmental factors. If we consider a soil at one locality, each soil property commonly will require a different amount of time to attain the steady-state condition (Fig. 6-2). The soil profile could be said to be in a steady state when its diagnostic properties are each in a steady state. Examples of curves that flatten with time are those for various weathering phenomena,[4] and those for soil–profile development indices.[6,16]

In the following chapters, weathering and soil formation will be discussed as a function of one main factor. Obviously, the other factors cannot be held strictly constant, and the latter factors will be discussed when they play a role in explaining a particular feature or features.

REFERENCES

1. Barshad, I., 1964, Chemistry of soil development, p. 1–70 *in* F.E. Bear, ed., Chemistry of the soil: Reinhold Publ. Corp., New York, 515 p.
2. Bunting, B.T., 1965, The geography of soil: Aldine Publ. Co., Chicago, 213 p.

3. Chesworth, W., 1973, The parent rock effect on the genesis of soil: Geoderma, v. 10, p. 215–225.

4. Colman, S.M., 1981, Rock-weathering rates as functions of time: Quaternary Res., v. 15, p. 250–264.

5. Crocker, R.L., 1952, Soil genesis and the pedogenic factors: Quat. Rev. Biology, v. 27, p. 139–168.

6. Harden, J.W., 1982, A quantitative index of soil development from field descriptions: Examples from a chronosequence in central California: Geoderma, v. 28, p. 1–28.

7. Harradine, F., and Jenny, H., 1958, Influence of parent material and climate on texture and nitrogen and carbon contents of virgin California soils: 1: Texture and nitrogen contents of soils: Soil Sci., v. 85, p. 235–243.

8. Harris, S.A., 1968, Comments on the validity of the law of soil zonality: 9th Internat. Cong. Soil Sci. Trans., v. 4, p. 585–593.

9. Jenny, H., 1941, Factors of soil formation: McGraw-Hill, New York, 281 p.

10. ———, 1946, Arrangement of soil series and types according to functions of soil-forming factors: Soil Sci., v. 61, p. 375–391.

11. ———, 1958, Role of the plant factor in the pedogenic functions: Ecology, v. 39, p. 5–16.

12. ———, 1961, Derivation of state factor equations of soils and ecosystems: Soil Sci. Soc. Amer. Proc., v. 25, p. 385–388.

13. ———, 1980, The soil resource—origin and behavior: Springer-Verlag, New York, 377 p.

14. Lavkulich, L.M., 1969, Soil dynamics in the interpretation of paleosols, p. 25–37 *in* S. Pawluk, ed., Pedology and Quaternary research: Univ. of Alberta Printing Dept., Edmonton, Alberta, 218 p.

15. Major, J., 1951, A functional, factorial approach to plant ecology: Ecology, v. 32, p. 392–412.

16. Meixner, R.E., and Singer, M.J., 1981, Use of a field morphology rating system to evaluate soil formation and discontinuities: Soil Sci., v. 131, p. 114–123.

17. Runge, E.C.A., 1973, Soil development sequences and energy models: Soil Sci., v. 115, p. 183–193.

18. Yaalon, D.H., 1975, Conceptual models in pedogenesis: Can soil-forming functions be solved?: Geoderma, v. 14, p. 189–205.

7

Influence of parent material on weathering and soil formation

Parent material influences many soil properties to varying degrees. Its influence is greatest in drier regions and in the initial stages of soil development. In wetter regions, and with time, other factors may overshadow the influence of parent material. Here we will deal with the effects of minerals, rocks, and unconsolidated deposits on weathering and soils. We also will explore briefly the genesis of soils in areas characterized by a slow rate of sedimentation.

MINERAL STABILITY

Minerals vary in their resistance to weathering; some weather quite rapidly (10^3 years), whereas others weather so slowly (10^5 to 10^6 years) that they persist through several sedimentary cycles. Minerals can be ranked by their resistance to weathering in several ways. One way is to study the soil or weathered material and compare the minerals there with minerals in the unweathered material on the basis of their respective depletion, evidence for alteration to clay, or etching of individual grains (Fig. 7-1). Goldich[26] established the mineral stability series for the common rock forming minerals in this way. His series is given in Table 7-1, modified somewhat by the work of Hay.[29] This latter work is added because it provides a cross-link between the weathering series of the fer-

Fig. 7-1 Photomicrographs of thin sections showing mineral weathering. (A) Two etched pyroxene grains in weathered ash. (Taken from Hay,[29] pl. 2, D, Jour. of Geology, © 1959, The University of Chicago.) *Left.* Hypersthene occupies a cavity that retains the original shape of the crystal. *Right.* Augite is surrounded by halloysite subsequently deposited in the cavity formed during grain weathering. Such sharp cockscomb terminations are characteristic of weathered pyrox-

(A) ├────0.5 mm────┤

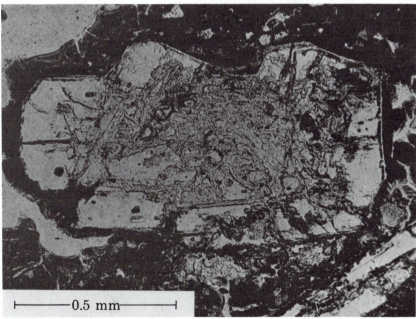

(B) ├────0.5 mm────┤

ene and hornblende grains in many areas. Lack of clay formation in place here might be related to insufficient aluminum in the parent grain. (B) Halloysite replacing interior of a zoned plagioclase. (Taken from Hay,[29] pl. 3, A, Jour. of Geology, © 1959. The University of Chicago.) The core of this grain probably was more calcic than the rim, thus the selective replacement.

Table 7-1
Ranking of Stability of Common Rock-forming Minerals, Their
Structural Classification, and Bonding Energies*

	Ferromagnesian minerals		Non-ferromagnesian minerals
Increased stability ↓	Olivine[1] (29)	Fe-rich Mg-rich	Ca^{2+}-plagioclase[4] (32) An$_{80-100}$ An$_{65-80}$
	Hypersthene[2a] Augite[2a] (31) Hornblende[2b] (32) Biotite[3] (30)		An$_{50-65}$
			Na^{+}-plagioclase[4] (34) K^{+}-feldspar[4] (34) Muscovite[3] Quartz[4] (37)

	Number of oxygens shared between adjacent silica tetrahedra	*Aluminosilicate mineral structures*
Increased sharing ↓	None	1. Nesosilicate (independent tetrahedrons)
	2	2a. Inosilicate (single chain)
	2 and 3	2b. Inosilicate (double chain)
	3	3. Phyllosilicate (sheet structure)
	4 (all)	4. Tectosilicate (interlinked network of tetrahedrons)

*Superscripts with mineral name are keyed to the numbered aluminosilicate structures.
Numbers in parentheses are bonding energies ($\times$ 1000 kg cal) for 24 oxygens in each mineral structure, calculated by Keller[34] (1954), and rounded to the nearest 1000. (Use granted by permission of the Mineralogical Society of America.)
(Taken from Goldich,[26] Table 18, Jour. of Geology, published by The Univ. of Chicago and Hay,[26] Fig. 8, Jour. Geology. © 1959, The Univ. of Chicago.)

romagnesian and non-ferromagnesian minerals, based on plagioclase composition. Although Hay did not find a difference between the weathering of hypersthene and augite, other stability studies have ranked augite as the more stable of the two, and that ranking is preserved in the table. Hay also found that iron-rich olivine grains weather more rapidly than magnesium-rich grains, and that the latter compare favorably with augite and hypersthene in resistance to weathering. This general sequence is in accord with many data, although the order may be changed somewhat under certain environmental conditions.

The stability of minerals also may be assessed by their occurrence in sedimentary rocks of varying ages; older rocks should have relatively greater amounts of the more resistant minerals. Pettijohn[47] has drawn up a list for the ferromagnesian and accessory minerals, and although it includes more minerals than does Table 7-1, the ranking of the common rock-forming minerals is similar. Other persistence schemes are compared with Pettijohn's by Brewer,[14] and these data indicate that rankings of mineral stability differ. Most studies, however, indicate that zircon and tourmaline compare with quartz in their high resistance to weathering, and therefore any of these three minerals can be used as the stable component in mineral-depletion weathering studies.[38] Raeside[48] points out, however, that some seemingly stable minerals may not always be stable for various mineralogical reasons.

Several factors account for the stability of the common rock-forming minerals. Some are related to the mineral itself, and some are related to the weathering environment. Only those related to the mineral will be considered here.

Mineral structure plays a major role in the resistance of a mineral to weathering. From the data in Table 7-1, it is noted that for some minerals the mineral-stability ranking parallels a ranking based on mineral structures: that is, the progressive increase in sharing of oxygens between adjacent silica tetrahedra correlates with increased resistance to weathering. This is especially true for the sequence olivine-pyroxene-hornblende-biotite-quartz. The energies of formation of the cation-oxygen bond seem to explain the above sequence. Keller[34] calculated the energies, and the data indicate that $Si-O$ is the strongest bond (> 3000 kg cal), $Al-O$ the next strongest bond (< 2000 kg cal), and the bond between the common base ions and oxygen is the weakest (< 1000 kg cal). Therefore, mineral resistance to weathering should be reflected in the structure, composition, and total energy of all the bonds in the mineral. Keller[34] calculated the energies of formation of all the bonds for a variety of minerals, based on a mineral unit with 24 oxygens (Table 7-1). The relationship between energies of the total bonds and the mineral-stability ranking is fairly good both for the non-ferromagnesian minerals and the ferromagnesian minerals, and the only mineral out of sequence is biotite. The correlation between the ferromagnesian and non-ferromagnesian minerals, however, does not correspond with that based on field data. From the bonding energy data, one could predict that volcanic glass should be highly susceptible to weathering, and that appears to be the case. Curtis[20] has taken another approach and calculated the total energy released by mineral breakdown; these data are then shown to correlate with the mineral persistence ranking.

Mineral breakdown during weathering probably is initiated at the site of the weakest bond, and the more sites that are exposed to weathering solutions, the more rapid should be the weathering.[3] Olivine weathers rapidly because the separate tetrahedra are linked together only by the fairly weak cation-oxygen bonds. The other minerals in the stability series contain tetrahedra linked together by the stronger $Si-O$ and $Al-O$ bonds and so are more resistant to weathering. In these minerals weathering is most effective if the weathering solutions have access to sites of the weak cation-oxygen bonds. Such sites are those that link the chains of the inosilicates together, and the sheets of the phyllosilicates together, or sites that offset the charge deficiency brought about by the substitution of Al^{3+} for Si^{4+} in the tectosilicates.

Other factors seem to explain the relative resistance of minerals classified in the same structural groups.[3] The plagioclases, for example, show a ranking from the more stable sodium-plagioclases to the least stable calcium-plagioclases. This sequence is due in large part to the substitution of Al^{3+} for Si^{4+} in the tetrahedral position as the plagioclase becomes more calcic; thus, stronger $Si-O$ bonds are replaced by weaker $Al-O$ bonds, and this is reflected in a decrease in the total bonding energies. Biotite and muscovite, although both phyllosilicates, differ in their weathering. Biotite is more susceptible to weathering, probably because the Fe^{2+} is readily oxidized to Fe^{3+}. This change in valency disrupts the electrical neutrality of the mineral, which, in turn, causes other cations to leave and thus weaken the crystal lattice. The tightness of packing of the oxygens might also be a factor with some minerals. In both of two mineral pairs, sodium-plagioclase and orthoclase, and olivine and zircon, the latter minerals have the tighter packing and are more resistant to weathering. This is especially important in the olivine-zircon pair, because it includes one of the least and one of the most resistant minerals.

SUSCEPTIBILITY OF ROCKS TO CHEMICAL WEATHERING

Rocks weather and erode at different rates, as can be seen in the variations in topographic relief that accompany variations in rock type. For rocks from widely spaced localities, however, factors other than rock type might influence the weathering variations observed. To be able to compare rocks of differing lithology under similar conditions of weathering, it might be best to study a sedimentary deposit, such as bouldery till or outwash, which includes a variety of rock types. Rocks from the same depth below the surface should have weathered under as similar conditions of weathering as one could hope to find.

Goldich's stability series for minerals can be used to predict igneous rock stability in the weathering environment. Rocks with a high content of more weatherable minerals should weather more rapidly than rocks with a high content of minerals resistant to weathering. To make a valid comparison, however, the rocks should be similar in crystallinity and grain size. For igneous rocks, therefore, resistance to weathering should increase in the order gabbro $\rightarrow$ granite or basalt $\rightarrow$ rhyolite. Clay production should follow these trends, and it commonly does.[2]

Grain size has an effect on the rate of weathering, for it is observed that coarser-grained igneous rocks commonly weather more rapidly than finer-grained rocks.[61] This is readily seen in many tills in the Cordilleran

Region. In the Sierra Nevada, for example, till of probable pre-Wisconsin age has the following variation in weathered clasts: coarse-grained granitic stones are weathered to grus, coarse-grained porphyritic andesitic pebbles and cobbles have thick weathering rinds or are weathered to the core, and fine-grained volcanic rocks of intermediate to basic composition have weathering rinds up to 4 mm thick.[7] Moreover, it is a common observation in large boulders that mafic inclusions, generally of finer grain size than the enclosing granitic rock, stand in relief above the latter due to weathering.[11,15]

Weathering rind studies help to quantify some of these relations. Weathering rinds are uniformly thick zones of chemical alteration around the periphery of clasts. A measure of chemical alteration is the thickness of the rind. Because rinds become thicker with time, rind data have been collected to date glacial deposits in various parts of the western United States (Table 7-2). Although there are differences in climate and specific rock type, basalt clasts collected from soils in the Rocky Mountain region have rinds rather similar to those for coarse-grained andesites in California, and the latter are usually thicker than those on fine-grained andesite. In contrast, rinds on granitic clasts at the ground surface are much thicker. Rinds on surface graywacke clasts in New Zealand are the thickest and reach about 6 mm thickness in about 9500 years.[17]

Table 7-2
Thicknesses of Weathering Rinds (mm) of Clasts in the Western United States

ESTIMATED AGE (YR)	BASALTS*		
	West Yellowstone, Mont.	*McCall, Idaho*	*Yakima Valley, Wash.*
10,000–20,000	0.10 ± 0.07	0.25 ± 0.14	0.25 ± 0.04
140,000	0.78 ± 0.19	1.61 ± 0.41	1.05 ± 0.17
	COARSE-GRAINED ANDESITES*		
	Truckee, Cal.	*Lassen Pk, Cal.*	
10,000–20,000	0.18 ± 0.13	0.18 ± 0.12	
140,000	0.97 ± 0.23	0.82 ± 0.25	
	GRANITIC ROCKS		
	*Mt. Sopris, Colo.***	*Central Sierra Nevada, Cal.****	
10,000–20,000	6–45	1.7	
140,000		2.9	

*Data for clasts in soils from Colman and Pierce.[19]
**Data for surface clasts from Birkeland.[9]
***Data for surface clasts from Burke and Birkeland.[15]

Rocks with high amounts of glass weather more rapidly than rocks with low glass contents, and rocks of finer grain size weather more slowly than rocks of coarser grain size. One reason for the latter may be that intergranular surface area increases with decrease in grain size; hence more energy probably would be required to disintegrate the finer-grained rock.

The weathering process that goes on in rocks containing biotite differs somewhat from that in biotite-free rocks. Biotite-bearing rocks commonly weather more rapidly, as seen in many outcrops of igneous and metamorphic rocks. The reason for this behavior was presented earlier (Ch. 3). Biotite, in addition to being the first mineral to weather in granitic rocks, forms alteration products that can occupy a greater volume than did the original biotite[42]; the result is mineral expansion, with numerous localized points of stress within the rock that eventually shatter the rock and form grus. This mechanism probably explains, in part, why the zone between fresh and weathered granitic rock usually is gradational over about 0.5 m or more. Basalt is an extreme case of rock lacking biotite. Weathering proceeds inward, grain by grain, and the boundary between fresh and chemically weathered rock can be quite sharp ($<$ 1 mm).[16] The point to be made is that the chemical alteration necessary to disaggregate a biotite-bearing granitic rock is not comparable to that necessary to weather basalt to a similar depth. Much of granitic rock weathering is mechanical shattering induced by slight chemical weathering, whereas basalt rock weathering is mainly chemical and, as noted, proceeds slowly inward. This, along with the variation in susceptibility of minerals to weathering, explains the common observation that soils formed from granitic rock are higher in sand and lower in clay than are soils formed from basalt.

The weathering of sedimentary rock differs from that of igneous rock, but there is no fast rule to determine which rocks are most susceptible to weathering. In some areas the evidence favors more rapid weathering of sedimentary rocks,[30] but this can be reversed in other areas with other rock types. Sedimentary rocks can have any combinations of nonclay and clay minerals, and the weathering of these rocks could proceed as predicted by the Goldich stability series or the Pettijohn persistence series. Rocks with a considerable clay-size component may break down more rapidly than rocks low in clays because clays by themselves do not bind minerals tightly together and they might expand and contract with variation in moisture content. The cementing agent can have a marked effect on the weathering of sedimentary rocks (Fig. 7-2). In pre-Wisconsin stream terrace deposits in the Colorado Piedmont area, for example, some clasts of silica-cemented arkoses in soils are intact, whereas adjacent granitic clasts of seemingly similar mineralogy are weathered to grus. Either the silica cement keeps weathering solutions from reaching the biotites or biotites were weathered prior to deposition or the cement is strong enough to withstand stresses set up by weathering biotites. In the same soils,

Fig. 7-2 Variation in boulder weathering as a function of rock type. Eagle River terrace gravel, western Colorado. Some granitic stones have weathered to grus (g), whereas other granitic stones (light tone) and sedimentary stones (dark tones) are unweathered.

clasts of silica-cemented quartz sandstones also are intact, but clasts of shale are weathered.

Limestones present a special case in the weathering of sedimentary rock. The rock weathers by solution of the readily dissolved carbonate minerals, and the rate of weathering is determined by carbonate equilibria.

Till deposits in Illinois contain a wide variety of rock types and thereby provide field sites for relative weathering rates. The data of Willman and others[63] suggest the following rock-stability ranking, from more stable to less stable:

Quartzite, chert > granite, basalt > sandstone, siltstone > dolomite, limestone

Although there are local variations in the ranking due to variations in specific lithologies, the overall trend seems reasonable.

INFLUENCE OF PARENT MATERIAL ON THE FORMATION OF CLAY MINERALS

The parent material, whether mineral or rock, exerts some control on the clay minerals that form because weathering releases constituents essential to the formation of the various clays. The specific ions, their concentration, and the molar $SiO_2:Al_2O_3$ ratio are all important aspects of the parent material. However, once weathering release has occurred, the micro- and macro-environments within the soil determine whether certain ions or other constituents are selectively removed or remain behind, and this then determines which clay mineral forms. In order to keep factors constant, samples should be collected from a certain depth within the soil, and comparisons should be made between the weathering products and the parent minerals or rocks. Most results substantiate the conditions that were discussed earlier as necessary for the formation of the clay minerals.

Even the clay mineral that forms may not be a stable end product, because it too can subsequently change to other products with changing conditions within the soil. There is a large literature on clay-mineral formation relative to the parent mineral or rock, and almost any product is possible, probably because of the large variety of environments that are possible. These will not be reviewed here because many of the relationships could involve parent material masked by other environmental factors. A review of some of this work is given in Loughnan.[36]

Rocks or minerals of varying composition weathering in the same environment can produce different clay minerals. Barnhisel and Rich,[1] for example, sampled boulders of different lithologies within the same weathered zone and analyzed the clays that formed. Granites and gneisses, rocks low in bases, produced kaolinite predominantely, whereas gabbro, a rock high in bases, produced mostly smectite. These same extremes in clay-mineral formation can be seen at the mineral level. In Hawaii, Bates[6] reports plagioclase altering to halloysite, whereas adjacent olivine grains alter to montmorillonite. Tardy and others[64] also note that the specific alteration product can vary with climate (Table 7-3). As a generalization,

Table 7-3
Variation in Clay-mineral Alteration Product from
Various Primary Minerals, as a Function of Climate

Climatic regime	Primary mineral	Clay mineral(s) formed
Arid	Biotite	Smectite
	Plagioclase	Kaolinite
Temperature	Biotite	Vermiculite
	Plagioclase	Smectite, kaolinite
Humid tropical	Biotite	Kaolinite
	K-feldspar	Kaolinite
	Plagioclase	Kaolinite
Very humid tropical	Biotite	Kaolinite
	K-feldspar	Kaolinite, gibbsite
	Plagioclase	Gibbsite

(Taken from Tardy and others.[64])

the parent mineral will greatly influence the mineralogy of the weathering by-product in those soils, or places within soils, characterized by low leaching, and the influence of the parent rock or mineral will gradually diminish with greater leaching.

Parent-material influences can also be studied by sampling over large regions. Barshad[4] has reported on such a study in California. Samples of the uppermost 15 cm of the soil were collected from a large part of the state. This assured representation of a large range in climate and in parent materials. Kaolinite, halloysite, montmorillonite, illite, vermiculite, and gibbsite are the major minerals present. In his data analysis, Barshad found that the main variation in clay minerals is a function of precipitation (Fig. 11-13) but that parent material affected some clay-precipitation relationships. The parent materials form two main groups: mafic and felsic igneous rocks. The influence of parent material was recorded in two ways. Illite is associated with felsic rocks only, probably because of their higher mica content as well as the availability of potassium. Montmorillonite, although abundant at low precipitation with both rock types, persists as a prominent clay mineral at higher precipitation in the mafic parent materials than in the felsic parent materials, probably because of the greater content of bases in the mafic rocks.

The eastern side of the Sierra Nevada offers another sampling region suitable for studying parent-material effects on clay mineralogy because tills and stream gravels with a variety of rock types are abundant.[8,10] Again, the major effects probably are climatic (Ch. 11), although parent-material influences are recognized in the drier areas. In soils formed from a parent material that is partly andesitic, montmorillonite can be present at a mean annual precipitation of 48 cm. In contrast, soils formed from materials of predominantly granitic lithology, at even lower rainfalls, contain illite and kaolinite, and montmorillonite only at depth. Again, the parent material helps determine the kind of clay that forms through availability and kind of bases.

INFLUENCE OF ORIGINAL TEXTURE OF UNCONSOLIDATED SEDIMENTS ON SOIL FORMATION

Much of the research undertaken by geomorphologists is with unconsolidated deposits of Quaternary age; hence, a brief treatment of the effect of textural variation of this material on soil formation seems in order. Although we shall try to keep lithology and mineralogy of parent material constant, this is not always possible in the field. When all other factors are equal, however, the texture of the parent material has a great influence on the course of soil formation.

Texture influences the rate and depth of leaching, and this is related to many soil properties. Textural influence can be such that soils that generally occur only in different climatic regions occur side by side, and yet each is stable for the prevailing site conditions. Figure 7-3 depicts some possible field conditions. The depth of leaching is governed by the texture, being greater in gravel, less in non-gravelly sand, and least in finer-textured material. For the same duration of soil formation, an argillic horizon could form in the finer-textured material but not in the sandy material, because it could only be a function of translocation of original clay-size material. If sufficient time has passed for clay formation to be noticeable, it probably would proceed more rapidly in the finer-textured material,[39] because weathering rate increases with greater surface area per unit volume due to the availability of more water for weathering over a longer period of time each year. Clay formation by precipitation from solutions is also enhanced by increased surface area, because surfaces promote retention of clay-forming constituents. In contrast, constituents released by weathering in the more gravelly materials may be leached from the soil before they have the opportunity to react to form clays. This observation may help explain the presence of well-expressed argillic horizons in late-Wisconsin tills and loesses in the midcontinent and the general lack or poor expression of such horizons in gravelly tills and outwashes of the same age in the Sierra Nevada and other parts of the Cordilleran Region (Fig. 8-10). In addition, there might be more release by weathering of Ca^{2+} in the finer-textured material, which, in combination with less leaching, might lead to formation of a Bk horizon. Greater leaching in the coarser-grained materials, especially during the wetter years, might tend to keep the Ca^{2+} content too low to form a Bk horizon. The clay minerals that form also follow these trends. Those in the finer-textured material might have a higher ratio of 2:1 layer clay to 1:1 layer clay than those in the more permeable material. These relationships can be seen in soil samples spaced closely together; in Hawaii, kaolinite can occur in freely drained parts of the soil, whereas montmorillonite can form in the same soil in local areas of slightly restricted drainage beneath stones.[57] Finally, organic

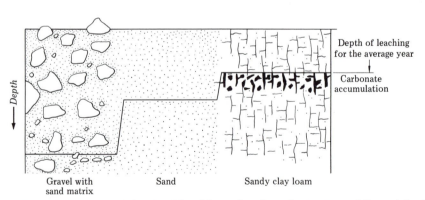

Fig. 7-3 Hypothetical depth of leaching related to the texture of the original parent material.

matter contents should be higher with the finer-textured material, a relationship that has been demonstrated in California[28] and other areas (Ch. 9).

It should be stressed that the above statements are generalizations and that some of the effects mentioned probably are not the same from one climatic region to another. Furthermore, although the quantitative classification scheme of soil-profile development of Harden[27] takes into account the original texture of the parent material, finer-textured soils should develop profile characteristics more rapidly. Obviously, these relationships are important to stratigraphic correlations based on soil development.

EXTREME EXAMPLES OF ROCK-TYPE CONTROL ON SOIL FORMATION

The composition of some rock types controls the direction of soil formation to an unusual extent. It is well known that podzolization takes place best in sandy materials poor in bases. Cline[18] reports on podzolized soils in New York formed in sandy, mainly crystalline parent materials. However, nearby till with high carbonate content does not exhibit podzolization characteristics, probably because the high base content keeps pH high and prevents translocation of Fe and Al, either inorganically or by chelating agents. As another example, it is commonly known that Bk horizons form in particular moisture regimes, but at the margins of areas with these regimes, the Ca content of the parent material could determine whether or not such horizons form.[32] Soils formed in limestone are still another example of bedrock control on soils, for a fine-grained soil with or without carbonate commonly overlies bedrock with sharp contact. The origin of the soil has been ascribed by some as the insoluble residue that accumulates at the surface upon removal by dissolution of the carbonate fraction[5,44,62]; hence, the properties of the insoluble residue of the limestone are reflected in the soil properties. Combined pedologic and stratigraphic studies, however, suggest that this model may be too simple. In places, there may not be enough limestone dissolved to produce the insoluble residue of fine grained material; the fines, therefore, are of eolian origin or have been deposited subaerially from higher terrain.[22,45,56,72]

Volcanic airfall deposits (ash, pumice, tuff, etc.) are parent materials that probably control soil formation more than any other parent materials.[25,35,49,65] Many of these soils have such similar properties, and these latter are so different from those of other parent materials, that it has been proposed that they be classified in a new order, termed Andisol. The diagnostic horizons allowed are histic, umbric, mollic, and ochric epipedons; cambic and placic horizons; and duripan. As usual, the definition is a bit complicated, but common properties are a low bulk density (<0.9) and a high content of alteration materials that are amorphous or of low crystallinity. A simple test for the presence of these latter materials is the NaF pH test[24]; the sample is placed in a solution of NaF, F replaces OH of the amorphous and low crystallinity materials, and the pH of the solution increases to 9.2 or 9.4 in two minutes. The glass of the parent materials alters quickly, and at one locality in New Zealand, the half-life of andesitic glass has been shown to be about 7000 years.[41] The alteration of glass seems to be first to allophane, imogolite, opaline silica, and Al- and Fe-humus complexes.[12] In time, a multitude of clay minerals can form, including gibbsite in a high leaching environment, halloysite, or if enough silica is available, the common 2:1 and 2:1:1 layer silicate minerals in less leaching environments with sufficient silica. In only some environments, however, the relatively amorphous intermediary products are precursors to layer-lattice clay minerals.[46]

TEST FOR UNIFORMITY OF PARENT MATERIAL

In any soil study, one should establish, beyond reasonable doubt, the identity of the parent material from which the soil formed. In some places this is no easy task, especially if the soil is quite weathered. If parent material cannot be identified in the field, several mineralogical tests for parent-material uniformity can be made.

In the field, one carefully examines the least weathered material, the C or R horizon, to determine if the soil could have formed from that material. For example, if the C or R horizon is granitic bedrock or colluvium, the soil should contain only that material; rounded gravel should be absent unless it can be shown that it resulted from weathering. Stone lines have been used to recognize unconformities in lateritic and other materials—unconformities that might otherwise be overlooked.[52] A textural B horizon, if present, should have a position and relationship to both overlying and underlying horizons that is pedogenic in character (see criteria for buried soils, Ch. 1). In stream gravel or till, uniformity would be suggested by a more or less constant gravel content to the surface. Common relationships to be watchful for are fine-grained overbank alluvium overlying high-energy mainstream alluvium, or if any sites are near valley walls, there probably will be colluvial or alluvial fan material overlying the main valley deposit of outwash or till. Actually, most soil forms from material of a size smaller than gravel, and so it is especially important to determine uniformity within those size fractions. Sands are usually the more resistant fraction, so one can carefully examine the sand fraction percent, normalized to 100 percent, for variation. Variation could be a function of original parent-material layering, or it could reflect subsequent weathering of grains from larger to smaller size fractions. However, even if the normalized sand fractions suggest uniform parent material, one has to be sure that this agrees with the vertical distribution of silt and clay, for the best evidence for layering could be in these fractions.

In many localities, even those far removed from sites of subaerial deposition, parent-material layering might be due to eolian deposition. Loess forms a thin surface layer in many soils,[13,69] and if it is not recognized as such, one might err in describing the soil-forming processes at a particular site. In the western United States, for example, we find loess to be an important component of the A and some B horizons of many gravelly mountain soils[58] (Fig. 7-4). Many of the soils in the basins between the ranges will have a surface layer of loess or loess mixed with the underlying parent material. Recently, it has been recognized in soils on California coastal islands.[40]

Microscopic examination of the soil and parent material might also indicate influx of foreign material either by eolian or subaerial processes. Contamination is recognized by a non-clay mineralogy of the soil that does not match that of the parent material or that cannot be explained by weathering of the original grains of the parent material.[38] In places, this kind of contamination has been shown to account for as much as one-half of the soil material.[37] More subtle eolian influxes can be determined by oxygen-isotope data on quartz in the soil and in the assumed parent material.[63,66]

Once the more obvious ways to ascertain uniformity have shown that the material is probably uniform, a final mineralogical check can be made. Within a specific sand or silt fraction, minerals that are relatively resistant to weathering should show a relative increase in abundance from the parent material to the surface, whereas more weatherable minerals should show a relative decrease in the same direction.[38] According to Brewer,[14]

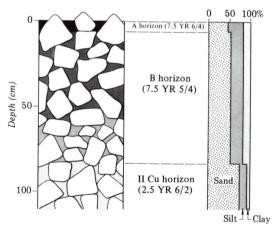

Fig. 7-4 Soil formed on rock-glacier deposit of probable late-Wisconsin age, Mt. Sopris, western Colorado.[9] Field and laboratory analysis of particle-size distribution indicates that the upper 83 cm is loess and that it overlies material derived from the rock glacier because the large sand grains in the lower material are made up of both quartz and feldspar, which could not have broken down to the sizes found in the upper layer in the time available.

plots such as these should change gradually and uniformly with depth; sharp inflections or reversals in trends might indicate lack of uniformity. Ratios of resistant mineral to nonresistant mineral also can be used to indicate the same trends (Fig. 8-4). Another technique is to determine the ratio of two resistant minerals; if the ratio remains nearly constant with depth and matches that of the parent material, uniformity is strongly suggested.[3,38] This may not be true in every case, however, because sedimentary layers of different textures could have similar ratios of resistant mineral to resistant mineral in the same size fraction.

CUMULATIVE SOIL PROFILES

Some soils receive influxes of parent material while soil formation is going on; that is, soil formation and deposition are concomitant at the same site. Nikiforoff[43] named these soils cumulative (Fig. 7-5). In such soils, the A horizon builds up with the accumulating parent material, and the material in the former A horizon can eventually become the B horizon. In contrast, other soils gradually lose material through surface erosion, so that the A horizon eventually forms from the former B horizon (Fig. 7-5). These soils were called non-cumulative by Nikiforoff.[43] Both kinds of soils are common along many slopes. Some examples of cumulative profiles will

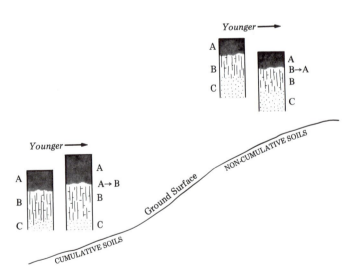

Fig. 7-5 Cumulative and non-cumulative soil profiles. If removal of material at the non-cumulative site surface is matched by soil-profile development, the profile characteristics do not vary with time. In contrast, the horizons in the cumulative profile commonly are thickened due to processes at the site and influx from upslope of organic matter, clay-size material, or other material. Because cumulative profiles are, at least in part depositional, subtle sedimentary layering might be discernible. See Ch. 1 for meaning of arrows.

be given here, and others are given in Chapter 9, where lateral soil variations with slope position are discussed.

Because cumulative soils have parent material continuously added to their surfaces, their features are partly sedimentologic and partly pedogenic. In a soil study, therefore, it is important that sedimentologic features are not ascribed to pedogenesis.

Some topographic positions are favorably situated for cumulative profile formation. They are especially common in colluvial and fan deposits at the base of hillslopes,[50] and along river floodplains.[33,59] Uplands receiving increments of loess during soil formation also can have cumulative profiles.

Cumulative profiles are sometimes recognized by properties that are not consistent with those of the soils in the surrounding region. A common property is an overthickened A horizon due to deposition from upslope of material rich in organic matter, to organic matter at the site being continually buried, or to a combination of both processes. At any rate, the gains in organic matter are not balanced by decompositional losses, and these gains, combined with deposition at the surface, produce thick A horizons. Cumulative profiles also may contain more clay to a greater depth than adjacent non-cumulative soils from which they derive their surface increments. This occurs because clay content in the cumulative soil is a function of clay formation at that site in addition to the clay delivered from upslope sites by erosion.

Some loess areas in the midcontinent are characterized by cumulative profiles. Smith,[60] in studying loess in Illinois, suggested an influence of parent material on the soil pattern. Many properties of the parent material loess are related to distance from the source of the loess, which is usually a river floodplain of glacial age. The loess is commonly thicker and coarser grained near the source and becomes thinner and finer grained downwind. The soils developed on the loess show a downwind increase in clay content, accompanied by the gradual loss of $CaCO_3$. Smith reasoned that loess deposition and soil formation were intimately involved in producing the soil pattern. Close to the source rivers, the loess was deposited too rapidly for soil formation to keep up, and mostly unweathered calcareous loess was deposited. At localities farther from the rivers, however, soil formation could keep up or even exceed the rate of deposition, so that carbonates, if originally present, would have been leached as rapidly as the material was being deposited. Moreover, weathering could have gone on for longer periods of time on loess materials away from the river where the supposed depositional rate was less. Thus, cumulative soil profiles could be found at some distance from the rivers, but not close to the rivers. The effect of parent material, therefore, is twofold. Downwind decrease in particle size results in finer-textured parent material and soils, and downwind decrease in rate of deposition results in cumulative soils in which carbonate leaching and other soil-forming processes can go on concomitant with deposition; thus, soil formation of longer duration is suggested for the downwind localities. Hutton[31] explains the loess-soils distribution in Iowa in a somewhat similar manner. In Alaska,[51] variation in Spodosol soil development with distance from the loess source is attributed to variations in the rate of loess deposition.

Detailed work by Ruhe[53,54,55] demonstrates the complexity of the loess landscape. In particular, in northeastern Kansas, the decrease in parent-material particle size away from the loess source is not the sole cause of regional variation in B-horizon clay contents. He showed that the age of the base of the loess in Iowa is younger away from the source area, and therefore it no longer can be assumed that the more distant soils in a loess landscape have weathered longer (Fig. 7-6). Moreover, both the surface and buried soils have more strongly expressed Bt horizons with distance from the source area, and the increase in development of this horizon takes place over a shorter distance for the Sangamon soil than it does for the surface soil (Fig. 7-7). Although part of the Bt-horizon development trend could be a function of clay formation and translocation, part of that for the surface soils is due to greater moisture in the progressively thinner

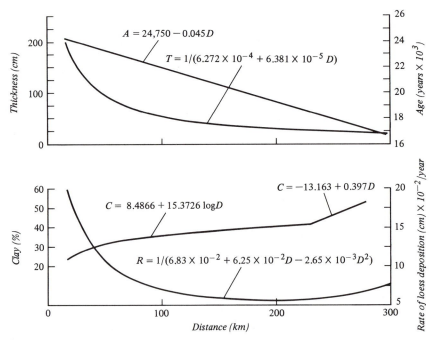

Fig. 7-6 Data for Wisconsin loess in Iowa. D, distance from source; A, age of A horizon of the basal buried soil in loess; T, thickness of loess; C, maximum clay content in surface soils; R, rate of loess deposition. (Taken from Ruhe,[54] Fig. 7.)

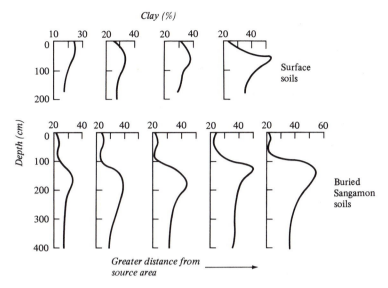

Fig. 7-7 Clay data for surface soils formed in Wisconsin loess and buried Sangamon soils formed in Illinoian loess, in Iowa. The changes in the surface soils take place in about 260 km from the source area; and those in the Sangamon soils, between about 18 and 33 km from the source area. (Taken from Ruhe,[54] Fig. 3.)

loess sheets because of a perched water table on the underlying buried soils. The explanation of the soil landscape as a result of this study involves many factors, both geological and pedologic, and parent material is only one of many factors.

An interesting transect demonstrating the influence of loess on pedogenesis is that from the Sinai Desert northward into Israel.[21,23,68,69,70] Fine-grained material is deposited by infrequent floods in the arroyos of the Sinai. Vegetation is scarce, and winds transport the dust northward into Israel, mainly from Sinai, but also from North Africa. The rate of dust accumulation and the mean size of the dust vary systematically northward. In parts of the more extreme desert, deposition is slow and the material is recognized as a fine-grained surface layer overlying gravelly materials, such as the ubiquitous alluvial-fan deposits. Where rainfall is sufficient, part of the loess is translocated into the soil. Still farther north, in the Negev Plains, the loess is thickest and most widespread, and the soils are formed entirely in loess. North of the Negev, the rate of deposition declines and the loess is more clay rich. In the coastal plain, there are sand dunes of various Quaternary ages in which the airborne clay has infiltrated to produce a soil-development sequence. Soils formed in young sand dune deposits have Bw horizons, but in time enough eolian clay has infiltrated to produce Bt horizons whose clay content increases with relative age. The end result is old soils that are so clay rich that Vertisols have formed. Here parent material in the form of dust accretion plays an important role in explaining the regional soils, but other factors also are important. These include rainfall, which influences both the location of maximum loess deposition and the infiltration of loess into the soil, and time, which helps explain soil development locally along many parts of the transect.

Alluvial fans also offer an opportunity to study the genesis of cumulative soils. One fan studied in southern California will be described here (Yerkes and Wentworth,[71] and unpublished work of the writer). The fan is at the mouth of a very small watershed underlain with sedimentary rocks, some of which are calcareous. The cumulative origin of the soil profile on the fan is suggested by the distribution of organic carbon with depth in the profile (Fig. 7-8). Organic carbon in the cumulative soil is higher in content, and the depth to a level that is less than 1 percent organic carbon is greater than it is in a nearby stable, non-cumulative soil profile. Radiocarbon dates suggest that deposition commenced about 10,000 years ago, and this agrees with stratigraphic and radiocarbon data from nearby areas. Textural variation with depth is rather uniform, implying that deposition has been rapid and continuous enough for sedimentation to mask any B-horizon development.

Carbonate in the fan deposit does not help identify it as a cumulative

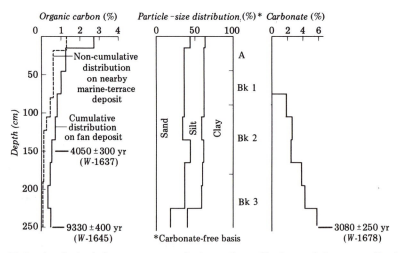

Fig. 7-8 Analytical data on a cumulative soil profile formed from an alluvial-fan deposit. The fan deposit overlies a coastal marine terrace west of Santa Monica at the mouth of Coral Canyon. Radiocarbon dates are not from this profile, but, rather, from a nearby profile with similar morphology; horizontal lines indicate the depth from which the dated samples were taken. Two samples are on organic carbon, one on carbonate. (Soil data from unpublished work of the writer. Radiocarbon dates by M. Rubin, U.S. Geological Survey.)

profile, however (Fig. 7-8). The soil carbonate originates from the weathering of rock fragments and calcareous sediments transported from the drainage basin. The radiocarbon date on the carbonate, although subject to error, is a minimum for sediment in that part of the profile, and it is consistent with the other dates because carbonate can undergo solution and re-precipitation (Ch. 5). Carbonate translocation may have gone on during sediment deposition, or the bulk of it may have been translocated at a later time. If carbonate translocation went on during deposition, thick fan deposits should have overthickened Bk horizons that extend to the base of the deposit. A deep artificial exposure in colluvium adjacent to the fan in similar parent material disclosed that the Bk horizon does not extend throughout the deposit, but has a base, below which is colluvium with a relatively low carbonate content. This can be taken to mean that carbonate did not accumulate with the buildup of sediment on the surface or that its absence at depth is explained by other factors, which may be paleoclimatic or lithologic. That some other factor is involved is supported by rough calculations on rates of sedimentation and carbonate translocation; these suggest that translocation can keep pace with sedimentation under the present climatic conditions.

These data demonstrate the intimate interrelationship between geology and pedology in fan and colluvial deposits that form at the bases of slopes or at the mouths of stream valleys. These and loess landscapes are some of the more difficult areas to work in because of the dynamic nature of the geomorphical and pedologic systems; they serve to demonstrate that it is difficult to isolate parent material from other soil-forming factors.

REFERENCES

1. Barnhisel, R.I., and Rich, C.I., 1967, Clay mineral formation in different rock types of a weathering boulder conglomerate: Soil Sci. Soc. Amer. Proc., v. 31, p. 627–631.
2. Barshad, I., 1958, Factors affecting clay formation: 6th Natl. Conf. Clays and Clay Minerals Proc., p. 110–132.
3. ———, 1964, Chemistry of soil development, p. 1–70 *in* F.E. Bear, ed., Chemistry of the soil: Reinhold Publ. Corp., New York, 515 p.
4. ———, 1966, The effect of a variation in precipitation on the nature of clay mineral formation in soils from acid and basic igneous rocks: 1966 Internat. Clay Conf. (Jerusalem), Proc. v. 1, 167–173.
5. Barshad, I., Halevy, E., Gold, H.A., and Hagin, J., 1956, Clay minerals in some limestone soils from Israel: Soil Sci., v. 81, p. 423–437.
6. Bates, T.F., 1962, Halloysite and gibbsite formation in Hawaii: 9th Natl. Conf. Clays and Clay Minerals Proc., p. 307–314.
7. Birkeland, P.W., 1964, Pleistocene glaciation of the northern Sierra Nevada, north of Lake Tahoe, California: Jour. Geol. v. 72, p. 810–825.

8. ———, 1969, Quarternary paleoclimatic implications of soil clay mineral distribution in a Sierra Nevada–Great Basin transect: Jour. Geol., v. 77, p. 289–302.

9. ———, 1973, Use of relative age dating methods in a stratigraphic study of rock glacier deposits, Mt. Sopris, Colorado: Arctic Alpine Res., v. 5, p. 401–416.

10. Birkeland, P.W., and Janda, R.J., 1971, Clay mineralogy of soils developed from Quaternary deposits of the eastern Sierra Nevada, California: Geol. Soc. Amer. Bull., v. 82, p. 2495–2514.

11. Blackwelder, E.B., 1931, Pleistocene glaciation in the Sierra Nevada and Basin Ranges: Geol. Soc. Amer., Bull., v. 42, p. 865–922.

12. Bleeker, P., and Parfitt, R.L., 1974, Volcanic ash and its clay mineralogy at Cape Hoskins, New Britain, Papua New Guinea: Geoderma, v. 11, p. 123–135.

13. Bouma, J., Hoeks, J., van der Plas, L., and Scherrenburg, B. van, 1969, Genesis and morphology of some alpine podzol profiles: Jour. Soil Sci., v. 20, p. 384–398.

14. Brewer, R., 1964, Fabric and mineral analysis of soils: John Wiley and Sons, New York, 470 p.

15. Burke, R.M., and Birkeland, P.W., 1979, Reevaluation of multiparameter relative dating techniques and their application to the glacial sequence along the eastern escarpment of the Sierra Nevada, California: Quaternary Res., v. 11, p. 21–51.

16. Cady, J.G., 1960, Mineral occurrence in relation to soil profile differentiation: 7th Internat. Cong. Soil Sci. Trans., v. 4, p. 418–424.

17. Chinn, T.H.J., 1981, Use of rock weathering-rind thickness for Holocene absolute age-dating in New Zealand: Arctic Alpine Res., v. 13, p. 33–45.

18. Cline, M.G., 1953, Major kinds of profiles and their relationships in New York: Soil Sci. Soc. Amer. Proc., v. 17, p. 123–127.

19. Colman, S.M., and Pierce, K.L., 1981, Weathering rinds on andesitic and basaltic stones as a Quaternary age indicator, western United States: U.S. Geol. Surv. Prof. Pap. 1210, 56 p.

20. Curtis, C.D., 1976, Stability of minerals in surface weathering reactions: A general thermochemical approach: Earth Surface Processes, v. 1, p. 63–70.

21. Dan, J., and Yaalon, D.H., 1966, Trends of soil development with time in the Mediterranean environments of Israel, p. 139–145 *in* Trans. of Conference on Mediterranean Soils (Madrid).

22. Dan, J., Yaalon, D.H., and Koyumdjisky, H., 1972, Catena soil relationships in Israel. 2. The Bet Guvrin catena on chalk and Nari limestone crust in the Shefala: Israel Jour. Earth Sci., v. 21, p. 99–114.

23. Dan, J., Gerson, R., Koyumdjisky, H., and Yaalon, D.H., 1981, Aridic soils of Israel—properties, genesis and management: Agric. Res. Organ. Spec. Publ. No. 190, The Volcani Center, Bet Dagan, Israel.

24. Fieldes, M., and Perrott, K.W., 1966, The nature of allophane in soils. Part 3. Rapid field and laboratory test for allophane: New Zealand Jour. Sci., v. 9, p. 623–629.

25. Flach, K.W., Holzhey, C.S., De Coninck, F., and Bartlett, R.J., 1980, Genesis and classification of Andepts and Spodosols, p. 411–426 *in* B.K.G. Theng, ed., Soils with variable charge: New Zealand Soc. Soil Sci., Lower Hutt.

26. Goldich, S.S., 1938, A Study in rock-weathering: Jour. Geol., v. 46, p. 17–58.
27. Harden, J.W., 1982, A quantitative index of soil development from field descriptions: Examples from a chronosequence in central California: Geoderma, v. 28, p. 1–28.
28. Harradine, F., and Jenny, H., 1958, Influence of parent material and climate on texture and nitrogen and carbon contents of virgin California soils. I. Texture and nitrogen contents of soils: Soil Sci., v. 85, p. 235–243.
29. Hay, R.L., 1959, Origin and weathering of late Pleistocene ash deposits on St. Vincent, B.W.I.: Jour. Geol. v. 67, p. 65–87.
30. Hembree, C.H., and Rainwater, F.H., 1961, Chemical degradation of opposite flanks of the Wind River Range, Wyoming: U.S. Geol. Surv. Water Supply Pap. 1535-E, 9 p.
31. Hutton, C.E., 1951, Studies of the chemical and physical characteristics of a chrono-litho-sequence of loess-derived prairie soils of southwestern Iowa: Soil Sci. Soc. Amer. Proc., v. 15, p. 318–324.
32. Jenny, H., 1941, Calcium in the soil: III. Pedologic relations: Soil Sci. Soc. Amer. Proc., v. 6, p. 27–35.
33. ———, 1962, Model of a rising nitrogen profile in Nile Valley alluvium, and its agronomic and pedogenic implications: Soil Sci. Soc. Amer. Proc., v. 26, p. 588–591.
34. Keller, W.D., 1954, Bonding energies of some silicate minerals: Amer. Mineralogist, v. 39, p. 783–793.
35. Leamy, M.L., Smith, G.D., Colmet-Daage, F., and Otowa, M., 1980, The morphological characteristics of Andisols, p. 17–34 *in* B.K.G. Theng, ed., Soils with variable charge: New Zealand Soc. Soil Sci., Lower Hutt.
36. Loughnan, F.C., 1969, Chemical weathering of the silicate minerals: American Elsevier Publ. Co., Inc., New York, 154 p.
37. Marchand, D.E., 1970, Soil contamination in the White Mountains, eastern California: Geol. Soc. Amer. Bull., v. 81, p. 2497–2506.
38. Marshall, C.E., 1977, The physical chemistry and mineralogy of soils, Vol. II: Soils in place: John Wiley and Sons, New York, 313 p.
39. Miles, R.J., and Franzmeier, D.P., 1981, A lithochronosequence of soils formed in dune sand: Soil Sci. Soc. Amer. Jour., v. 45, p. 362–367.
40. Muhs, D.R., 1983, Airborne dust fall on the California Channel Islands, U.S.A.: Jour. Arid Environments, v. 6, p. 222–238.
41. Neall, V.E., 1977, Genesis and weathering of Andosols in Taranaki, New Zealand: Soil Sci., v. 123, p. 400–408.
42. Nettleton, W.D., Flach, K.W., and Nelson, R.E., 1970, Pedogenic weathering of tonalite in southern California: Geoderma, v. 4, p. 387–402.
43. Nikiforoff, C.C., 1949, Weathering and soil evolution: Soil Sci., v. 67, p. 219–223.
44. Norrish, K., and Rogers, L.E.R., 1956, The mineralogy of some terra rossa and rendzinas in South Australia: Jour. Soil Sci., v. 7, p. 294–301.
45. Olson, C.G., Ruhe, R.V., and Mausbach, M.J., 1980, The terra rossa limestone contact phenomena in karst, southern Indiana: Soil Sci. Amer. Jour., v. 44, p. 1075–1079.
46. Parfitt, R.L., Russell, M., and Orbell, G.E., 1983, Weathering sequence of soils from volcanic ash involving allophane and halloysite, New Zealand: Geoderma, v. 29, p. 41–57.

47. Pettijohn, F.J., 1941, Persistence of heavy minerals and geologic age: Jour. Geol., v. 49, p. 610–625.

48. Raeside, J.D., 1959, Stability of index minerals in soils with particular reference to quartz, zircon, and garnet: Jour. Sed. Petrol., v. 29, p. 493–502.

49. Read, N.E., ed., 1974, Soil groups of New Zealand. Part 1. Yellow-brown pumice soils: New Zealand Soc. Soil Sci., Lower Hutt, 251 p.

50. Riecken, F.F., and Poetsch, E., 1960, Genesis and classification considerations of some prairie-formed soil profiles from local alluvium in Adair County, Iowa: Iowa Acad. Sci., v. 67, p. 268–276.

51. Rieger, S., and Juve, R.L., 1961, Soil development in Recent loess in the Matanuska Valley, Alaska: Soil Sci. Soc. Amer. Proc., v. 25, p. 243–248.

52. Ruhe, R.V., 1959, Stone lines in soils: Soil Sci., v. 87, p. 223–231.

53. ———, 1969a, Quaternary landscapes in Iowa: Iowa State Univ. Press, Ames, 255p.

54. ———, 1969b, Application of pedology to Quaternary Research, p. 1–23 *in* S. Pawluk, ed., Pedology and Quaternary Research: Univ. of Alberta Printing Dept., Edmonton, Alberta, 218 p.

55. ———, 1973, Backgrounds of model for loess-derived soils in the upper Mississippi River Basin: Soil Sci., v. 115, p. 250–253.

56. Ruhe, R.V., Cady, J.G., and Gomez, R.S., 1961, Paleosols of Bermuda: Geol. Soc. Amer. Bull., v. 72, p. 1121–1142.

57. Sherman, G.D., and Uehara, G., 1956, The weathering of olivine basalt in Hawaii and its pedogenic significance: Soil Sci. Soc. Proc., v. 20, p. 337–340.

58. Shroba, R.R., and Birkeland, P.W., 1983, Trends in late-Quaternary soil development in the Rocky Mountains and Sierra Nevada of the western United States, p. 145–156, *in* S.C. Porter, ed., Late-Quaternary environments of the United States. Vol. I. The late-Pleistocene: University of Minnesota Press, Minneapolis, Minn.

59. Skully, R.W., and Arnold, R.W., 1981, Holocene alluvial stratigraphy in the Upper Susquehanna River Basin, New York: Quaternary Res., v. 15, p. 327–344.

60. Smith, G.D., 1942, Illinois loess—variations in its properties and distribution: A pedologic interpretation: Univ. Illinois Agri. Exp. Sta. Bull. 490, p. 137–184.

61. Smith, W.W., 1962, Weathering of some Scottish basic igneous rocks with reference to soil formation: Jour. Soil Sci., v. 13, p. 202–215.

62. Stace, H.C.T., 1956, Chemical characteristics of terra rossa and rendzinas in South Australia: Jour. Soil Sci., v. 7., p. 280–293.

63. Syers, J.K., Jackson, M.L., Berkheiser, V.E., Clayton, R.N., and Rex, R.W., 1969, Eolian sediment influence on pedogenesis during the Quaternary: Soil Sci., v. 107, p. 421–427.

64. Tardy, Y., Bocquier, G., Paquet, H., and Millot, G., 1973, Formation of clay from granite and its distribution in relation to climate and topography: Geoderma, v. 10, p. 271–284.

65. Wada, K., 1980, Mineralogical characteristics of Andisols, p. 89–107 *in* B.K.G. Theng, ed., Soils with variable charge: New Zealand Soc. Soil Sci., Lower Hutt.

66. Wilding, L.P., Smeck, N.E., and Dress, L.R., 1977, Silica in soils: Quartz, cristobalite, tridymite, and opal, p. 471–552 *in* J.B. Dixon and S.B. Weed, eds., Minerals in soil environments: Soil Sci. Amer., Madison, Wisconsin.

67. Willman, H.B., Glass, H.D., and Frye, J.C., 1966, Mineralogy of glacial tills and their weathering profile in Illinois II. Weathering profiles: Illinois State Geol. Surv. Circ. 400, 76 p.

68. Yaalon, D.H., and Dan, J., 1974, Accumulation and distribution of loess-derived deposits in the semi-desert and desert fringe areas of Israel: Zeitscht. Geomorphol., Suppl. Bd. 20, p. 91–105.

69. Yaalon, D.H., and Ganor, E., 1973, The influence of dust on soils during the Quaternary: Soil Sci., v. 116, p. 146–155.

70. Yaalon, D.H., and Ganor, E., 1979, East Mediterranean trajectories of dust-carrying storms from the Sahara and Sinai, p. 187–193 *in* C. Morales, ed., Saharan dust—mobilization, transport, deposition: John Wiley and Sons, New York.

71. Yerkes, R.F., and Wentworth, C.M., 1965, Structure, Quaternary history, and general geology of the Corral Canyon area, Los Angeles County, California: U.S. Geol. Surv. open-file report, 215 p.

72. Zaidenberg, R., Dan, J., and Koyumdjisky, H., 1982, The influence of parent material, relief and exposure on soil formation in the arid region of eastern Samaria: Catena Supplement 1, p. 117–137.

8

Weathering and soil development with time

Rock and mineral weathering and the development of prominent soil features are time-dependent. The time necessary to produce various weathering and soil features varies, however; those soil properties associated with organic matter buildup develop rapidly, but those associated with the weathering of the primary minerals develop rather slowly.[141] In this chapter, data are presented on the rates of development of various weathering and soil features, soil orders, and clay-mineral alteration products in various environments.

THE TIME SCALE

In order to compare soil data on an even semiquantitative basis, a time scale must be adopted. The dating of the deposits and soils discussed in this chapter, and in others, will vary in accuracy. Some deposits are dated directly by radiometric methods, others are bracketed by radiometric dates, still others by relative dating methods[17] or by correlation with dated deposits elsewhere. Although there is no consensus on the absolute ages for the deposits I will discuss, the ages presented by Colman and Pierce[39] and some correlations of Birkeland and others[16] will be used as a basic framework for the Cordilleran Region. Correlation of Cordilleran deposits with those of the midcontinent is speculative beyond the range of radiocarbon dating (about 40,000 years).[27,48] Where time-stratigraphic names are used, the following *approximate* ages in years before the present are suggested:

> Holocene (0–10,000)
> Late Wisconsin (10,000–25,000)
> Middle Wisconsin (25,000–50,000)
> Early Wisconsin (50,000–70,000)
> Sangamon (70,000–135,000)
> Illinoian (135,000–145,000)

Not all workers would accept these ages, but they are reasonable estimations. These ages will be used to plot data for the figures. The alternative, using local geological names, would be confusing to readers not familiar with the local successions; these latter can be found in the references cited.

RATE OF ROCK AND MINERAL WEATHERING

As discussed in Chapter 7, rocks and minerals weather at different rates. Here I will present some quantitative data on the rate of weathering. Some data on initial rates of weathering come from the study of tombstones and other cultural features, but the majority of the data on longer durations of weathering for which the time factor is reasonably well known come from the study of Quaternary unconsolidated deposits.

Tombstones or other manmade structures are good indicators of the rate at which weathering can proceed above the ground surface. Data are not too numerous, however; they cover a variety of rock types in a variety of climates, and they are commonly referred to in most physical geology textbooks, as well as in other books on weathering.[73,100,140] At any rate, rocks that weather quite rapidly, such as some limestones, can lose their tombstone inscriptions in as little as 100 years in a humid climate. With more resistant rock, such as some sandstones and igneous rocks, it may take several centuries before the tombstone shows distinct signs of weathering.[108] It would seem, therefore, that several centuries are sufficient for weathering to be visible on almost any rock type in a humid climate. Arid climate weathering proceeds at much slower rates.[5] The recent study by Meierding[87] on the weathering of marble indicates a methodology that could be used in future quantitative studies of tombstone weathering.

Studies of glacial tills and outwash deposits are ideally suited to the determination of weathering of different lithologies as a function of time because all factors, except paleoclimate, can be kept constant. Because quantitative weathering studies usually are made in conjunction with stratigraphic studies, there are data from which fairly sound conclusions can be drawn. Most of these methods were pioneered by Blackwelder[21] and have been continued by other workers. Tills with deep exposures are best for these comparisons because, in places where only surface boulders are present, one cannot be certain that any lithologic variations that are recognized are due solely to weathering subsequent to deposition. Some variation in the percentage of fresh boulders, and in boulder and lithologic frequencies at the surface, for example, could be a function of the weathered nature of the terrain over which the glaciers advanced. If the terrain had been highly weathered, only the more resistant lithologies could be picked up, transported, and deposited as boulders. If the terrain had not been weathered, boulders of all lithologies might be plentiful. Little is known of the amount of weathering of landscapes between glacial advances. In a study in Idaho, however, concentration of thorium-bearing minerals as placer deposits in outwash seems to be correlated with the duration of the interglaciation immediately prior to the glacial advance that produced the outwash.[119]

In weathering studies of glacial tills, one can compare the amounts of subaerial versus subsurface weathering with time (Fig. 8-1). Many tills in the Cordilleran Region display the weathering succession shown in Figure 8-2, at least for granitic rock types. Weathering of late-Wisconsin tills has taken place on the exposed surfaces of all stones so that they show signs of oxidation and have some relief due to weathering. Weathering rinds are present on surface clasts and reported maximum thicknesses are 3 mm for till in the Sierra Nevada[33] and 45 mm for rock-glacier deposits in Colorado.[13] The same rock types in glacial till at depth are less weathered and retain glacial polish and striations. Hence, for that period of time subaerial weathering exceeds subsurface weathering. In contrast, tills of Illinoian age contain granitic stones weathered at the surface with somewhat thicker rinds, and they are partly or wholly weathered to grus at depth. Therefore, for this period of time, weathering of granitic clasts at depth is equal to or exceeds that at the surface. Under extremely arid conditions, salt weathering can be such that subsurface weathering exceeds surface weathering, even in the younger deposits.[38]

(A)   (B)

Fig. 8-1   Variation in stone weathering in the Rocky Mountains with age of till and position with respect to surface. (A) Rotated boulder on late-Wisconsin till; the part exposed to surface weathering (left) is quite weathered; the part beneath the surface (right) is virtually unweathered. (B) Rotated boulder on Illinoian till; weathering is extensive both on parts of the boulder that have remained at the surface (right) and on parts that have been beneath the surface (left).

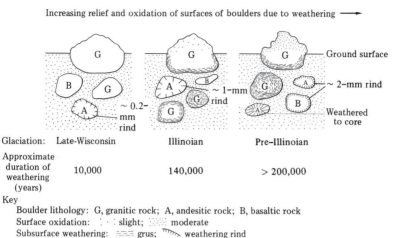

Increasing relief and oxidation of surfaces of boulders due to weathering ⟶

Glaciation: Late-Wisconsin   Illinoian   Pre-Illinoian

Approximate
duration of   10,000   140,000   > 200,000
weathering
(years)

Key
  Boulder lithology: G, granitic rock; A, andesitic rock; B, basaltic rock
  Surface oxidation: ⠂⠂ slight; ▦ moderate
  Subsurface weathering: ≡≡ grus; ⌢⌢ weathering rind

Fig. 8-2 Variation in subaerial and subsurface weathering as a function of time, Sierra Nevada, California.[10,19,33,39]

Weathering rinds on volcanic clasts also show a progressive increase in thickness with age of till.[107] The most extensive study is that of Colman and Pierce,[39] in which clasts were collected from soils to avoid the problem of surface spalling with occasional fire. Logarithmic curves fit the data best, and surprisingly, show a gross similarity in shape for basalts and andesites (Fig. 8-3). A logarithmic curve also has been derived for rind thickness vs. time for graywacke in New Zealand,[37] and the rate for that rock type could be one of the fastest in the world (6 mm in 9500 years).

Rinds offer a means of dating soils in places where other methods may not work well. Material for adequate absolute dating of parent materials seldom is available, and there are problems with contamination and resolution with almost all methods. Rinds collected at the soil surface or within the soil offer a satisfactory means for dating soils provided the relationship between rind thickness and age is fairly well known.

To summarize, initial weathering or granitic clasts is characterized by rates of subaerial weathering that exceed those of subsurface weathering; with time, however, subsurface rates are greater than subaerial rates. Thus, in old tills, most granitic stones at the surface have been at the surface for a long time; they are not lag gravel from depth. This difference in weathering with position relative to the surface is basic to an understanding of the origin of topographic features in granitic terrain.[135] Furthermore, it is important to remember that in the time necessary to alter a sound granitic boulder to grus in a soil, dense volcanic rocks may show only thin weathering rinds.

The rate of subsurface weathering of granitic stones east of the Rocky Mountains is somewhat similar to that reported above. Granitic clasts in outwash in Ohio show progressively greater alteration with age, but only in possible pre-Illinoian outwash are they thoroughly decomposed.[75]

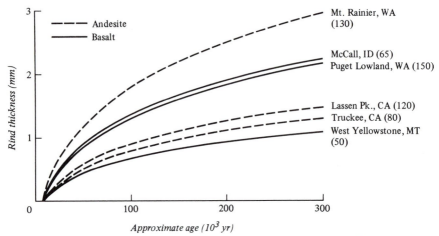

Fig. 8-3 Curves for weathering-rind thickness of andesite and basalt versus time, for till localities in the western United States. (Taken from Colman and Pierce,[39] Fig. 19.) Numbers in parentheses are the mean annual precipitation (cm). The curves for the Puget Lowland and Mt. Rainier are the more rapid rate of the two calculated.

Surface-weathering studies can be used to indicate the amount of granitic-rock breakdown with time. One commonly measured feature is the depth of near-circular weathering pits. Along the base of the eastern escarpment of the Sierra Nevada, these reach a maximum of 75 mm on late Wisconsin boulders and 370 mm on Illinoian boulders.[33] Weathering at high altitude above treeline in the same range is even more rapid; early Holocene clasts have pits 130 mm deep.[142] Another measured feature is the percent of weathered boulders on a surface, such as a moraine. An adequate definition of a weathered granitic clast is that, on over one-half of the exposed surface, weathering penetrates to the depth of the average grain diameter.[20] These surfaces are characterized by loose minerals or minerals that can be readily loosened with a light hammer blow. It should be noted that the weathered boulders may have spalled subsequent to deposition and still be considered fresh; the only requirement is the condition of the mineral grains on the present surface. Another measurement is the number of surface boulders per unit area; this is termed surface-boulder frequency. Data collected from widely separated areas give somewhat comparable results (Table 8-1); the scatter in the data could be real or be partly due to different definitions used by the various workers. We can conclude that at least 10^5 years are necessary for most stones to obtain weathered surfaces, and over 10^5 years to decompose all surface granitic stones. In many places, however, surface granitic stones persist for much longer periods of time, up to 0.5 million years or more.

There are few data on surface weathering of non-granitic clasts, but several workers have published data on the ratios of granitic to non-granitic lithologies of surface stones. In most cases, the granitic stones are the

Table 8-1
Weathering Data for Tills in Various Mountain Ranges, Western
United States

Age	Fresh granitic boulders (%)	Surface-boulder frequency (number in 186 m^2)	Area
Late-Wisconsin	88–95	96–180	Colorado[91]
	91–98	—	Wyoming[82]
	78–88	38–53	California[32,33] *
	50–90	180–300	California[121] *
	80–98	130–280	Washington[106]
Illinoian	40–66	62–95	Colorado[91]
	65–68	—	Wyoming[82]
	18–38	33–66	California[32,33] *
	5–20	60–115	California[121] *
	18–60	0–100	Washington[106]

*Data are from the same set of moraines.

more readily decomposed. For example, in one area in the eastern Sierra Nevada, the granitic to metamorphic clast ratio is 80 to 20 for late-Wisconsin till and 30 to 70 for Illinoian till.[121] In northeastern Oregon, tills are composed of granitic, basaltic, and metamorphic clasts.[32] Late Wisconsin moraines have 33 to 88 percent granitic boulders on the surface, whereas Illinoian moraines have a maximum of 4 percent; most localities, however, have no granitic clasts. Thus, resistant non-granitic rocks persist at the surface long after the granitic rocks have been reduced to grus.

The rate of mineral weathering can be determined in two ways. In one, the uniformity of the parent material can be established, and then the ratio of resistant to nonresistant minerals for dated surfaces gives the rates at which the more weatherable minerals are depleted. Ruhe[111,113] has done this for an area in Iowa in which erosion surfaces of varying age are cut on Kansan till (Fig. 8-4). Soils that formed over a longer period of time have higher ratios of resistant to nonresistant minerals. The duration of weathering is difficult to estimate because the landscape has undergone burial by loess, and in some places, it has been stripped by erosion. Nevertheless, Ruhe[111] estimates that the soil related to the Wisconsin surface may have weathered 6800 years, the soil related to the late-Sangamon surface no less than 13,000 years, and probably much longer, and the soil related to the Yarmouth-Sangamon surface 10^5 years or more. The change with time is considerable, and 10^4 years or more are necessary to record detectable variations in the ratios. The ratios also show that the surface horizons are the most strongly weathered and that weathering extends into the C horizons of the older soils. Brophy[31] analyzed Sangamon soils in Illinois in a similar way and described the influence of parent material

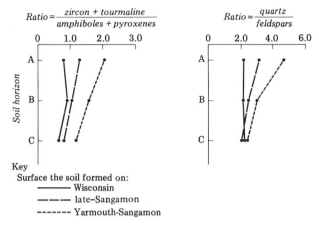

Fig. 8-4 Average ratios of resistant to nonresistant minerals for soils developed on surfaces of different age from Kansan Till in Iowa. (Taken from Ruhe,[111] Figs. 5 and 6, © 1956, The Williams & Wilkins Co., Baltimore, and Ruhe,[115] Table 3.1, reprinted by permission from *Quaternary Landscapes in Iowa,* © 1969 by the Iowa State University Press, Ames.)

texture on weathering in addition to that of age (Fig. 8-5). Outwash has lost 90 percent of the original hornblende compared to a 60 percent loss in till. The reason for the differences with texture is that outwash deposits are more permeable than tills, and therefore, depletion in hornblende can proceed at a more rapid rate than it can in till. Tills in Indiana also show some mineral depletion with time.[7] Heavy minerals in soils on Wisconsin tills display little alteration, whereas those on Illinoian and Kansan tills display a lot; these two older tills cannot be differentiated by their weathering alone, however. Guccione[62] has conducted similar studies in Missouri with similar results, but she cautions workers that reduction of grain size from one size fraction to another by weathering can confound the results of these kinds of studies.

Another way to study rates of mineral weathering is to examine individual mineral grains for signs of weathering, such as cockscomb terminations (Fig. 8-6). This method may be more sensitive than the ratio method in showing change, at least in the earlier stages of weathering, because minerals can show signs of weathering and yet not be depleted. Data on individual grains are not always reported, so little information is available. Data from several sources (Fig. 8-7), however, suggests age limits for mineral weathering. The weathering in soil in the British West Indies is one of rapid alteration and depletion, whereas that in soils in till and outwash deposits in the western United States proceeds at a fairly slow rate. Locke[77] quantified etching studies by measuring the depth of etching of the cockscomb terminations and calculating the mean maximum etching depth. Measuring this for hornblende grains in cold, dry arc-

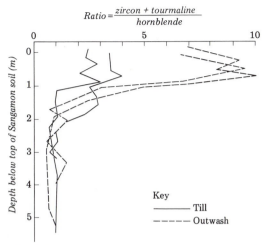

$$Ratio = \frac{zircon + tourmaline}{hornblende}$$

Key

——————— Till

- - - - - Outwash

Fig. 8-5 Relationship of resistant to nonresistant mineral ratios with depth and with texture of the parent material. (Taken from Brophy,[31] Fig. 11.) The ratio is for the indicated depth relative to that for the three lowest samples. Clay contents in the assumed parent materials are 14 to 25 percent for the two tills and 3 to 10 percent for the two outwash deposits. If the data of Ruhe for the oldest soils (Fig. 8-4) are recalculated on the same basis as the ratios here, the ratios for till in both areas are comparable.

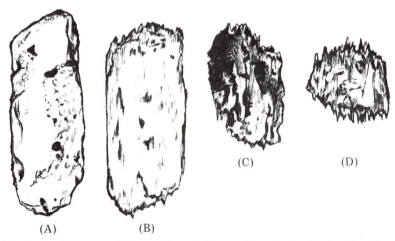

(C)   (D)

(A)   (B)

Fig. 8-6 Variation in etching of hypersthene grains in the Sierra Nevada tills described by Birkeland.[10] J. G. LaFleur (unpub. report, 1972) has studied the relationship between degree of grain etching and age of soil to find (A) unetched or (B) slightly etched grains that are characteristic of soils formed in late-Wisconsin deposits; (C) and (D) are highly etched forms that are common in soils formed in Illinoian till.

tic soils, he found etching to be greatest at the surface and to diminish with depth; the maximum values for mean etching depth were approximately 2 μ in about 8500 years, 2.7 μ in about 100,000 years and 5 μ in still older deposits. Comparing his diagram of etching scale with that used by other workers, etching rates in the Arctic could be the lowest in the world. Hay[68] makes the point that, in comparing the rate of weathering of two profiles by mineral alteration, the depth through which weathering is noticeable should be considered. That is, intensive weathering to a shallow depth may be equal in the total amount of material altered by less intense weathering to a greater depth.

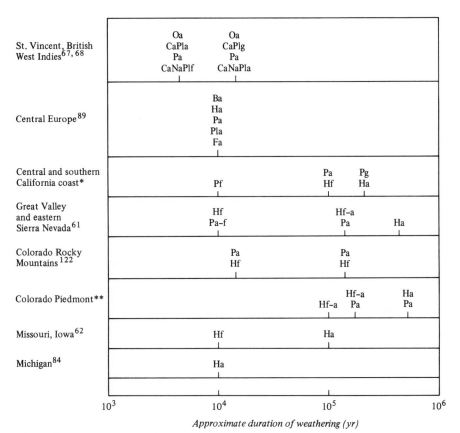

Approximate duration of weathering (yr)

Fig. 8-7 Some data on the length of time necessary to alter or deplete various minerals. Asterisk, data from Bradley[28,29] and Birkeland (unpub.); double asterisk, data from A. J. Crone (unpub.) The data from the two latter areas are from projects for a 1972 soils course at the University of Colorado and should be considered tentative. Mineral key: O, olivine; P, pyroxene; H, hornblende; B, biotite; F, K-feldspar; Pl, plagioclase (Ca, calcic; CaNa, less calcic). Condition-of-mineral key: f, fresh; a, altered (etched); g, gone or nearly gone.

SOIL MORPHOLOGY AND TIME (CHRONOSEQUENCES AND CHRONOFUNCTIONS)

Many of the prominent properties of soil profiles require a fairly long time to form. The time it takes to acquire particular soil features varies with the feature and is of considerable interest to pedologists who work on the soil-forming processes and to geomorphologists who might use the data to date deposits or surfaces.[125]

Classification of Chronosequences

Vreeken[133] recognizes four kinds of chronosequences, based on the relationship between soils and deposits. Post-incisive chronosequences are those where there is a sequence of deposits of different age, and the soils related to each deposit have formed from the end of the time of deposition to the present. In contrast, a pre-incisive chronosequence is a sequence where soil began to form on a particular deposit, but subsequent burial of the soil took place at different times. A time-transgressive chronosequence without historical overlap is a vertical stacking of sediments and buried soils, so that the latter record times of non-deposition. Finally, a time-transgressive chronosequence with historical overlap is the most compli-cated sequence, and incorporates aspects of the other three. Each kind of chronosequence has its attributes for study of soil development with time. Vreeken argues that the time-transgressive sequence with historical over-lap is the most valuable for soil chronosequence studies, but it could be argued that it is not always a very pure chronosequence because some properties might be as much related to post-burial diagenesis or to slope position as they are to time. Although he rightly notes the many problems with post-incisive chronosequences, they are the most common, and we have the most data on them, so most of the examples given here are of that kind. In any such study, one must carefully choose the described sites to avoid subsequent erosion or deposition. Daniels and others[44] make the point well that older soils in a chronosequence may not have gone through the same history as younger soils; for example, climate could have changed or water-table history could differ. Only by detailed study of the soils, the surficial geology, and the paleoclimate indicators can we even hope to put the soils into the context of Quaternary events.

A-horizon Organic Matter and Associated Properties

Organic matter probably reaches a steady state more rapidly than any other property of the soil. Data are given in Figure 8-8 for widely spaced localities, different parent materials, different properties of the A horizon, and different sampling depths. Nevertheless, these and other data[63,98,102,103,118] suggest that the time to achieve steady state may range from as little as 200 to perhaps 10,000 years. The data can also be used to suggest that the A horizon can reach a new steady state quite rapidly, if

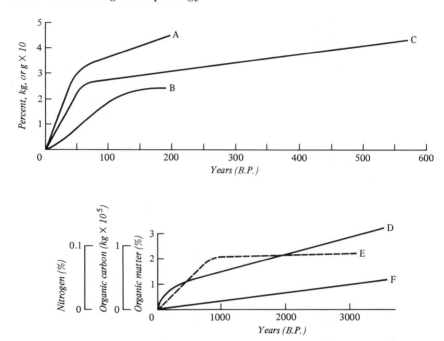

Fig. 8-8 Variation in the properties of the A horizon with time. Key for plotted data: (A) percent organic matter in the surface layer, Storglaciär moraines, Sweden[126]; (B) organic carbon (kg/m^2) for the top 46 cm of the soil, Glacier Bay moraines, Alaska[41]; (C) organic carbon (g/46 cm^2) for the top 91.5 cm of the soil, Mt. Shasta mudflows, California[45]; (D) organic carbon (kg/hectare) for the top meter of the soil, New Zealand sand dunes[128] (note that this curve increases to about 2.1 at 10,000 years; thus the rate of increase diminishes beyond the last plotted point); (E) nitrogen in the top 10 cm of the soil, Lake Michigan sand dunes[101]; (F) nitrogen on surface A horizon, Front Range moraines, Colorado (unpubl. data of author and co-workers; Mahaney[81] has a plot for organic matter).

conditions change. Hence, it is one soil property that is amenable to the functional use of factors because it can be quantified, and polygenesis is not an insolvable problem. The A horizon properties of most soils are probably in a steady state with prevailing conditions and therefore are of limited use in stratigraphic studies except for very young deposits or surfaces.

Several other soil properties develop as rapidly as the trends of the organic-matter constituents of the A horizon, and they probably respond to these trends. These changes in properties are reported in many of the articles referenced above, in Figure 8-8, as well in Crocker and Dickson.[42] Soil pH commonly becomes more acid quickly and could reach a steady state at about the same time as does the A-horizon organic matter content (Fig. 8-9). Carbonates, if present, will influence the pH unless they are

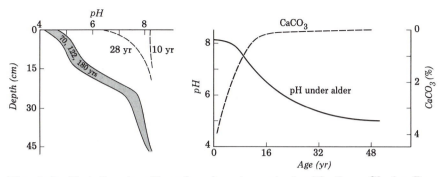

Fig. 8-9 Variation in pH and carbonate content with time, Glacier Bay, Alaska. (Taken from Crocker and Major,[41] Figs. 4 and 11.) Data in the right-hand figure are for the top 5.1 cm of the soil.

leached as rapidly as the organic constituents build up. The rate of leaching will depend on the local climate. The eventual pH at the steady state will depend on some combination of climatic, vegetational, and parent-material influences.

Bt (Argillic) Horizon Development

The formation of a Bt horizon that meets argillic criteria is time-dependent because eolian influx, weathering, clay formation, and translocation are relatively slow processes. Although the best approach here would be to calculate the quantities of clay formed and translocated as a function of time, quantitative data on rates are quite rare,[68] because bulk density data are seldom collected. Another approach would be to calculate the clay accumulation index,[76] which is the clay content in the Bt horizon minus that in the C horizon multiplied by the thickness of the Bt horizon. In lieu of these, I will take the stratigraphic approach and compare clay-content profiles of variously aged deposits from different regions. Another important facet of B horizon formation is the progressive development of microscopic features seen in thin section.[30]

A plot of clay content against depth usually shows an increase in the amount and in the thickness of B-horizon clay with time. Data on soils from a variety of regions are given in Figure 8-10; these soils are thought to be representative of the rate of clay buildup for the respective region and for the indicated parent material. Several things are quite evident in the plots. One is that the rate of clay buildup varies remarkably from region to region. This is important in soil-stratigraphic studies because clay content is one of the basic properties used in the correlation of unconsolidated deposits. One should, therefore, be aware of the regional differences in the rates of clay buildup, or else erroneous correlations might result.

There are several reasons for the regional differences in rates of Bt-

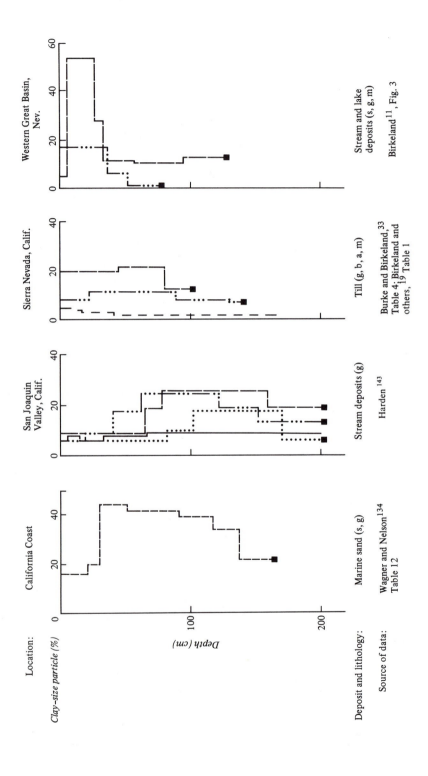

Location: California Coast | San Joaquin Valley, Calif. | Sierra Nevada, Calif. | Western Great Basin, Nev.

Clay-size particle (%)

Depth (cm)

Deposit and lithology: Marine sand (s, g) | Stream deposits (g) | Till (g, b, a, m) | Stream and lake deposits (s, g, m)

Source of data: Wagner and Nelson[134] Table 12 | Harden[143] | Burke and Birkeland,[33] Table 4; Birkeland and others,[19] Table 1 | Birkeland[11], Fig. 3

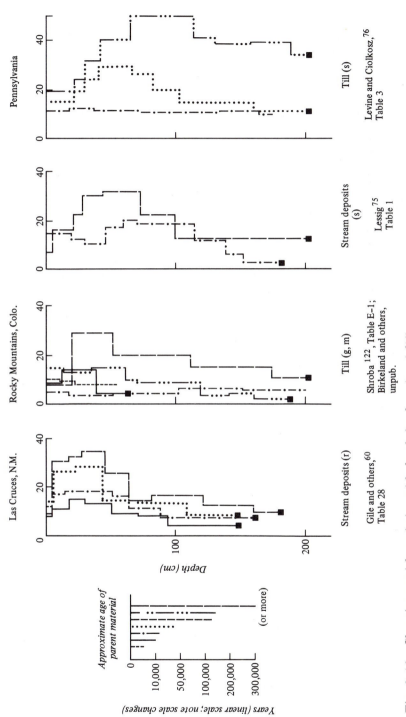

Fig. 8-10 Clay-size-particle variation with depth for deposits of different age from the California coast to Pennsylvania. The indicated age is for the parent material; the age of the soil could be younger because of erosion between the end of deposition and the beginning of soil formation or burial of the soil by younger deposits.

horizon development. More work needs to be done on the California coast because the example chosen may actually be younger fan material overlying marine sands. Nevertheless, a well-differentiated profile has formed, due perhaps to high original clay content; this material would have a high water-holding capacity to promote more rapid clay formation, and also have clay available for translocation. High Na^+ associated with coastal environments (Ch. 5) could help accelerate the translocation process. A soil chronosequence on marine terraces of an offshore California island suggests a rapid rate of clay accumulation (weight per profile) with time, but the amounts of clay are so high (>50 percent) that a significant eolian contribution is suspected.[95,96] Relatively low soil moisture, combined probably with relatively low dust influx, could limit the rate of Bt development in the San Joaquin Valley[3,64] and on the flanks of the Sierra Nevada and the Rocky Mountains. Of interest for stratigraphic studies is the lack of a Bt horizon in soils 10,000 to 15,000 years old in these localities, as well as in soils of similar age in Pennsylvania and the Puget Lowland of Washington,[74] and the gross similarity between clay accumulation plots for the Sierra Nevada and Rocky Mountains for soils about 140,000 years old. The Bt horizon for the early Holocene soil in the Rocky Mountains is above treeline and probably formed primarily from mechanically infiltrated loess. In contrast to the surrounding mountain flanks, soils in parts of the Great Basin and the Las Cruces area display rapid Bt-horizon development and they even form in Holocene soils in the latter area. Much of the clay could be of eolian origin. Part of the reason for rapid development is that the translocated clays are concentrated in a relatively thin zone, relative to the other areas. The Las Cruces area soils also demonstrate the influence of gravel content on the rate of Bt-horizon formation. The Bt horizons form more rapidly in the gravelly materials in the late Holocene, because there is less non-gravelley material between clasts for the eolian fine materials to infiltrate and accumulate. If, however, carbonate content is sufficiently high, Bt-horizon formation is inhibited and may require all of Holocene time to form.[55] In Southern California, McFadden[85] reports Bt horizons in early Holocene soils in a semiarid climate and states that formation of these horizons takes longer than the duration of the Holocene in arid environments. Perhaps dust influx is less there than in New Mexico. Maximum time for the formation of a Bt horizon is 40,000 years for the San Joaquin Valley and Pennsylvania and about 140,000 years for the Sierra Nevada and Rocky Mountains; it could form sooner, but deposits of younger ages have not been recognized. In Pennsylvania, for example, Bilzi and Ciolkosz[9] cite other work to suggest that occasionally Bt horizons are present in soils formed from alluvium about 12,000 years old.

Finally, Bt horizons form rapidly in some midcontinent deposits and environments. Soils formed on the late Wisconsin loess have such well-differentiated clay profiles (Fig. 7-7) that Bt horizons are expected to form within the Holocene.[115,116] Perhaps this rate results from some combination of high-clay-content parent materials and a leaching soil-moisture regime. They also form rapidly in stream deposits in Ohio. In contrast, Bt horizons require at least Holocene time to form in sand-dune deposits.[90]

A few studies have reported Bt-horizon formation within the latter part of the Holocene. Clay translocation is recognized in a soil formed in loess spoil about 100 years old in Iowa, but the amount falls short of argillic criteria.[63] One of the more rapidly formed Bt horizons formed in about 350 years near Lubbock, Texas[69]; recall, however, that this is an area where mud rains are a contributing factor (Ch. 5). In the Nevada desert, a Bt horizon has formed in a soil that might be less than 2000 years old, and this is attributed to both dust and the dispersive effects of Na^+.[105]

In contrast to the above, Bt horizons are not common in exceptionally dry environments. In the cold desert of Antarctica, with less than 1 cm of annual water equivalent precipitation, Bt horizons are not present in soils several million years old.[26] They also are not present in southern Israel or in Sinai soils that are several hundred thousand years old; present precipitation in these latter areas is less than 5 cm year.[43]

The development of Bt horizons with time is somewhat different in very sandy materials, such as dune sand.[56,90] Discrete clay bands are the first indication of translocated clays. With time, these increase in number, lateral continuity, and thickness, until they eventually merge into the common type of Bt horizons.

Development of Red Color

A common observation in many areas is that older soils are redder than younger soils.[112,114] This is readily shown by the color of the B horizon, or the Cox horizon if a B is not present, and by the dry rubification values of Harden[64] (Table 8-2). The offshore California colors show little progression with time, probably because the soils are clay-rich, and the older ones are Vertisols and thus mix. For many of the other areas, data are shown for the parent-material colors ($t = 0$ yr), so one can better appreciate both the color change with time and the value of obtaining a color closest to the parent material as possible. Redness seems to increase more rapidly in the San Joaquin Valley relative to the Sierra Nevada, and colors in the Rocky Mountains have a rate that is the same or somewhat lower than those in the Sierra Nevada. The Colorado Piedmont has a rate that could be slightly higher than that for the San Joaquin Valley. Some of the highest rates are those in arid areas, such as Las Cruces and the Lower Colorado River. Also of interest is the relatively rapid increase in redness of alpine soils, especially those in New Zealand and the Rocky Mountains; other alpine areas in the Rocky Mountains give similar rates as those in Table 8-2.[92] The rubification values more or less parallel the redness rates, but for any detailed study, all colors should be taken in the same manner by one observer. Although data are not given in the table,

Table 8-2

Color of the B Horizon, or Cox Horizon if no B Horizon is Present, for Soils at Various Localities, and Rubification Values (in Parentheses)

Location	Parent material	APPROXIMATE AGE (YR)*					Moist (m) or dry (d)
		0	10^3	10^4	10^5	10^6	
1. California offshore island[95]	Eolian over marine sediment	—			10YR 3/3	10YR 3/3	m
2. San Joaquin Valley, Cal.[64]	River sediment	10YR 7/2	10YR 5/4 (20)	10YR 5/4 10YR 6/6 (40)	5YR 4/4 2.5YR 3/6 (40) (70)	10R 3/6 (80)	d
3. Sierra Nevada, Cal.[19,33]	Till**	5Y 7/2		10YR 5/4 (40)	7.5YR 5/4 5YR 4/4 (50) (60)		d
	Till***	5Y 7/1		10YR 5/3 (40)			d
4. Lower Colorado River, Cal.[85]	River sediment	10YR 7/2	10YR 7/2	7.5YR 6/6 5YR 5/8 (50) (80)	5YR 5/6 2.5YR 5/6 (60) (70)		d
5. Las Cruces, New Mexico[60]	River sediment	—	5YR 3/4	5YR 5/4 5YR 5/5	2.5YR 4/6		d
6. Rocky Mountains, Colo.[122]	Till	2.5Y 7/2		7.5YR 6/3 (30)	7.5YR 5/4 7.5YR 6/6 (40) (60)		d
	Till†	7.5Y 6/1	10YR 4/4 (60)	7.5YR 4/5 (80)			d
7. Southern Alps, New Zealand	Till‡	5Y 6/1	10YR 6/4 (50)	7.5YR 4/8 (100)			d
8. Colorado Piedmont[79]	River sediment	10YR 3/2		7.5YR 4/6 (50)	5YR 5/5 (50)	10R 4/6 (80)	m,d

*Scale is arithmetic between tick marks.
**Includes unpubl. data of author or co-workers.
***Tills in alpine environment (unpubl. data of author, R.M. Burke, and J.C. Yount).
†Tills in alpine environment (unpubl. data of author, R.M. Burke, and J.W. Harden).
‡Tills in alpine environment (unpubl. data of author).

the rate of increase in redness is very slow in cold, dry arctic conditions[14] and even slower in the Antarctic.[23,38]

One problem in linking color development with age of parent material is the role of paleoclimate in producing redness. It is shown above that red soils form fairly rapidly in areas characterized by high temperatures. Hence, if soil redness increases with age of deposit, one or both of two factors may be responsible. One factor is time, and the other is warmer past climates. It is difficult to separate the effects of these two factors.[112] Paleoclimatic interpretation of soils, therefore, should not rest solely on color, but also on other soil properties.

Trends in Iron, Aluminum, and Phosphorus

The amounts of free Fe and Al, and the P fractions in soils, are all time-dependent, but the rate of accumulation or depletion is also related to bioclimate. As an example of these relationships, various co-workers and I have analyzed chronosequences of soils formed in tills in a variety of alpine and arctic environments.[18] We have data on all the soils of the chronosequences, but here (Fig. 8-11) I give data only on the parent material and soils for about 10,000 years old. Expected trends are accumulation of oxalate-extractable Fe and Al (Fe_o, Al_o), and depletion of acid-extractable P (P_a). In most of the areas studied, these trends are seen. The relatively wet and warm environment of New Zealand results in the greatest accumulations of Fe_o and Al_o, and the near total depletion of P_a. The Mt. Everest area and the Wind River Mountains show somewhat similar trends, with the former area having greater accumulations of Fe_o and Al_o, as well as greater depletion of P_a. In contrast, the cold, dry arctic environment of Baffin Island displays no trends that can be solely ascribed to pedogenesis; Evans and Cameron,[47] however, found such trends in a nearby area. In this latter area, at least 100,000 years of pedogenesis has had little impact on trends in these elements. It should be emphasized that some of these trends are fairly remarkable, given the fairly harsh environment in which some of the soils have formed and the relative youth of the soils.

Other studies support the above general trends with time. Soil chronosequences spanning 1 to 3 my in the extremely cold and arid Antarctic show no to a slight increase in dithionite-extractable Fe (Fe_d) with time.[23,104] For the warm arid to semiarid area near Las Cruces, New Mexico, Gile and others[60] did not obtain meaningful trends with time. McFadden[85] studied chronosequences in Southern California in arid, semiarid, and xeric climatic regimes. The semiarid chronosequence could span 300,000 years and the others approximately 0.9 to 1 my. Accumulation trends in Fe_o and Fe_d are slight in the arid area, increase slightly under more moist conditions, and increase more for Fe_d than for Fe_o. None of the Fe_o values approach those of Figure 8-11, exclusive of Baffin Island. One problem with relying on Fe_o for trends with age is that with

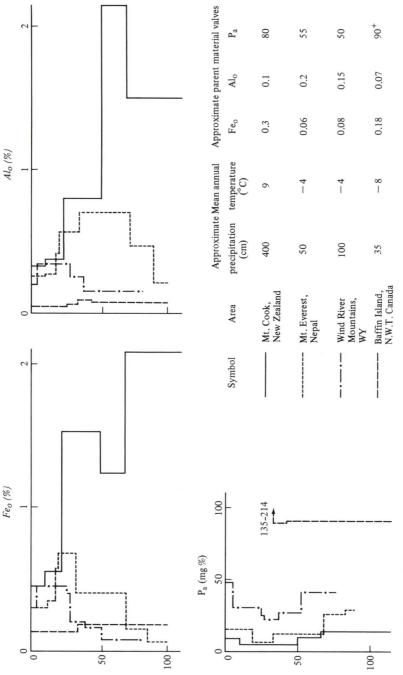

Symbol	Area	Approximate precipitation (cm)	Approximate Mean annual temperature (°C)	Approximate parent material values		
				Fe_o	Al_o	P_a
——	Mt. Cook, New Zealand	400	9	0.3	0.1	80
------	Mt. Everest, Nepal	50	−4	0.06	0.2	55
—·—·	Wind River Mountains, WY	100	−4	0.08	0.15	50
— —	Baffin Island, N.W.T. Canada	35	−8	0.18	0.07	90+

Fig. 8-11 Vertical distribution of oxalate-extractable Fe and Al (Fe_o and Al_o) and acid-extractable P (P_a) for soils about 10,000 years old in various alpine and arctic localities. (Taken from unpublished data of P. W. Birkeland, R. M. Burke, A. L. Walker, F. B. Fox, and J. B. Benedict.[18]

time it can convert to Fe_d and actually decline in value; this seems to be more common under high temperatures.[85] Other semiarid environments give mixed results. In the Sierra Nevada, California, soils developed in Late Wisconsin, Illinoian, and pre-Illinoian tills show few trends in Fe and Al extracts that are age-dependent.[19,34] Alexander[1] also obtained few trends for Fe extracts for a soil chronosequence in stream deposits in Nevada, east of the Sierra Nevada. In contrast, in a xeric climate in Spain, Torrent and others[130] obtained good trends in Fe_d for one stream-deposit chronosequence, but not another, probably because the stream deposits in the latter area were pre-weathered in the source area. In contrast to these marginal results is the well-known rapid accumulation of Al and Fe extracts in more humid environments,[120] especially those in which Spodosols form.[35,50,110,132]

Other studies support the data in Figure 8-11 that variation in P fractions can be attributed to both time and bioclimate. In New Zealand, trends are marked in chronosequences with annual precipitation ranging from 65 to 509 cm, and areas of highest precipitation exhibit the most rapid change.[138] In contrast, weathering intensity is low in dry environments and few P trends with age are discernible.[19,34,117]

In much of this work on chemical trends, it is important to know the size fraction in which the trends are taking place. In work on P fractions in New Zealand, it was shown that the finer fractions exhibit the greatest changes with time.[127] Such changes may be difficult to decipher in soils if analyses are on the total <2 mm fraction, especially in sandy soils because the bulk of the material is in the coarse fraction that exhibits little change. One can enhance trends of any extracts by analyzing just the finer fractions.[47]

Soil CaCO₃ Buildup and Removal

The rate at which carbonate builds up in soil depends on the process of formation as well as the rate of leaching in the soil. If the necessary Ca^{2+} comes from mineral weathering, buildup will be controlled by the rate of weathering of the calcium-bearing minerals. If $CaCO_3$ was originally present in the parent material, the rate of buildup in the soil is a function of the rate at which it can be translocated by leaching waters in the soil profile. If, however, the $CaCO_3$ is of atmospheric origin, buildup is a function of the Ca^{2+} in the rainfall, the carbonate content of dust, and the amount of rainfall to both move Ca^{2+} into the soil and to dissolve carbonate in dust, and to translocate it to some depth where precipitation occurs. Machette and others[80] discuss some of these factors as they pertain to carbonate in soils in the southwestern United States. In areas they term rainfall limited, there could be a high influx of Ca^{2+} from all sources relative to annual precipitation, but not all the Ca^{2+} is carried into the soil. Other areas are considered to be influx limited; these are areas of low Ca^{2+} influx relative to precipitation. Here the Ca^{2+} is leached into the soil, but some can also be leached beyond the base of the carbonate horizon. In both of these extreme cases, the eventual amount of total Ca^{2+} residing in the soil is less than that delivered to the soil surface. One should be aware of this when evaluating dust-trap data.

Provided the $CaCO_3$ can be shown to have been distributed uniformly in the soil parent material, one can roughly calculate how long it might take to redistribute the $CaCO_3$ by solution, translocation, and re-precipitation. Arkley[2] has presented a method for these calculations, using estimations on the volume of water passing various levels in the soil, as determined from soil and climatic data, and data on $CaCO_3$ solubility. One such calculation is shown in Figure 8-12, and, although it and other ages he calculated are fairly rough and only a minimum, the calculated ages are reasonably consistent with the geological evidence on age. McFadden[85] has suggested an improvement on the Arkley model that takes into account such things as $CaCO_3$ solubility as a function of the partial pressure of CO_2, solution ionic strength, and the rate of atmospheric $CaCO_3$ influx into the soil.

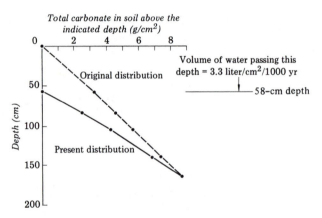

Fig. 8-12 A calculation of the approximate time necessary to redistribute carbonate in a soil.[2] Carbonate is assumed to have been uniformly distributed in the soil, subsequently removed from the uppermost 58 cm, and then precipitated in the 58 to 165 cm interval. It is calculated that about 3.3 liters of water/cm²/1000 yr move by the 58-cm-depth level in the soil, and from Fig. 5-7, a soil carbonate solubility of 0.1 g/liter is estimated for the prevailing physical and chemical conditions.

$$\frac{\text{Time for carbonate}}{\text{translocation}} = \frac{\substack{\text{Amount of carbonate translocated from}\\ \text{uppermost 58 cm of soil}}}{\substack{\text{Volume of water passing}\\ \text{through the .58-cm depth} \times \text{Carbonate solubility}\\ \text{level of soil}}}$$

$$= \frac{3.24 \text{ g}}{(3.3 \text{ liter/1000 yr}) \times (0.1 \text{ g/liter})} = 9800 \text{ years}$$

Several studies have been done on the rate of buildup of pedogenic CaCO$_3$ with time. The classical area for this is near Las Cruces, New Mexico, where Gile and co-workers[57,58] defined the carbonate morphology stages and showed changes in both carbonate stage and amount with time (Fig. 8-13). They[60] also showed that lower amounts of total carbonate per unit volume are needed to reach a particular morphological stage in gravelly materials relative to non-gravelly ones. This is because there is a lower volume of non-gravelly matrix in the gravelly parent materials, and the morphology is expressed in the non-gravelly materials. For this reason, in an area of both gravelly and non-gravelly parent materials, higher stages are reached in much less time in the gravelly materials.

Regional soil-stratigraphic studies by Machette and others[80] provide data on regional rates of buildup for the various carbonate stages (Fig. 8-14). The rates vary widely, from the low rates of the Colorado Mountains (upper Arkansas Valley), where stage III is the maximum stage reached, to the highest rates in the Roswell-Carlsbad area of New Mexico. Buildup in this latter area is rapid partly because of a local source of carbonate dust. Stage VI carbonate morphology is only reached in three of the areas studied, and in only one area is it reached within the Pleistocene.

Another way to rank areas is by the total carbonate in the soil for soils of various ages. This is the preferred method because gravel content is taken into account. Amounts are calculated using the following equation:[80]

$$Cs = (C_3P_3 - C_1P_1)d$$

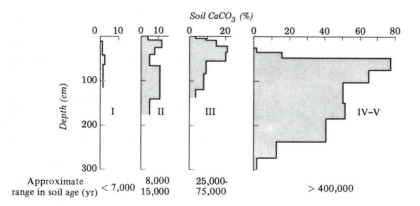

Fig. 8-13 Carbonate distribution with depth for soils of different approximate ages, Las Cruces area, New Mexico.[54,59,60] The data are for nongravelly parent materials; Roman numerals correspond to carbonate buildup stages (Appendix 1).

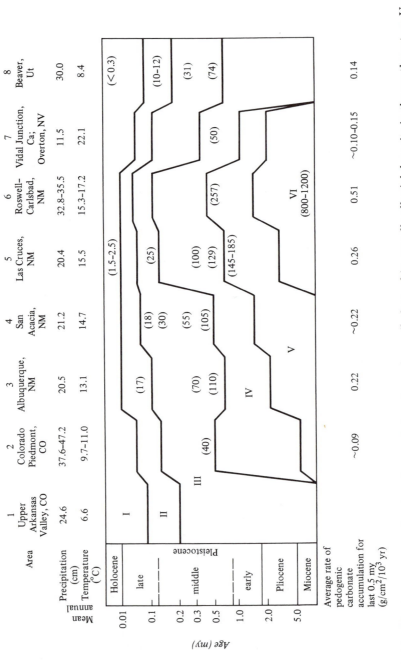

Fig. 8-14 Maximum stages of carbonate morphology in surface soils formed in gravelly alluvial deposits in the southwestern United States. Numbers in parentheses are the average grams of $CaCO_3/cm^2$/soil column in soils of the indicated ages. (Taken from Machette and others,[80] Table 2; Machette,[78] Table 1; M. N. Machette, written communication, 1983; and McFadden.[85])

where

Cs is the weight of secondary carbonate in a soil horizon per unit area (g/cm^2)

C_3 is the present total $CaCO_3$ content (g $CaCO_3$/100 g oven-dry soil)

C_1 is the initial (primary) $CaCO_3$ content

P_3 is the present oven-dry bulk density (g/cm^3)

P_1 is the initial oven-dry bulk density

d is the thickness (cm) of the sampled interval

Here C_1 and P_1 have to be estimated either from Cu horizons at depth or from Cu horizons of younger soils in the area. The Cs values for each sampled interval can be summed to give CS, the total secondary carbonate for the soil. This is expressed as grams of $CaCO_3$ per square centimeter of surface area for the calcareous portion of the soil. For the areas studied, the ranking based on CS values is the same as that based on the rate of buildup of the various morphological stages (Fig. 8-14).

One way to estimate ages of calcareous soils is to divide the present-day influx rate into the CS value, but few data are available on the influx rate. However, airborne-sediment traps have been in place in the Las Cruces area for several years, and data from them, as well as chemical data on rainfall, indicate that much of the soil $CaCO_3$ is of atmospheric origin.[60] The yearly influx ranges from 9.3 to 26.3 $\times$ 10^{-4} g/cm^2. Silt commonly exceeds clay and the latter makes up about 20 to 40 percent of the sample; organic carbon is 3 to 7 percent and $CaCO_3$ is 1 to 3 percent. The Ca^{2+} in the precipitation is considered to contribute much more to the pedogenic carbonate than does the carbonate in the dry dust, perhaps by a factor of two to three. A rough estimate is that Ca^{2+} from all sources is sufficient to form about 0.2 g pedogenic $CaCO_3$/cm^2 every 1000 years assuming all the available Ca^{2+} enters the soil and is deposited as pedogenic carbonate.

It is difficult to date soils by the above method because the system is very complex. Some complications are (a) part of the water runs off and therefore does not translocate $CaCO_3$ in the soil; (b) some of the eolian $CaCO_3$ may be blown away before the next rainfall can move into the soil; and (c) the Ca^{2+} that goes to making additional $CaCO_3$ may come from other sources, such as calcium-bearing salts and the Ca^{2+} on the exchangeable sites of the colloidal fraction.

In lieu of placing dust traps throughout the southwestern United States to get regional data on influx rates of $CaCO_3$, Machette and others[80] estimated the long-term mean rate of pedogenic carbonate accumulation from pedologic and geological data. Soils about 0.5 my old are recognized in several areas; mean rates of accumulation can be calculated by dividing the CS of these soils by their age. The Roswell-Carlsbad area has the most rapid rate. In most cases, the ranking of areas by rate of accumulation equals that by rate of attainment of specific carbonate stages. The data for Las Cruces are not too different from that calculated above for the present day (Fig. 8-14). In the first edition of this book, it was shown that the potential influx of carbonate fell short of the rate estimated from soil data and the assumed ages of the soils. It now appears that the assumed ages were too young, and therefore, the measured and estimated present-day rate is quite similar to the calculated long-term rate.

Parent materials in some humid regions contain original $CaCO_3$, and because of the prevailing climate this material is removed slowly in solution by the percolating waters. Some figures on the rate of $CaCO_3$ depletion from soils are as follows: removal from the top 5 cm of the soil in less than 50 years at Glacier Bay, Alaska,[41] and from the top 2 m in 1000 years in sand dunes along Lake Michigan.[101] Stratigraphic studies in the mid-continent have tended to use the depth of $CaCO_3$ leaching as one criterion for the approximate age of the parent material. Thorp[129] points out, for example, that Wisconsin drifts in northern Illinois are leached to a 0.5 to 1.5 m depth, whereas Illinoian drifts in the southern part of the state are leached to a 2.5 to 3.5 m depth. Thus for any region with a uniform parent material, there might be a depth of $CaCO_3$ leaching corresponding to the age of the parent material. Many workers, however, warn that this criterion for age should be used with caution because many variables can alter the rate of $CaCO_3$ removal. Furthermore, even if environmental conditions remain constant for long periods of time, the rate of leaching would decrease with depth in the soil because less water is available for leaching at greater depths in the soil, and the water there could be saturated with respect to $CaCO_3$. Hence, one would not expect a linear relationship between depth of leaching and duration of soil formation.

CLAY MINERALOGY

The clay mineralogy of a soil may change with time, and if it does, this may confound efforts to use clay minerals as paleoclimatic indicators. Two cases will be explored. In one, the initial parent material contains little inherited clay, and clay formation from alteration of non-clay minerals predominates. In the other, the parent material contains clay minerals of diverse types, and many of the clays that form in the soil do so by alteration of preexisting clays. Whether or not a clay mineral will alter is a function of its stability in the soil solution as shown by stability diagrams (Figs. 4-5 and 4-6).

Clay minerals that form from the weathering of non-clay minerals should be stable in that environment, or else they would not have formed. If, however, the properties of a soil change with time so that the internal environment of the soil changes, the originally formed clays may alter. For example, the buildup of clay-size particles in a soil may change the leaching environment enough that (a) existing clay minerals are unstable or (b) clay minerals that form in the future differ from those that formed in the past, yet both were in equilibrium with the water chemistry at the time their respective formation. Another example of a pure time control on clay-mineral alteration would be if the weathering of the non-clay minerals was so intense that certain minerals that delivered weathering products and ions to the solution were depleted and that their depletion so altered the water chemistry that the previously formed clays then became unstable. This may happen in areas of intense weathering, such as the tropics, areas of podzolization, or stable landscapes in which weathering dates back to the early Quaternary and Tertiary. In contrast, some areas, such as a Sierra Nevada–Great Basin transect, show little age control on soil clay mineralogy.[12,15]

In places where a variety of clay minerals are present in the original parent material, changes in mineral species may take place for any mineral not stable in the environment as shown by the stability diagrams. Jackson and others[71] ranked the clay minerals by relative resistance to weathering and assigned a number to each mineral characteristic of a weathering stage (Fig. 8-15). Three of the more stable minerals are not given in Fig. 8-15; they are allophane (stage 11), hematite (stage 12), and anatase (stage 13). For most of the minerals, a decrease in iron, magnesium, and silicon is associated with an increase in mineral stability. The arrows in the figure indicate some possible paths for alteration from one mineral to another. In soils with a combination of original clay minerals in the parent material, any alteration due to weathering will be regulated by the water-mineral equilibria. The extent of the alteration, or in other words, the weathering stage reached, will be a function of time because the alterations involve slow reactions and water-mineral equilibria; that is, alteration takes place until an equilibrium mineral assemblage is formed, after which time further alteration probably does not occur.

The midcontinent is a good region to study the effect of time on clay-mineral alteration. Tills and loesses of several ages are present in many environments, and parent materials contain a variety of clay minerals derived from bedrock units over which the glaciers advanced. Clay-mineral transformations are judged from the vertical sequence of clay minerals in a soil and by comparison of soils of different ages. Many data have

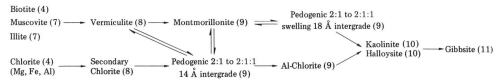

Fig. 8-15 Various clay-mineral reactions in soil. Parenthetical numbers are weathering stability index numbers for the clay minerals; mineral stability increases with number. (From Jackson,[70] p. 124 *in* Chemistry of the Soil by F. E. Bear, ed., © 1964 by Litton Educational Publishing, Inc. Reprinted by permission of Van Nostrand Reinhold Company.)

been accumulated on the region, but only two example areas will be discussed.

Clay minerals formed in soils from tills in Illinois demonstrate the changes possible with advancing age (Fig. 8-16). The soil on early-Wisconsin till shows only an alteration of chlorite to vermiculite-chlorite in the B and leached Cox horizons. Some soils on late-Wisconsin till display similar alteration.[51,53] This similarity between clay-mineral alteration in soils formed from early- and late-Wisconsin tills does not always hold true. In some places, illite depletion is recognized in soils on early-Wisconsin tills, but not in soils on late-Wisconsin tills;[139] in addition, some of the early-Wisconsin tills with original montmorillonite show an increase in montmorillonite in the soil. In contrast to the soils on early-Wisconsin tills, Sangamon soils have more altered clays, and the alteration goes to greater depth[31,139] (Fig. 8-16). Chlorite is altered through a vermiculite-chlorite stage to material termed "heterogeneous-swelling material" that probably is either montmorillonite or a mixed-layer mineral.[139] Illite alters to illite-montmorillonite and finally to montmorillonite. Older soils seem to be altered to about the same stage.

Soil formed from late-Wisconsin loess in Illinois shows slight clay-mineral alteration at most.[52] The major alteration is that of montmorillonite to the heterogeneous-swelling material. Other changes are slight depletion of illite, modification of chlorite, and possibly formation of vermiculite and kaolinite.

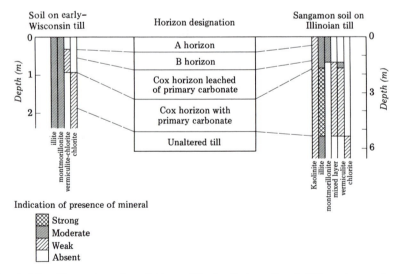

Fig. 8-16 Clay-mineral variation with depth in soils of different ages formed from till, Illinois. (Taken from Frye and others,[51] Fig. 3.) The expandable vermiculite in the original publication has subsequently been shown to be heterogeneous-swelling material[139]; it is here shown as montmorillonite in the Sangamon soil. Note the difference in vertical scale for the two profiles.

In Illinois, therefore, the stable clay mineral seems to be stage 9 montmorillonite, but of the heterogeneous-swelling material variety. This material can form in the time span represented by the weathering of late-Wisconsin deposits. More intense alteration to great depths, however, requires weathering durations on the order of Sangamon time. Forsyth[49] used similar relationships to help date a buried soil in Ohio. A different clay-mineral pattern is seen in the accretion-gley deposits, and this is discussed in Chapter 9.

Soils in Indiana show somewhat similar trends.[6] Illite and chlorite are present in the parent tills. The alteration goes to vermiculite and illite-montmorillonite in post-Wisconsin soils (between stages 8 and 9) and to montmorillonite (stage 9) in Sangamon soils. Kaolinite (stage 10) is found in the parent material of older soils, but also some seems to have formed in the soils.

The fibrous clay minerals also show some time dependency for formation. In carbonate soils, montomorillonite transforms to palygorskite in about 120,000 years, and it takes more than 300,000 years for palygorskite to dominate.[4] Sepiolite then forms by conversion from palygorskite.

It should be kept in mind that the rapidity of the clay-material transformation depends on bioclimate, just as do certain chemical trends as discussed above. Greater leaching with warmer temperatures should increase the rate of change. Bockheim[24] reviews his data and that of others for the Arctic and Antarctic and concludes that little or no change takes place in the Antarctic, even though some chronosequences extend beyond 1 my. If changes do occur in the Antarctic, they probably are slow and seem to be

$$\text{muscovite} \rightarrow \text{hydrated} \rightarrow \text{vermiculite} \rightarrow \text{smectite}$$
$$\text{mica} \qquad \text{mica}$$

In contrast, in the Arctic, the alteration goes to vermiculite, and if leaching is sufficient, to kaolinite. Alteration is more rapid in the alpine environment of the Colorado Rocky Mountains, where within 10^4 years mica is still present, but has a subdued X-ray peak, and mixed layer (10–18 Å) clays are the main transformation product.[122,123] At slightly lower altitude within 100 m above the present treeline, mica alteration has been more intense, and vermiculite is present in addition to the mixed-layer clay. These latter tundra sites could have been forested earlier. In a still wetter and leaching alpine environment in New Zealand, unpublished data of the author suggest the nearly total elimination of original mica and chlorite in 10^4 years, and their transformation to mica-vermiculite, mica-chlorite, and mica-smectite. Other New Zealand clay-mineral studies at lower altitude support rapid clay-mineral change in areas of high rainfall.[35,93] In addition to bioclimate, parent material will exert some control on the rate of clay-mineral formation, for halloysite can form from volcanic glass in several hundred to several thousand years.[22]

In summary, it seems that the following generalizations are valid. Clay minerals that form by weathering of primary grains in soils from parent materials low in clay content probably form mineral assemblages stable in that environment, and therefore variation in the minerals with age may not be found. Soils formed from parent materials high in clay content, and with a variety of clay minerals, are another matter. Some clays in the assemblage may be unstable and gradually change over to more stable forms. The change will progress from the surface downward, and complete change to a stable assemblage may take considerable time. In these cases, the presence of clay-mineral assemblages may aid in identifying either the age of a surface soil or the occurrence of a period of weathering and/or soil formation within a thick section of sediments. In addition to time, the rate of such change depends on bioclimate.

TRENDS IN DEVELOPMENT INDICES

The goal of the development indices (Ch. 1) is mainly to condense complex field data to numbers that can be plotted against time to determine if trends exist. Indices have been calculated and plotted against time for soils at localities in the eastern United States,[8,9] western United States,[64,85,88] regional comparisons across the United States,[65] and in Australia.[136,137] All these show quite significant and useful trends with time, and some have important ramifications in considering the impact of climate on pedogenesis (Ch. 10). If carefully done, these trends can be as good as, or better than, trends based on time-consuming chemical analyses. Workers should seriously consider using them as a first step in using soils and soil data as a stratigraphic correlation tool.

QUANTIFICATION OF CHRONOSEQUENCES TO PRODUCE CHRONOFUNCTIONS

Much of the data in this chapter is intended to show and explain expected trends; for only a few trends have equations been derived. In a recent review, Bockheim[25] demonstrates what more workers should be doing with chronosequence data. Of prime importance is fairly good age control. He selected properties of chronosequences from the literature for 27 areas. Each property was correlated with time using regression techniques and three models:

$$Y = a + bX$$
$$Y = a + (b \log X)$$
$$\log Y = a + (b \log X)$$

where Y is the soil property and X is time. Correlation coefficients were calculated, as were their statistical significance. The model $Y = a + (b$

log X) yielded the largest correlation coefficients, and 85 percent of these coefficients were statistically significant. Because the chronosequences varied in their parent materials and climate, Bockheim also was able to demonstrate the influence of these factors on soil properties. Although more data for more areas are needed, he showed for the chronosequences he studied that (a) rates of decrease in pH and in base saturation are similar and are not controlled by parent material and climate, (b) the rate of increase in B-horizon clay content and in solum thickness are positively correlated with the clay content of the parent material, (c) rates of increases of solum thickness, depth of oxidation, salt content, and B-horizon clay content are positively correlated with mean annual temperature, and (d) the rate of change in C:N ratio does not seem to correlate with either climate or parent material. To summarize, Bockheim's analysis is not only good for the derivation of chronofunctions, but also gives insight into the influence of the other factors.

A sophisticated statistical analysis, using multivariate techniques, was performed by Sondheim and others[124] to evaluate a 550-year-old chronosequence on the west coast of Canada. A principal component analysis showed that most of the variance in the data is explained by two significant components. Loadings on the first component reflect pedogenic processes, loadings on the second the influence of sea spray. They were also able to derive an equation for soil development and predict both the shape of the soil-development curve and when the steady state might be reached.

RELATIONSHIP BETWEEN SOIL ORDERS AND TIME

There is a fairly good correlation between the soil orders and the age of the underlying deposits or landscapes in some regions. A striking example in the United States is the widespread occurrence of Ultisols in the midcontinent and in the east (Fig. 2-4). These soils have formed on deposits and landscapes of pre-Wisconsin age. Although it is admittedly difficult to decide whether the main factor involved in their formation is primarily time or paleoclimate,[112] at least some investigators feel that the soils required long periods of time to form.[99] Another example is the band of Ultisols along the west slope of the Sierra Nevada and in the Cascade Range of California[66] (Fig. 2-4). Detailed stratigraphic studies in the region indicate that these soils are primarily restricted to pre-Wisconsin deposits.[40,72] In northwestern Oregon, Trimble[131] suggests that 10^6 years or more were required to form some of the deeply weathered Ultisols.

The Oxisols of the world also seem to have required long periods of time to form. Maignien[83] reviews most of the data on them and shows that many oxisols date from the Tertiary or the early Quaternary. Although the climate in some regions, such as Australia, may have been different during the time of Oxisol formation, these soils are so highly weathered that formation times of hundreds of thousands of years or more do not seem unreasonable.

In contrast to soil orders that seem to require a long time to form, some can form in a short time. Entisols are so little developed that material would be classified in that order shortly after exposure to soil-forming processes; perhaps a century is all that is required. Others, such as Histosols and Vertisols, require so little pedogenesis that formation under appropriate conditions could be very rapid, surely less than 10^3 years. Inceptisols could form in that time, or slightly longer. Podzolization occurs very rapidly, as shown and reviewed by Ellis,[46] but McKeague and others[86] believe that soils that qualify as Spodosols probably require several thousand years to form. Both Mollisols and Aridisols would require variable times to form because they include both cambic and argillic horizons; those with cambic horizons probably could form in 1,000 to 2,000 years, those with argillic horizons about 10^4 years. Alfisols, because an argillic horizon is required, probably also requires about 10^4 years or more.

From this it can be seen that soil maps at the order level can be used to suggest broad groupings of deposits or landscapes on the basis of age.

TIME NECESSARY TO ATTAIN THE STEADY-STATE CONDITION

Many soil properties reach a steady-state condition. Curves for the buildup of most properties initially are fairly steep, but after some time the curves flatten, indicating little visible change thereafter with time.[124] Because of this, earlier studies concluded that it was difficult to differentiate older soils of the major interglacials from the Sangamon soil on most field criteria.[94,109]

The time necessary to reach the steady state will vary with the soil property being studied, the parent material, erosion, and the particular kind of soil profile that forms in a particular environment.[141] Thus, A-horizon properties form rapidly whereas many B-horizon properties form rather slowly (Fig. 8-17, A). The variation in rates of formation of the various kinds of B horizons is a function of the processes responsible for each kind of B horizon. Because a soil profile is the sum total of many soil properties, a profile can be said to be in a steady state only when most of its diagnostic properties are in a steady state. Hence, soil orders will vary in the time necessary to reach a steady state (Fig. 8-17, B).

The steady-state condition is reached for different properties by different processes. For example, at the steady state for organic matter in the A horizon, the gain of undecomposed organic matter to the soil is balanced by the loss of organic matter due to decomposition. The ongoing processes at the steady state for the Bt, Bk and K horizons are another matter. I doubt if the growth of clay minerals is balanced by their chemical destruction, or that the precipitation of $CaCO_3$ is balanced by its solution and removal. Rather, these steady states might involve a delicate bal-

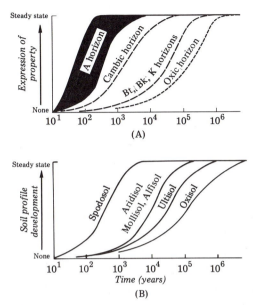

Fig. 8-17 Schematic diagrams showing the variations in time to attain the steady state for (A) various soil properties and (B) various soil orders.

ance between removal of surficial material by erosion and the slow extension of the soil profile at depth into unweathered material.[97] With oxic horizons, the steady state might be reached only when most of the weatherable minerals have been depleted by weathering, thus rendering further change unlikely.

The above statements on steady state are intended only as broad guidelines. In detail, for specific areas, they may not be always acceptable. Recent plots of soil data demonstrate that some curves have not flattened off even after long durations of pedogenesis (10^6 years).[25,95] Furthermore, plots of the Harden soil development index for central California have not flattened at the age of the oldest soil (3×10^6 years).[64] At any rate, the flattening of the curves means that for that time span where the curve is relatively flat, correlation of deposits on soil data will be less precise. Finally, the time at which the curve flattens will vary with the soil-forming factors.

REFERENCES

1. Alexander, E.B., 1974, Extractable iron in relation to soil age on terraces along the Truckee River, Nevada: Soil Sci. Soc. Amer. Proc., v. 38, p. 121–124.
2. Arkley, R.J., 1963, Calculation of carbonate and water movement in soil from climatic data: Soil Sci., v. 96, p. 239–248.

3. ———, 1964, Soil survey of the eastern Stanislaus area, California: U.S. Dept. Agri., Soil surv. series 1957, no. 20, 160 p.

4. Bachman, G.O., and Machette, M.N., 1977, Calcic soils and calcretes in the southwestern United States: U.S. Geol. Surv. Open-File Report 77-794, 163 p.

5. Barton, D.C., 1916, The disintegration of granite in Egypt: Jour. Geol., v. 24, p. 382–393.

6. Bhattacharya, N., 1962, Weathering of glacial tills in Indiana: I. Clay minerals: Geol. Soc. Amer. Bull., v. 73, p. 1007–1020.

7. ———, 1963, Weathering of glacial tills in Indiana: II. Heavy minerals: Jour. Sed. Petrol., v. 33, p. 789–794.

8. Bilzi, A.F., and Ciolkosz, E.J., 1977, A field morphology rating scale for evaluating pedological development: Soil Sci., v. 124, p. 45–48.

9. Bilzi, A.F., and Ciolkosz, E.J., 1977, Time as a factor in the genesis of four soils developed in recent alluvium in Pennsylvania: Soil Sci. Soc. Amer. Jour., v. 41, p. 122–127.

10. Birkeland, P.W., 1964, Pleistocene glaciation of the northern Sierra Nevada north of Lake Tahoe, California: Jour. Geol., v. 72, p. 810–825.

11. ———, 1968, Correlation of Quaternary stratigraphy of the Sierra Nevada with that of the Lake Lahontan area, p. 469–500 *in* R.B. Morrison and H. E. Wright, Jr., eds., Means of correlation of Quaternary successions: Internat. Assoc. Quaternary Res., VII Cong., Proc. v. 8, 631 p.

12. ———, 1969, Quaternary paleoclimatic implications of soil clay mineral distribution in a Sierra Nevada–Great Basin transect: Jour. Geol., v. 77, p. 289–302.

13. ———, 1973, Use of relative age-dating methods in a stratigraphic study of rock glacier deposits, Mt. Sopris, Colorado: Arctic Alp. Res., v. 5, p. 401–416.

14. ———, 1978, Soil development as an indication of relative age of Quaternary deposits, Baffin Island, N.W.T., Canada: Arctic Alp. Res., v. 10, p. 733–747.

15. Birkeland, P.W., and Janda, R.J., 1971, Clay mineralogy of soils developed from Quaternary deposits of the eastern Sierra Nevada, California: Geol. Soc. Amer. Bull., v. 82, p. 2495–2514.

16. Birkeland, P.W., Crandell, D.R., and Richmond, G.M., 1971, Status of correlation of Quaternary stratigraphic units in the western conterminous United States: Quaternary Res., v. 1, p. 208–227.

17. Birkeland, P.W., Colman, S.M., Burke, R.M., Shroba, R.R., and Meierding, T.C., 1979, Nomenclature of alpine glacial deposits, or, what's in a name: Geology, v. 7, p. 532–536.

18. Birkeland, P.W., Walker, A.L., Benedict, J.B., and Fox, F.B., 1979, Morphological and chemical trends in soil chronosequences: Alpine and arctic environments: Agron. Absts., p. 188.

19. Birkeland, P.W., Burke, R.M., and Walker, A.L., 1980, Soils and subsurface rock-weathering features of Sherwin and pre-Sherwin glacial deposits, eastern Sierra Nevada, California: Geol. Soc. Amer. Bull., v. 91, p. 238–244.

20. Birman, J.H., 1964, Glacial geology across the crest of the Sierra Nevada, California: Geol. Soc. Amer. Spec. Pap. 75, 80 p.

21. Blackwelder, E.B., 1931, Pleistocene glaciation of the Sierra Nevada and Basin Ranges: Geol. Soc. Amer. Bull., v. 42, p. 865–922.

22. Bleeker, P., and Parfitt, R.L., 1974, Volcanic ash and its clay mineralogy at Cape Hoskins, New Britain, Papua New Guinea: Geoderma, v. 11, p. 123–135.

23. Bockheim, J.G., 1979, Relative age and origin of soils in eastern Wright Valley, Antarctica: Soil Sci., v. 128, p. 142–152.

24. ———, 1980, Properties and classification of some desert soils in coarse-textured glacial drift in the Arctic and Antarctic: Geoderma, v. 24, p. 45–69.

25. ———, 1980, Solution and use of chronofunctions in studying soil development: Geoderma, v. 24, p. 71–85.

26. ———, 1982, Properties of a chronosequence of ultraxerous soils in the Trans-Antarctic Mountains: Geoderma, v. 28, p. 239–255.

27. Bowen, D.Q., 1978, Quaternary geology: Pergamon Press, New York, 221 p.

28. Bradley, W.C., 1957, Origin of marine-terrace deposits in the Santa Cruz area, California: Geol. Soc. Amer. Bull., v. 68, p. 421–444.

29. ———, 1965, Marine terraces on Ben Lomond Mountain, California, p. 148–150 *in* Guidebook for Field Conf. I, Northern Great Basin and California: Internat. Assoc. Quaternary Res., VII Cong., Nebraska Acad. Sci., Lincoln.

30. Brewer, R., and Sleeman, J.R., 1969, The arrangement of constituents in Quaternary soils: Soil Sci., v. 107, p. 435–441.

31. Brophy, J.A., 1959, Heavy mineral ratios of Sangamon weathering profiles in Illinois: Illinois State Geol. Surv. Circ. 273, 22 p.

32. Burke, R.M., 1979, Multiparameter relative dating (RD) techniques applied to morainal sequences along the eastern Sierra Nevada, California and Wallowa Lake area, Oregon: Ph.D. thesis, Univ. of Colorado, Boulder, 166 p.

33. Burke, R.M., and Birkeland, P.W., 1979, Reevaluation of multiparameter relative dating techniques and their application to the glacial sequence along the eastern escarpment of the Sierra Nevada, California: Quaternary Res., v. 11, p. 21–51.

34. Burke, R.M., Walker, A.L., and Birkeland, P.W., 1979, Preliminary remarks on chemical data for soils formed in post-Sherwin glacial deposits, eastern Sierra Nevada, California, p. 111–113 *in* Burke, R.M., and Birkeland, P.W., eds., Field guide to relative dating methods applied to glacial deposits in the third and fourth recesses and along the eastern Sierra Nevada, California, with supplementary notes on other Sierra Nevada localities: Friends of the Pleistocene, Pacific Cell, Guidebook.

35. Campbell, A.S., 1975, Chemical and mineralogical properties of a sequence of terrace soils near Reefton, New Zealand: Ph.D. thesis, Lincoln College, Canterbury, New Zealand, 477 p.

36. Campbell, I.B., and Claridge, G.G.C., 1975, Morphology and age relationships of Antarctic soils: Royal Soc. New Zealand, Bulletin 13, p. 83–88.

37. Chinn, T.J.H., 1981, Use of rock weathering-rind thickness for Holocene absolute age-dating in New Zealand: Arctic Alp. Res., v. 13, p. 33–45.

38. Claridge, G.G.C., and Campbell, I.B., 1968, Some features of antarctic soils and their relation to other desert soils: 9th Internat. Cong. Soil Sci. Trans., v. 4, p. 541–549.

39. Colman, S.M., and Pierce, K.L., 1981, Weathering rinds on andesitic and basaltic stones as a Quaternary age indicator, western United States: U.S. Geol. Surv. Prof. Pap. 1210, 56 p.

40. Crandell, D.R., 1972, Glaciation near Lassen Peak, northern California: U.S. Geol. Surv. Prof. Pap. 800-C, p. C181–C190.
41. Crocker, R.L., and Major, J., 1955, Soil development in relation to vegetation and surface age at Glacier Bay, Alaska: Jour. Ecology, v. 43, p. 427–448.
42. Crocker, R.L., and Dickson, B.A., 1957, Soil development on the recessional moraines of the Herbert and Mendenhall glaciers, southeastern Alaska: Jour. Ecology, v. 45, p. 169–185.
43. Dan, J., Yaalon, D.H., Moshe, R., and Nissim, S., 1982, Evolution of Reg soils in southern Israel and Sinai: Geoderma, v. 28, p. 173–202.
44. Daniels, R.B., Gamble, E.E., and Cady, J.C., 1971, The relation between geomorphology and soil morphology and genesis: Adv. in Agron., v. 23, p. 51–88.
45. Dickson, B.A., and Crocker, R.L., 1953, A chronosequence of soils and vegetation near Mt. Shasta, California. II. The development of the forest floors and the carbon and nitrogen profiles of the soils: Jour. Soil Sci., v. 4, R. p. 142–154.
46. Ellis, S., 1980, Physical and chemical characteristics of a podzolic soil formed in Neoglacial till, Okstindan, northern Norway: Arctic Alp. Res., v. 12, p. 65–72.
47. Evans, L. J., and Cameron, B.H., 1979, A chronosequence of soils developed from granitic morainal material, Baffin Island, N.W.T.: Canad. Jour. Soil Sci., v. 59, p. 203–210.
48. Flint, R.F., 1971, Glacial and Quaternary geology: John Wiley and Sons, New York, 892 p.
49. Forsyth, J.L., 1965, Age of the buried soil in the Sidney, Ohio, area: Amer. Jour. Sci., v. 263, p. 571–597.
50. Franzmeier, D.P., and Whiteside, E.P., 1963, A chronosequence of podzols in northern Michigan. I. Physical and chemical properties: Mich. State Univ. Agri. Exp. Sta. Quat. Bull., v. 46, p. 21–36.
51. Frye, J.C., Willman, H.B., and Glass, H.D., 1960, Gumbotil, accretion-gley, and the weathering profile: Illinois State Geol. Surv. Circ. 295, 39 p.
52. Frye, J.C., Glass, H.D., and Willman, H.B., 1968, Mineral zonation of Woodfordian loesses of Illinois: Illinois State Geol. Surv. Circ. 427, 44 p.
53. Frye, J.C., Glass, H.D., Kempton, J.P., and Willman, H.B., 1969, Glacial tills of northwestern Illinois; Illinois State Geol. Surv. Circ. 437, 47 p.
54. Gile, L.H., 1970, Soils of the Rio Grande Valley border in southern New Mexico: Soil Sci. Soc. Amer. Proc., v. 34, p. 465–472.
55. ———, 1975, Holocene soils and soil-geomorphic relations in an arid region of southern New Mexico: Quaternary Res., v. 5, p. 321–360.
56. ———, 1979, Holocene soils in eolian sediments of Bailey County, Texas: Soil Sci. Soc. Amer. Jour., v. 43, p. 994–1003.
57. Gile, L.H., Peterson, F.F., and Grossman, R.B., 1965, The K horizon—a master soil horizon of carbonate accumulation: Soil Sci., v. 99, p. 74–82.
58. Gile, L.H., Peterson, F.F., and Grossman, R.B., 1966, Morphological and genetic sequences of carbonate accumulation in desert soils: Soil Sci., v. 101, p. 347–360.
59. Gile, L.H., Hawley, J.W., and Grossman, R.B., 1971, The identification, occurrence and genesis of soils in an arid region of southern New Mexico:

Training sessions, desert soil-geomorphology project, Dona Ana County, N.M., Soil Conservation Service, 177 p.

60. Gile, L.H., Hawley, J.W., and Grossman, R.B., 1981, Soils and geomorphology in the Basin and Range area of southern New Mexico—Guidebook to the Desert Project: New Mexico Bur. Mines and Mineral Resources Memoir 39, 222 p.

61. Gillam, M.L., Ensey, C., Page, W.D., and Blum, R.L., 1977, Heavy mineral etching in soils from the Merced and Truckee areas, California, p. 230–246 *in* M.J. Singer, ed., Soil development, geomorphology, and Cenozoic history of the northeastern San Joaquin Valley and adjacent areas, California: Guidebook for joint field session of the Amer. Soc. Agron., Soil Sci. Soc. Amer., and Geol. Soc. Amer.

62. Guccione, M.J.W., 1982, Stratigraphy, soil development and mineral weathering of Quaternary deposits, midcontinent, U.S.A.: Ph.D. thesis, Univ. of Colorado, Boulder, 302 p.

63. Hallberg, G.R., Wollenhaupt, N.C., and Miller, G.A., 1978, A century of soil development in spoil derived from loess in Iowa: Soil Sci. Soc. Amer. Jour., v. 42, p. 339–343.

64. Harden, J.W., 1982, A quantitative index of soil development from field descriptions: Examples from a chronosequence in central California: Geoderma, v. 28, p. 1–28.

65. Harden, J.W., and Taylor, E.M., in press, A quantitative comparison of soil development in four climatic regimes: Quaternary Res.

66. Harradine, F., 1966, Comparative morphology of lateritic and podzolic soils in California: Soil Sci., v. 101, p. 142–151.

67. Hay, R.L., 1959, Origin and weathering of late Pleistocene ash deposits on St. Vincent. B.W.I.: Jour. Geol., v. 67, p. 65–87.

68. ———, 1960, Rate of clay formation and mineral alteration in a 4000-year-old volcanic ash soil on St. Vincent, B.W.I.: Amer. Jour. Sci., v. 258, p. 354–368.

69. Holliday, V.T., 1982, Morphological and chemical trends in Holocene soils at the Lubbock Lake Archeological Site, Texas: Ph.D. thesis, University of Colorado, Boulder, 285 p.

70. Jackson, M.L., 1964, Chemical composition of the soil, p. 71–141 *in* F.E. Bear, ed., Chemistry of the soil: Reinhold Publ. Corp., New York, 515 p.

71. Jackson, M.L., Tyler, S.A., Willis, A.L., Bourbeau, G.A., and Pennington, R.P., 1948, Weathering sequence of clay-size minerals in soils and sediments. I. Fundamental generalizations: Jour. Phys. Colloid. Chem., v. 52, p. 1237–1260.

72. Janda, R.J., 1966, Pleistocene history and hydrology of the upper San Joaquin River, California: Ph.D. thesis, Univ. Calif. (Berkeley), 425 p.

73. Jenny, H., 1941, Factors of soil formation: McGraw-Hill, New York, 281 p.

74. Lea, P.D., 1983, Glacial history of the southern margin of the Puget Lowland, Washington: M.S. thesis, Univ. of Washington, Seattle.

75. Lessig, H.D., 1961, The soils developed on Wisconsin and Illinoian-age glacial outwash terraces along Little Beaver Creek and the adjoining upper Ohio Valley, Columbiana County, Ohio: Ohio Jour. Sci., v. 6, p. 286–294.

76. Levine, E.L., and Ciolkosz, E.J., 1983, Soil development in till of various ages in northeastern Pennsylvania: Quaternary Res., v. 19, p. 85–99.

77. Locke, W.W., III, 1979, Etching of hornblende grains in arctic soils: An indicator of relative age and paleoclimate: Quaternary Res., v. 11, p. 197–212.

78. Machette, M.N., 1982, Guidebook to the late Cenozoic geology of the Beaver Basin, south-central Utah: U.S. Geol. Surv. Open-file Rept. 82–850, 42 p.

79. Machette, M.N., Birkeland, P.W., Burke, R.M., Guccione, M.J., Kihl, R., and Markos, G., 1976, Field descriptions and laboratory data for a Quaternary soil sequence in the Golden-Boulder portion of the Colorado Piedmont: U.S. Geol. Surv. Open-File Report 76-804, 21 p.

80. Machette, M.N., Harper-Tervet, J., and Timbel, N.R., in press, Calcic soils and calcretes of the southwestern United States: Geol. Soc. Amer. Spec. Pap.

81. Mahaney, W.C., 1974, Soil stratigraphy and genesis of neoglacial deposits in the Arapaho and Henderson cirques, central Colorado Front Range, p. 197–240 *in* W.C. Mahaney, ed., Quaternary environments: York University (Toronto, Canada) Geogr. Mono. No. 5.

82. ―――, 1978, Late-Quaternary stratigraphy and soils in the Wind River Mountains, western Wyoming, p. 223–264 *in* W.C. Mahaney, ed., Quaternary soils: Geo Abstracts Ltd., Univ. of East Anglia, Norwich, England.

83. Maignien, R., 1966, Review of research on laterites: UNESCO, natural resources research, IV, 148 p.

84. Matelski, R.P., and Turk, L.M., 1947, Heavy minerals in some podzol soil profiles in Michigan: Soil Sci., v. 64, p. 469–487.

85. McFadden, L.D., 1982, The impacts of temporal and spatial climatic changes on alluvial soils genesis in Southern California: Ph.D. thesis, University of Arizona, Tucson, 430 p.

86. McKeague, J.A., Ross, G.J., and Gamble, D.S., 1978, Properties, criteria of classification, and concepts of genesis of podzolic soils in Canada, p. 27–60 *in* W.C. Mahaney, ed., Quaternary soils: Geo Abstracts, Univ. of East Anglia, Norwich, England.

87. Meierding, T.C., 1981, Marble tombstone weathering rates: A transect of the United States: Physical Geography, v. 2, p. 1–18.

88. Meixner, R.E., and Singer, M.J., 1981, Use of a field morphology rating system to evaluate soil formation and discontinuities: Soil Sci., v. 131, p. 114–123.

89. Meyer, B., and Kalk, E., 1964, Verwitterungs-mikromorphologie der mineralspezies in mittel-Europäischen Holozän-böden aus Pleistozänen und Holozänen lockersedimenten, p. 109–129 *in* A. Jongerius, ed., Soil micromorphology: Elsevier Publ. Co., New York, 540 p.

90. Miles, R.J., and Franzmeier, D.P., 1981, A lithochronosequence of soils formed in dune sand: Soil Sci. Soc. Amer. Jour., v. 45, p. 362–367.

91. Miller, C.D., 1979, A statistical method for relative-age dating of moraines in the Sawatch Range, Colorado: Geol. Soc. Amer. Bull., v. 90, p. 1153–1164.

92. Miller, C.D., and Birkeland, P.W., 1974, Probably pre-Neoglacial age of the type Temple Lake moraine, Wyoming: Discussion and additional relative-age data: Arctic Alpine Res., v. 6, p. 301–306.

93. Mokma, D.L., Jackson, M.L., Syers, J.K., and Stevens, P.R., 1973, Mineralogy of a chronosequence of soils from graywacke and mica-schist alluvium, Westland, New Zealand: New Zealand Jour. Sci., v. 16, p. 769–797.

94. Morrison, R.B., 1964, Lake Lahontan: Geology of the southern Carson Desert: U.S. Geol. Surv. Prof. Pap. 401, 156 p.

95. Muhs, D.R., 1982, A soil chronosequence on Quaternary marine terraces, San Clemente Island, California: Geoderma, v. 28, p. 257–283.

96. ———, 1983, Airborne dust fall on the California Channel Islands, U.S.A.: Jour. of Arid Environments, v. 6, p. 222–238.

97. Nikiforoff, C.C., 1949, Weathering and soil evolution: Soil Sci. v. 67, p. 219–230.

98. Nørnberg, P., 1977, Soil profile development in sands of varying age in Vendsyssel, Denmark: Catena, v. 4, p. 165–179.

99. Novak, R.J., Motto, H.L., and Douglas, L.A., 1971, The effect of time and particle size on mineral alteration in several Quaternary soils in New Jersey and Pennsylvania, U.S.A., p. 211–224 *in* D.H. Yaalon, ed., Paleopedology: Israel Univ. Press, Jerusalem, 350 p.

100. Ollier, C.D., 1969, Weathering: Oliver and Boyd, Edinburgh, 304 p.

101. Olson, J.S., 1958, Rates of succession and soil changes on southern Lake Michigan sand dunes: Botanical Gazette, v. 119, p. 125–170.

102. Parsons, R.B., Scholtes, W.H., and Riecken, F.F., 1962, Soils on Indian mounds in northeastern Iowa as benchmarks for studies of soil genesis: Soil Sci. Soc. Amer. Proc., v. 26, p. 491–496.

103. Parsons, R.B., Balster, C.A., and Ness, A.O., 1970, Soil development and geomorphic surfaces, Willamette Valley, Oregon: Soil Sci. Soc. Amer. Proc., v. 34, p. 485–491.

104. Pastor, J., and Bockheim, J.G., 1980, Soil development on moraines of Taylor glacier, Lower Taylor Valley, Antarctica: Soil Sci. Soc. Amer. Jour., v. 44, p. 341–348.

105. Peterson, F.F., 1980, Holocene desert soil formation under sodium salt influence in a playa-margin environment: Quaternary Res., v. 13, p. 172–186.

106. Porter, S.C., 1969, Pleistocene geology of the east-central Cascade Range, Washington: Guidebook for third Pacific Coast Friends of the Pleistocene field conference, Sept. 27–28, 1969, 54 p.

107. ———, 1975, Weathering rinds as a relative-age criterion: Application to subdivision of glacial deposits in the Cascade Range: Geology, v. 3, p. 101–104.

108. Rahn, P.H., 1971, The weathering of tombstones and its relationship to the topography of New England: Jour. Geol. Educ., v. 19, p. 112–118.

109. Richmond, G.M., 1962, Quaternary stratigraphy of the La Sal Mountains, Utah: U.S. Geol. Surv. Prof. Pap, 324, 135 p.

110. Ross, C. W., Mew, G., and Searle, P.L., 1977, Soil sequences on two terrace systems in the North Westland area, New Zealand: New Zealand Jour. Sci., v. 20, p. 231–244.

111. Ruhe, R.V., 1956, Geomorphic surfaces and the nature of soils: Soil Sci., v. 82, p. 441–455.

112. ———, 1965, Quaternary peleopedology, p. 755–764 *in* H.E. Wright, Jr., and D.G. Frey, eds., The Quaternary of the United States: Princeton Univ. Press, Princeton, 922 p.

113. ———, 1967, Geomorphology of parts of the Greenfield quadrangle, Adair County, Iowa: U.S. Dept. Agri. Tech. Bull. 1349, p. 93–161.

114. ———, 1968, Identification of paleosols in loess deposits in the United States, p. 49–65 *in* C.B. Schultz and J.C. Frye, eds., Loess and related eolian deposits of the world: Internat. Assoc. Quaternary Res., VII Cong., Proc. v. 12, 369 p.

115. ———, 1969, Quaternary landscapes in Iowa: Iowa St. Univ. Press, Ames, 225 p.

116. ———, 1983, Aspects of Holocene pedology in the United States, *in* H.E. Wright, Jr., ed., Late-Quaternary environments of the United States, Vol. 2. The Holocene: Univ. of Minnesota Press, Minneapolis.

117. Ryan, J., and Zghard, M.A., 1980, Phosphorus transformations with age in a calcareous soil chronosequence: Soil Sci. Soc. Amer. Jour., v. 44, p. 168–169.

118. Schafer, W.M., Nielsen, G.A., and Nettleton, W.D., 1980, Minesoil genesis and morphology in a spoil chronosequence in Montana: Soil Sci. Soc. Amer. Jour., v. 44, p. 802–807.

119. Schmidt, D.L., and Mackin, J.H., 1970, Quaternary geology of the Long and Bear valleys, west-central Idaho: U.S. Geol. Surv. Bull. 1311-A, 22 p.

120. Scott, W.E., 1977, Quaternary glaciation and volcanism, Metolius River area, Oregon: Geol. Soc. Amer. Bull., v. 88, p. 113–124.

121. Sharp, R.P., and Birman, J.H., 1963, Additions to classical sequence of Pleistocene glaciations, Sierra Nevada, California: Geol. Soc. Amer. Bull., v. 74, p. 1079–1086.

122. Shroba, R.R., 1977, Soil development in Quaternary tills, rock-glacier deposits, and taluses, southern and central Rocky Mountains: Unpub. Ph.D. thesis, Univ. of Colorado, Boulder, 424 p.

123. Shroba, R.R., and Birkeland, P.W., 1983, Trends in the late-Quaternary soil development in the Rocky Mountains and Sierra Nevada of the western United States, p. 145–156 *in* S.C. Porter, ed., Late-Quaternary environments of the United States, Vol. 1. The late Pleistocene: Univ. of Minnesota Press, Minneapolis.

124. Sondheim, M.W., Singleton, G.A., and Lavkulich, L.M., 1981, Numerical analysis of a chronosequence, including the development of a chronofunction: Soil Sci. Soc. Amer. Jour., v. 45, p. 558–563.

125. Stevens, P.R., and Walker, T.W., 1970, The chronosequence concept and soil formation: Quat. Rev. Biol., v. 45, p. 333–350.

126. Stork, A., 1963, Plant immigration in front of retreating glaciers, with examples from the Kebnekajse area, northern Sweden: Geografiska Annaler, v. XLV, p. 1–22.

127. Syers, J.K., Shah, R., and Walker, T.W., 1969, Fractionation of phosphorus in two alluvial soils and particle-size separates: Soil Sci., v. 108, p. 283–289.

128. Syers, J.K., Adams, J.A., and Walker, T.W., 1970, Accumulation of organic matter in a chronosequence of soils developed on wind-blown sand in New Zealand: Jour. Soil Sci., v. 21, p. 146–153.

129. Thorp, J., 1968, The soil—a reflection of Quaternary environments in Illinois, p. 48–55 *in* R.E. Bergstrom, ed., The Quaternary of Illinois: Univ. of Illinois, College of Agri., Spec. Publ. no. 14, 179 p.

130. Torrent, J., Schwertmann, U., and Schulze, D.G., 1980, Iron oxide mineralogy of some soils of two river terrace sequences in Spain: Geoderma, v. 23, p. 191–208.

131. Trimble, D.E., 1963, Geology of Portland, Oregon and adjacent areas: U.S. Geol. Surv. Bull. 1119, 119 p.

132. Ugolini, F.C., 1966, Soils, p. 29–72 *in* A. Mirsky, ed., Soil development and ecological succession in a deglaciated area of Muir Inlet, southeast Alaska: Inst. Polar Studies, Ohio State Univ., Rept. No. 20.

133. Vreeken, W.J., 1975, Principal kinds of chronosequences and their significance in soil history: Jour. Soil Sci., v. 26, p. 378–394.

134. Wagner, R.J., and Nelson, R.E., 1961, Soil survey of the San Mateo area, California: U.S. Dept. Agri., Soil surv. series 1954, no. 13, 111 p.

135. Wahrhaftig, C., 1965, Stepped topography of the southern Sierra Nevada, California: Geol. Soc. Amer. Bull., v. 76, p. 1165–1190.

136. Walker, P.H., and Coventry, R.J., 1976, Soil profile development in some alluvial deposits of eastern South Wales: Aust. Jour. Soil Res., v. 14, p. 305–317.

137. Walker, P.H., and Green, P., 1976, Soil trends in two valley fill sequences: Aust. Jour. Soil Res., v. 14, p. 291–303.

138. Walker, T.W., and Syers, J.K., 1976, The fate of phosphorus during pedogenesis: Geoderma, v. 15, p. 1–19.

139. Willman, H.B., Glass, H.D., and Frye, J.C., 1966, Mineralogy of glacial tills and their weathering profiles in Illinois: Part II. Weathering profiles: Illinois State Geol. Surv. Circ. 400, 76 p.

140. Winkler, E.M., 1975, Stone: Properties, durability in man's environment: Springer-Verlag, New York, 230 p.

141. Yaalon, D.H., 1971, Soil-forming processes in time and space, p. 29–39 *in* D.H. Yaalon, ed., Paleopedology: Israel Univ. Press, Jerusalem, 350 p.

142. Yount, J.C., Birkeland, P.W., and Burke, R.M., 1982, Holocene glaciation: Mono Creek, central Sierra Nevada, California: Geol. Soc. Amer. Absts. with Prog., v. 14, no. 4, p. 246.

143. Harden, J.W., 1982, A study of soil development using the geochronology of Merced River deposits, California: Ph.D. thesis, Univ. of California, Berkeley, 237 p.

9

Topography-soil relationships

Topography, or local relief, controls much of the distribution of soils in the landscape, to such an extent that soils of markedly contrasting morphologies and properties can merge laterally with one another and yet be in equilibrium under existing local conditions (Fig. 9-1). Many of the differences in soils that vary with topography are due to some combination of microclimate, pedogenesis, and geological surficial processes, and the sorting out of the effects of each on soil distribution is difficult. The fields of pedology and geomorphology probably overlap here more than with any other pedologic factor, as discussed by Gerrard.[19]

Soil properties vary laterally with topography. One reason for this is the orientation of the hillslopes on which soils form; this affects the microclimate and, hence, the soil. Another is the steepness of the slope; this affects soil properties because the rates of surface-water runoff and erosion vary with slope. In areas of rolling terrain, soil properties vary because lower areas are likely to be areas of accumulation of water runoff and sediment derived from surrounding higher-lying areas. Also, low areas might be influenced by a high water table, which could have a considerable effect on the soil. These relationships are important for the correct identification of buried soils.[55] In this chapter, I will discuss examples of these processes. Of course, we assume other factors to be nearly constant in this analysis; this might be true for some examples, but perhaps not for all.

INFLUENCE OF SLOPE ORIENTATION ON PEDOGENESIS

Slope orientation results in microclimatic and vegetation differences, and thus in soil differences. Jenny[27,28] argues that topography is the primary factor in explaining soil variation in these field situations. Here we

Fig. 9-1 Variation in soils with topographic position, in Indiana. Poorly drained, dark-colored Aquolls occur in the lowlands, and somewhat poorly drained Aqualfs occur in the higher parts of the landscape. These latter soils are light colored here because erosion has exposed the light-colored B horizon. (Photograph from Marbut Memorial Slide Collection, prepared and published by the Soil Science Society of America, Madison, Wisconsin, in 1968.)

are concerned with topographic relief of meters, or tens of meters, and therefore regional climate can be considered a constant.

An instructive example of the effect of slope orientation on soil properties is the study of Finney and others[15] in southeastern Ohio (Fig. 9-2). Several NW–SE-trending valleys were studied. The parent material is mostly colluvium derived from sandstone. The microclimate varies quite markedly with orientation, with the SW-facing slopes displaying higher temperatures on the leaf litter, as well as greater annual fluctuations. Temperatures beneath the leaf litter show the same trends with orientation, but the differences between maximum and minimum values are less. Soil-moisture values follow the temperature differences, with soils on NE-facing slopes generally being more moist than those on SW-facing slopes. Vegetation correlates with the moisture-temperature trends. A mixed-oak association is dominant on the SW-facing slopes, whereas a mixed mesophytic plant association is dominant on the NE-facing slopes. The microclimatic-vegetation differences produce soil suborder differences. The NE-facing slopes are characterized by Ochrepts with thick A horizons on slopes with gradients over 40 percent and Udalfs with relatively thick A horizons on the rest of the slopes. However, Udalfs with thin A horizons predominate on the SW-facing slopes, and Ochrepts with thin A horizons are present locally (H.R. Finney, written commun., 1972).

Slope orientation greatly affects soil organic carbon distribution with depth, the presence or absence of an E horizon, pH, and percent exchangeable bases (Fig. 9-2). Organic matter differences probably result from greater moisture and vegetation cover on the NE-facing slopes, combined with greater organic matter decomposition rates on the SW-facing slopes, along with some loss due to surface runoff. E horizons are more common on SW-facing slopes, and the reason for this is not clear. If anything, one would predict that they would be more common on the moister, NE-facing slopes. Base saturation and pH trends are somewhat parallel with depth, and these can be explained in terms of vegetation differences and fire. The vegetation assemblage on the NE-facing slopes has a higher base content, and bases are returned to the soil surface as litterfall. However, the mixed-oak association has a lower base content and probably a higher incidence of fire. Fire not only burns off some of the soil organic matter, but also brings about the loss of bases released from the organic matter by either runoff or deep leaching.

A somewhat similar relationship of soils to slope orientation is seen in the rolling hills of eastern Washington.[34] Parent material is Wisconsin loess deposited in NW–SE-trending ridges that have about 65 m of relief (Fig. 9-3). Although the regional precipitation is 53 cm/year, most of which occurs in the winter, distinct microclimates are produced by the orientation of the hillslopes. It is estimated that the S-facing slopes receive close to the average annual precipitation. However, because about one-fifth of the precipitation is snow, the moisture can be easily redistributed in the landscape. It is estimated that the hilltops might receive only 25 cm of precipitation, with losses attributed to the removal of snow by wind as well as high evaporation rates on the exposed ridges. In contrast, the N-facing slopes have more effective moisture than do the S-facing slopes due to a combination of accumulated drifting snow and lower evaporation rates. The virgin vegetation follows these trends, with the more mesic shrubs, grasses, and forbs on the N-facing slopes and the more xeric forbs and grasses on the S-facing slopes.

Soils are closely related to the microclimate (Fig. 9-3). The S-facing slopes are characterized by noncalcareous Mollisols that are thought to be normal for the region. In contrast, the N-facing slopes, largely because of the additional moisture, have more strongly developed Mollisol profiles and greater amounts of organic matter in the A horizon. The soils on the ridgetops are thin Mollisols, characterized by weak B-horizon development and carbonate accumulation at depth, both properties that are clearly the result of a more arid microenvironment. Thus, it is seen that

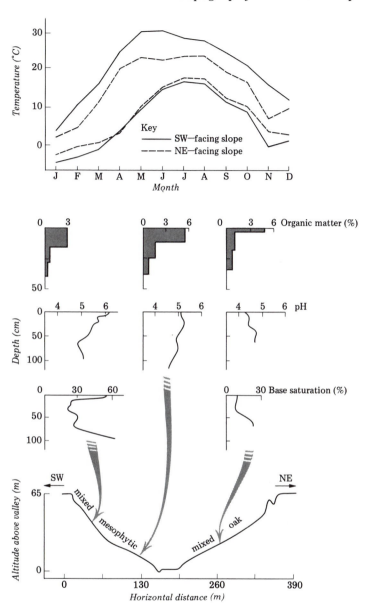

Fig. 9-2 Topographic, vegetative, soil, and microclimatic data for slopes of different orientation, southeastern Ohio. (Taken from Finnery and others,[15] Figs. 1, 3, 5, 6, and 7.) Temperatures are maximum and minimum monthly averages on leaf litter, 1956.

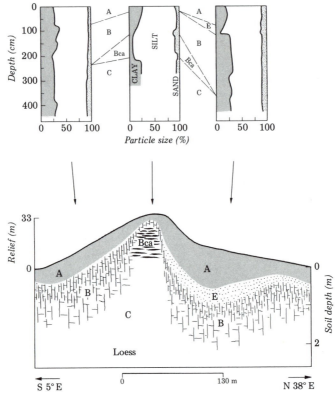

Fig. 9-3 Variation in soil properties with slope orientation in loess hills, eastern Washington. (Taken from Lotspeich and Smith,[34] Figs. 2 and 3, © 1953, The Williams & Wilkins Co., Baltimore.)

fairly large differences in soil properties and profiles can be closely associated with each other because of microclimatic differences associated with slope orientation.

SOIL CATENAS

Numerous studies have shown that many soil properties are related to the gradient of the slope as well as to the particular position of the soil on a slope. Milne[38,39] proposed the term catena to describe this lateral variability on a hillslope and emphasized that each soil along a slope bears a distinct relationship to the soils above and below it, for a variety of geomorphical and pedologic reasons. These also are called toposequences. Yaalon[61] has reviewed many of the topofunctions derived from various areas in the world.

Several models have been proposed to describe the landscape that soil catenas are related to, and each has its attributes. Conacher and Dalrymple[7] discuss soil-slope relations in considerable detail and propose a nine-unit landsurface model defined on process and response. For many studies, however, it is sufficient to use a simple five-unit model based on slope form.[50] Going from the top to the base of a slope, these are the flat summit, the convex portion or shoulder, the more or less uniform slope of the backslope, and the concave portion at the base, which is divided into an upper footslope position and a lower toeslope position. Soil catenas can be further divided into open and closed systems.[50,58] In the former, drainage is such that sediment can leave the area; the latter is characterized by a closed depression and all sediment is trapped in a topographic low.

Although most soil catenas are depicted as a two-dimensional cross section, Huggett[24] suggests that a better approach would be to consider a three-dimensional model, such as a small watershed. The boundary is the divide with the adjacent watershed. He then makes the point that properties of soil catenas in such a setting are partly explained in terms of lateral flow of soil waters, called throughflow water, from the higher parts of the watershed to the lower parts.[60] Referring to the three-dimensional soil chemistry that might result, one can identify, in the upper reaches, an eluvial zone in which soils may show a net loss, and along the lower parts of the watershed, an illuvial zone where the same constituents might accumulate; separating these is a transluvial zone. The degree to which the soil properties match the zones will be mainly a function of the climatic regime. Clay particles also are considered to move laterally by throughflow waters.[25]

It is common to find markedly different soils in juxtaposition in rolling topography in which rounded hills rise above very gentle slopes or closed depressions. Soils on the uplands commonly are well drained, whereas those in the depressions are poorly drained and rich in clay and organic matter, with signs of various degrees of gleying. In dry climates, saline and alkaline soils occupy the depressions, better-leached soils the slopes, and the less-leached soils the summits. The differences in soil properties with position could be due to pedogenesis in place, resulting from differences in moisture, leaching, and vegetation over the rolling landscape. In this case, the various parts of the landscape are assumed to be approximately the same age, and soil differences are attributed to the topographic factor. Recent work, however, has cast some doubt on this simple model, because it fails to take into account the fact that some material in the depressions could be derived from erosion of the landscape that slopes into the depression. With this model it is unlikely that the parent material of the soils formed on the slopes is the same as that in the depressions. The differences, however, have their primary origin in topographic position.

In addition to the strictly pedologic redistribution of materials on a slope, one also must consider the geomorphic processes operative on slopes that cause erosion in some places and deposition in others. Important processes include rainsplash, overland flow of water, and such mass movements as creep.[19]

The K-cycle model of Butler[5] uses much of the above, and attempts to place slope events in a time framework. Start with a simple slope, such as the one mentioned above. For a variety of reasons, the steeper, upper part of the slope might be unstable and erode fairly rapidly; material of this event can be deposited in the footslope-toeslope positions. Stability might follow and be accompanied by soil formation along the entire slope. One K-cycle encompasses the erosional-depositional interval, as well as the soils that subsequently form over the entire slope. A subsequent period of instability could result in a second K-cycle. The aerial position and extent of both the erosion and deposition zone can change with time. The end result could be that erosion keeps soils relatively poorly developed in the upper parts of the slopes, and a sequence of buried soils is present in the lower parts of the slopes. Some areas of the slope can be beyond the areas of maximum erosion and deposition. If, instead of the periodicity of K-cycles, erosion and deposition are more gradual through time, the cumulative and non-cumulative profiles depicted in Figure 7-5 could be produced.

Catenas have been shown to vary with climate,[44] so the following examples of soil catenas are organized according to major climatic regimes.

Catenas in Arctic and Alpine Areas

Soil catena relationships are well expressed in arctic areas.[52] Extreme climatic conditions retard weathering and soil development, and shallow ice permafrost influences both the water regime in the soil and the vegetation pattern. My experience on Baffin Island, in the eastern Canadian Arctic, has been that well-drained topographic highs, such as sandy moraines, are windswept and dry and have sparse lichen vegetation. In contrast, tundra vegetation is common on the backslopes, footslopes, and toeslopes. Permafrost is deep enough not to influence summit soils, but is progressively closer to the surface going down the slope; hence, turbation of the soil by freeze-thaw processes does not seem to affect the summit soils, although it has a progressively greater influence on the wetter soils in the lower parts of the landscape. Because of these influences soils exhibit well-drained and oxidized profiles at the summit and progressively more reduced profiles in the lower parts of the landscape. Tedrow's[52] data for northern Canada show a fourfold increase in organic matter in the surface horizon in going from the summit (0.9 percent) to the footslope (3.8 percent). The lateral movement of ions may be restricted except in the wetter, lower parts of the landscape.

Marked geochemical trends are apparent in humid subarctic areas of northern Norway.[20] Horizon thicknesses increase downslope, as do humus and Al_2O_3 contents in the Bh horizon; in contrast, Fe_2O_3 content changes little with slope position (Table 9-1). The chemistry of these soils seems to follow the direction of movement and the amount of throughflow water, with possible organic matter complexes moving Al in preference to Fe.

Alpine soil catenas also can show marked profile contrast with slope position. Those in the extremely cold and windswept environment above treeline in the Colorado Rocky Mountains display morphological variation with position on the rolling upland terrain.[4] Topography controls soil moisture by limiting much of the snowfall to the lee sides of hills; these same sites also trap loess. Soil distribution follows trends in snowcover duration and in thickness of loess. Summits and shoulders have little snow cover, and the soils are Cryochrepts; there is no loess at the most wind-blown sites, and patches of loess are up to 8 cm thick at more protected sites. Soils lower down on the slopes show a relationship to duration of snow cover. In late winter, snow covers much of the area. With spring and summer melting, the soils become exposed, with those higher on the slopes first, and those lower down last. Below the shoulder are what are called the minimal snow-cover sites, characterized by the most developed soils—Cryumbrepts with a pergelic temperature regime. These are characterized by thick organic matter-rich A horizons formed in loess. Next downslope are Cryumbrepts, where the snowbanks melt early. Loess is not as thick as in the minimal snow-cover soils, but the A horizons are the thickest of all. Soils in areas where the snowbanks melt late are poorly developed Cryochrepts with no loess. Finally, at the slope bases are the poorly drained, mottled, and gleyed soils, mainly Cryaquepts, characterized by high silt and clay transported during snowmelt from upslope positions.

Table 9-1
Properties of Soils in a Catena in Northern Norway*

Soil property	Landscape position		
	Summit/shoulder	Backslope	Footslope
A horizon thickness (cm)	15	25	30
Bh horizon thickness (cm)	10	30	55
Humus in Bh (%)	2.1	2.2	4.4
Fulvic acid:humic acid in Bh	1.8	—	9.3
Fe_2O_3 in Bh (%)	6.9	6.9	7.8
Al_2O_3 in Bh (%)	18.4	22.2	33.4

*The author is not clear on the extraction procedure used for Fe and Al.
(Taken from Glazovskaya,[20] Fig. 2.)

Soil Catenas under Humid Conditions

With an increase in the amount of water leaching through the system, aided perhaps in places by organic matter complexes, quite marked pedologic changes take place over short distances. McKeague[37] presents data on a catena involving Spodosols in eastern Canada and includes information on both the depth to water table and the redox conditions. Cryorthods are present in the summit, shoulder, and backslope positions. At these latter slope positions, the top of the water table is always below about 1 m upslope, but is at the surface for several weeks every spring downslope; measured redox conditions are more oxidizing upslope than downslope, where gleyed soil properties are present. In contrast, the footslope soils have the water table within 30 cm of the surface, and redox conditions vary from high to low Eh in the shallow zone of a fluctuating water table. The resulting soils are Cryaquents. Extractable Fe in the B horizon doubles from the summit (1.6 percent) to the footslope (3.2 percent). Clay minerals above the C horizon are smectite and chloritized vermiculite in the upslope soils and vermiculite, smectite, and mixed-layer clays in the footslope soils.

New Zealand workers have obtained soil catena data somewhat similar to that in relatively humid areas above, and they have compared them to catenas in somewhat drier conditions.[53,63] Under about 100 cm of precipitation, the sequence is Inceptisols at and above the backslope position and Aquepts at the footslope position (Fig. 9-4). In contrast, under about 200 cm of precipitation, the system is more leached, with Spodosols at the summit, Aquods at the backslope, and Histosols at the footslope. As expected, the more humid catena displays a greater elemental loss relative to the drier catena. Clay-mineral alteration is more advanced in the more humid catena, and for both catenas, alteration is more advanced in the more leached upslope positions than for the less leached downslope positions. The geochemistry of the waters at the downslope sites, specifically the relatively high accumulation of bases and Al, seems to inhibit higher clay-mineral alteration stages (Fig. 8-15).

Many catenas have been investigated in the midcontinent, in both loess and till.[36,56] Wisconsin loess of uniform particle size is the parent material for a soil catena in southeastern Nebraska.[1] Soils formed in the downslope positions are more strongly developed than are soils in upslope positions and on interfluves (Fig. 9-5). Resistant to non-resistant mineral ratios also vary markedly with slope position, and this suggests that much of the clay content variation is due to mineral weathering within each profile. Soil-moisture values in the C horizon for the summer of 1966 averaged 21 percent (by volume) at site A compared with 25 percent at site B. Although this slight difference in moisture might explain the differences in weathering and clay formation, it is possible that past conditions of greater moisture in the downslope soils, suggested by gleyed features, may

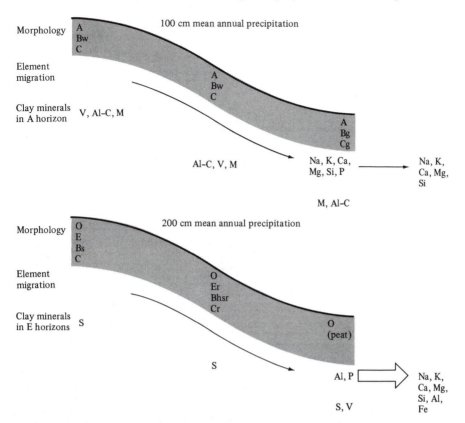

Fig. 9-4 Schematic diagrams showing soil morphology, element migration, and surface soil clay mineralogy for two catenas in New Zealand, in different precipitation regimes. Key to clay-mineral symbols: V, vermiculite; Al-C, Al-chlorite; M, mica; and S, smectite; "r," reduced horizons. For element migration, the width of the arrow is proportional to the amount of element moving through that portion of the catena in through flow water. (Taken from Tonkin and others,[53] Fig. 1, and Young and others,[63] Fig. 1 and Table 2.)

have accelerated the trends. The quartz to feldspars ratio with depth can be compared with those of Ruhe (Fig. 8-4). The ratio at site A is typical for Wisconsin-age deposits, but that at site B compares with those obtained from soils formed on Yarmouth-Sangamon surfaces. Thus in this case, there seems to be a severalfold acceleration of weathering and clay formation with slope position.

One problem with studying soils on slopes is that the slopes may differ in age with position, and therefore the soils may not be the same age from place to place. Ruhe and Walker[50] point out such a case for soils formed

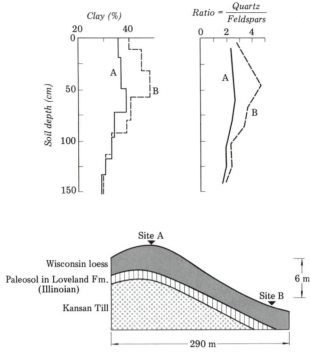

Fig. 9-5 Relationship between topography, stratigraphy, and soils, southeastern Nebraska. (Taken from Al-Janabi and Drew,[1] Figs. 3 and 4, Table 2.)

on rolling loess topography in Iowa (Fig. 9-6). Many soil properties correlate well with the slope steepness. In particular, with increasing steepness the soils are thinner, and there is less organic matter in the A horizon; also the depths to a pH of 6, to base saturation of 80 percent, and to a greater than 1 percent carbonate all decrease. Thus, the soils are shallower and less well developed on steeper slopes.

These soil trends with slope gradient may result, in part, from surface erosion, because, as Ruhe and Walker[50] and Walker and Ruhe[58] point out, the slopes are not all the same age. The summit surfaces, for example, are more than 14,000 years old, whereas the slopes adjacent to the summit areas have undergone more recent erosion and are less than 6800 years old. Age of surface, therefore, a factor not always appreciated in soil-slope studies, could explain many of the soil differences in a comparison of summit and slope areas.

Slopes are difficult to date. However, alluvium derived from slope erosion commonly has been deposited at the foot of slopes. Therefore, by radiocarbon dating of the alluvium one obtains dates on the times of stability and erosion of the slopes from which the alluvium came.

The till landscapes of the midcontinent are good examples of the interplay of pedologic, sedimentologic, and erosional processes in areas char-

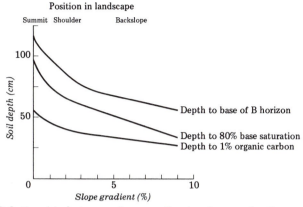

Fig. 9-6 Relationship between slope gradient and several soil properties, Iowa. (Data from Ruhe and Walker,[50] Table 1.)

acterized by closed depressions. We will first consider processes operative on the present landscape and then discuss the well-known problem of gumbotil.

Walker[57] has made a detailed study of closed depressions and associated hillslopes in Iowa. The parent material is till dated at about 13,000 BP. The soils vary from Udolls in well-drained positions to Aquolls in poorly drained, closed depressions. Major differences in the soils are increase in clay and organic matter toward the center of the depression, along with a decrease in the gravel content (Fig. 9-7).

The stratigraphy and radiocarbon dating of several depressions indicate that the sedimentation rate there and the erosion rate of the surrounding slopes have varied in the past. The uppermost 60 cm of material in the depression shown in Fig. 9-7 is rich in organic matter and was deposited during a period of relatively slow slope erosion dating from about 3000 BP to the present. Material below 60 cm, however, dates back to 8000 BP; it is relatively low in organic matter and was deposited during an interval of relatively rapid slope erosion. Older deposits, indicating additional periods of slope erosion and stability, are found deeper in the depression.

The depression, therefore, contains mostly postglacial sediment, and its properties directly influence the soil properties. These sediments were derived from the adjacent slopes, and the operative processes brought about the lateral separation of size fractions, so that gravel sizes were not moved into the centers of the depressions, whereas clay sizes were. Thus, the major lateral differences in soil-particle sizes are sedimentologic, not pedologic, in origin. Add to this primary textural control the variation in soil moisture due to variation in internal drainage, and most of the lateral variation in soils is adequately explained. Finally, the stratigraphic relationships indicate that most of the soils in the area formed over the past

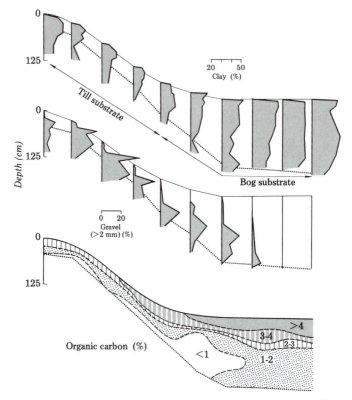

Fig. 9-7 Relationship of particle size and organic carbon with topographic position in a closed depression and on surrounding hillslopes, Iowa. (Taken from Walker,[57] Figs. 19, 30, and 31.) The dotted line separates hillslope sediments from till on the hillslopes, and younger from older deposits in the low-lying area.

3000 years under grassland vegetation. Hillslope soils that formed over a previous period under postglacial forest vegetation were essentially removed by erosion before 3000 BP.

The formation of the well-known gumbotils of the midcontinental Pleistocene record probably results from a combination of the topographic factor in an undulating landscape and the time factor. Kay[30] defined gumbotil as "a gray to dark-colored, thoroughly leached, non-laminated, deoxidized clay, very sticky and breaking with a starchlike fracture when wet, very hard and tenaceous when dry, and ... chiefly, the result of weathering of drift." Kay[30,31] and Kay and Pearce[32] considered gumbotil to be primarily the result of prolonged chemical weathering of pre-Wisconsin tills on flat, poorly drained plains; it is a gleyed soil. That chemical weathering had occurred was shown by chemical analyses of the gumbotil, compared with the unweathered till, and by the concentration in the gumbotil of siliceous pebbles that are resistant to weathering.

A lively debate on what was included and what was excluded in the original definition of gumbotil has developed in recent years. These arguments are reviewed by many workers,[17,18,33,49,54] and the details will not be repeated here. Basically, it is agreed that there are two contrasting origins for gumbotil. One origin is weathering in place under conditions of poor internal drainage; the other origin is that they are deposits laid down in depressions on the original till surface. Although Kay considered chemical weathering to be the main factor in gumbotil origin, he did agree that other processes, such as slope wash, were operative. Ruhe,[48] in his comparison of gumbotils located on swells and in swales in slightly undulating topography (Fig. 9-8), also recognized both processes of gumbotil formation. Ruhe thought that either weathering and sedimentation could go on contemporaneously in the swales, or that the two processes could be separated in time, and demonstrated from mineral ratios that material in the swales is more weathered than is material on the swells.

Frye and others[17] have called attention to these two different origins of gumbotil and have suggested that, if the term is kept, it be restricted to profiles that have weathered in place from till. They suggest that gleyed material that accumulates in depressions be called accretion gley. It is important to recognize both of these materials and probably to use separate terms for both because the profiles formed in place attain their main characteristics by pedogenesis in poorly drained areas, whereas accretion gleys attain their main characteristics by sedimentary processes, as well as pedogenesis, in poorly drained areas. Some criteria found to be helpful in differentiating gumbotil from accretion gley are listed in Table 9-2.

The relationship of the gumbotils to the adjacent accretion gleys is

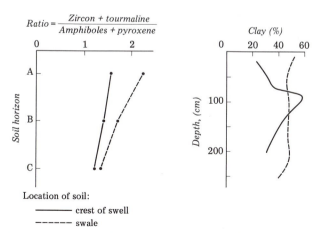

Fig. 9-8 Comparison of pre-Wisconsin soils formed in different parts of an undulating landscape, Iowa. (Taken from Ruhe,[48] Figs. 4 and 5, Table 1, © 1965, The Williams & Wilkins Co., Baltimore.) Parts of both profiles would be classed as gumbotil.

Table 9-2
Some Criteria for Differentiating Gumbotil from Accretion Gley[17,18,54]

Criterion	Gumbotil (fomed in place by weathering)	Accretion gley (deposited in lows in the landscape)
Stratification	Massive, not stratified, leached of carbonates	Stratification can be seen as humus-rich layers toward the top of the deposit and as layers of contrasting particle size
Contact with fresh till	Gradational contact with fresh till includes a leached and oxidized zone over an unleached and oxidized zone	Contact with underlying fresh till can be sharp; configuration of the contact might suggest that the deposit occupies a depression on an old paleo-landscape
Distribution of resistant pebbles	Pebbles of resistant lithologies show an orderly increase in relative abundance upward	Some deposits may have few resistant pebbles, relative to the till, due to sorting on adjacent hillslopes
Silicate mineral weathering	Silicate minerals are decomposed in order of weatherability; decomposition shows an orderly increase toward surface	Weatherable silicate minerals may be depleted; depletion toward the surface need not be systematic
Clay-mineral alteration	Clay minerals show an orderly arrangement of weathering stage with position in the profile	Clay minerals are a mixture of minerals of different stages of weathering; mechanical mixing is suggested

analogous to that of deposits and soils with topography described by Walker,[57] mentioned above. This relationship has been pointed out by Ruhe,[49] and the lateral variation in gumbotils and accretion gleys appears to be due to the topographic factor. Time is also involved here, however, because gumbotils are found only on pre-Wisconsin surfaces. Thus, given enough time, the materials and soils described by Walker might progress to the gumbotil and accretion gley stage of development (compare Figs. 9-7 and 9-8 and Table 9-2).

Follmer[16] incorporates many of the above concepts in a detailed discussion of the Sangamon soil in southern Illinois. The soil varies with topographic position on the Illinoian till plain, and in places its recognition as a separate entity is obscured by the overlying younger deposits. Clues are given on how to correctly interpret these soils and stratigraphic relations.

Catenas in which all soil-forming factors are reasonably well known may be more difficult to locate in strongly developed soils of warm, humid climates because the landscapes are so old and their histories more difficult to decipher. Ollier[44] and Gerrard[19] discuss some of these catenas. In contrast, many of the catenas previously mentioned have formed on landscapes formed of young glacial or loess materials, and therefore, their histories are easier to interpret.

Despite the problems, there are trends common to many soil catenas in warm humid climates.[26,29,40,43,45,47,59,62] Internal soil drainage is a major factor responsible for many of the trends. Soils in the higher positions of the landscape are well drained and oxidized red. As one progresses to the lower parts of the landscape, the soils are mottled where under the influence of a fluctuating water table, and those at the base of the slope are usually gleyed. Iron and manganese concentrations and concretions can be present in the more poorly drained soils. As an example, soils of catenas in Sri Lanka and Rhodesia have hues of 10R and 2.5-5YR with high chromas at well-drained sites, whereas colors commonly are 10YR 6-7/1-2 at poorly drained sites. Kaolinite is the common silicate clay mineral in the well-drained soils, smectite in the poorly drained ones. Both minerals form as a function of the geochemical environment of their respective soils, which is related to the leaching and groundwater conditions. If kaolinite is eroded and transported to the topographic lows, however, there has to be some mechanism for its conversion to smectite. In one place, however, kaolinite is common and smectite not present in the gleyed soils,[59] so either such a kaolinite → smectite alteration does not take place, the geochemical environment inhibits the alteration, or it is time dependent and more time is needed for the alteration to occur.

Finally, a still wetter environment is a swamp on the coastal plain of northern Florida.[8] Sandy Aquents are present on the better-drained higher parts of the area. On the more poorly drained slopes are Aquods with Bh horizons; those on the lower parts of the slopes can be submerged for parts of the year. Finally, humus-rich Aquepts are present in the lowest parts of the topography. Trends from topographic high to low position are clay content increasing from less than 10 percent to over 20 percent; organic carbon in the surface increasing from 0.5 percent to 37 percent, and staying generally above 10 percent in the upper 81 cm of the Aquept; pH decreasing from near 5 to 3–4; and extractable cations showing a general increase in the surface horizons, from 0.3 meq/100 g to 9.7, but since the environment is sufficiently leaching, most soil horizons have no bases.

Daniels and co-workers[11,12,13,14] recognize some complications in soil-slope relations in the Ultisols of the flat Coastal Plain of the eastern United States. These are old soils, with some beginning development during the early Pleistocene; in others, development was still earlier. Water table has an important influence on the soil distribution. Where divides are wide (about 3 km), the water table is high and within 50 cm of the surface about one-half of the time. Subtle depressions (relief <2 m) within these exceptionally flat areas have the usual gleyed soils in contrast to the yellowish brown soils of the slightly better drained surrounding terrain. In the better-drained soils at the upper edge of the depressions there is a marked increase in texture between the A and argillic horizon (approximately 20 to 40 percent clay). Soils in the topographic lows have about 20 percent clay, again with an abrupt increase at the A/B boundary. Between these two locations, the soils have much less clay and are loamy sands or sandy loams. The interpretation is that the clays have been eluviated from these sandy soils because the water table fluctuates there more than at the other two sites.

Soils adjacent to the major terrace escarpments in the coastal plain also display the influence of the water table, because the water table is deeper at the edge of the escarpment than it is a short distance inland from the scarp. Where the water table is relatively high, B-horizon colors are 10YR 5/6. Toward the escarpment, the water table is deep enough not to influence the soils much, and the colors can change to 7.5YR 5/6 and to 5YR 6/5 within several meters laterally. Greater amounts of free Fe accompany the increase in redness. Other consistent trends as one approaches the edge are that the E horizon thickens and tongues into the Bt horizon, the E-horizon clay content decreases, and the B-horizon clay content increases. In the areas of fluctuating water table, small patches of clay eluviation occur within the Bt horizons. Thus, a major change in these old highly weathered soils comes only when scarps retreat and the drainage of the soils at the edge improves, thus paving the way for a marked morphological change through eluvial-illuvial processes if a sufficiently long time has elapsed. They also point out that soils in the middle of the divide will be influenced by a high water table until the divide becomes less than about 0.6 km wide. Finally, with greater geomorphic age and widespread landscape dissection, the water table lowers and this will have a marked effect on the morphological trends over the region. Prior to dissection, the soils, except for those at the scarp edges, underwent little change with time.

Topographic position controls the formation of some Oxisols, for they have been shown to differ between upland sites and sites located on slopes.[22,35] Oxisols form by the usual weathering reactions and soil-forming processes in the uplands. The slopes, however, are characterized by fairly large quantities of laterally moving soil moisture derived mostly from upslope. Mobile constituents released by weathering can move with these waters (Fig. 9-9). Because iron and manganese are more mobile than

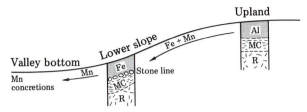

Fig. 9-9 Variation in morphology and chemistry of Oxisols with topographic position; the arrows depict general movement of soil moisture, iron, and manganese. (Taken from Hamilton,[22] Fig. 1.) Key: Al, bauxitic horizon; Fe, ferruginous horizon; MC, mottled clay; R, parent material.

aluminum, one sees a lateral segregation of these elements; aluminum is relatively concentrated in the upland areas, iron in the downslope areas, and manganese, because it is more mobile than iron, moves farthest downslope. Oxisols on the slopes, therefore, owe much of their iron content to precipitation as crystalline and amorphous iron compounds in voids in the soils or in the colluvial deposits on the slopes.

With dissection of the landscape, the previously deposited iron can be remobilized and moved into younger soils and deposits downslope. For example, if the landscape in Fig. 9-9 were to be dissected, a younger landscape would form downslope from the previously formed iron-rich Oxisol. Because of dissection, the iron-rich soil would be relatively well drained and waters from it could move laterally. The iron could, therefore, if remobilized, move to lower parts of the landscape and cement those soils and deposits. This process could continue with each new cycle of downcutting in the landscape.

Catenas in Dry Climates

Catenas in drier environments contrast with those described above in that the amount of soil moisture available for lateral transfer is less, and the geochemical environment for transfer of elements is different. Muhs[41] reviews these for Mediterranean climates, and discusses one catena in an arid Mediterranean climate. Here I will give other examples from Mediterranean climates, as well as from semiarid and arid climates.

Nettleton and others[42] have described soils formed from tonalite at various positions on a slope in southern California (Fig. 9-10). They assume that the three soils have formed on slopes of about the same geomorphic age, and therefore the soil differences can be attributed to topographic position. Because of the warm climate, with precipitation mainly in the winter and early spring (38 cm/year), the soils have ochric epipedons. However, B-horizon properties vary markedly downslope. The Vista soil has only a cambic horizon, whereas the Fallbrook has an argillic horizon, and the Bonsall a natric horizon. The Cr horizons are tonalite grus.

Soil-moisture measurements were taken at various times of the year,

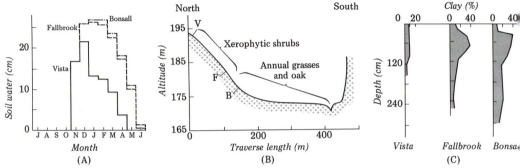

Fig. 9-10 Data for a toposequence of soils formed from tonalite, southern California. (A) estimated soil water for 1965–66; (B) topographic and vegetative relationships [at the sampled sites the slopes are 8% for Vista (V), 12% for Fallbrook (F), and 8% for Bonsall (B)]; (C) distribution of clay with depth. (Taken from Nettleton and others,[42] Figs. 1 and 12, and Table 3.)

and from these data and the precipitation data for 1965–66, the moisture in the soils was estimated on a monthly basis (Fig. 9-10). These data show that the soils that lie downslope (Fallbrook and Bonsall) receive more soil moisture than the upslope soil (Vista), and that they retain their moisture longer.

A marked feature of these soils is that the clay content increases downslope (Fig. 9-10). Weathering of the primary minerals is greater downslope, and thus most of the clay present can be attributed to weathering of the underlying rock at that site. Weathering and clay-formation differences downslope are most likely due to soil moisture as determined by slope position. Soils in lower slope positions can receive more moisture than those in upslope positions because of lateral movement, either at the surface or within the soil. Furthermore, once clay formation begins, the soil has a higher water-holding capacity that results in accelerated clay formation as compared to soils in which clay contents are low.

We have extended our study of soils on moraines of different ages in the Sierra Nevada to include those on the moraine slopes.[3] The advantage of this is that we can follow the development of catenas through time. Soils at all slope positions of the 20,000-year-old moraines have A/Bw/Cox profiles with maximum clay increase into the B horizon of 2 to 3 percent, and thick, gravel-free cumulic profiles in the footslope-toeslope positions. In contrast, soils formed on the 140,000-year-old moraines are much better developed, with Bt horizons at all positions, and an increase in clay into the B horizon of 10 to 18 percent. Not uncommonly, the most strongly developed of these profiles is in the footslope position, and the A horizons there are not cumulic. Our tentative conclusions are that moraine slopes take perhaps 20,000 years to adjust to a stable configuration and that they remain relatively stable beyond that time so that strongly developed soils are able to form at downslope positions.

Sand dunes along the Mediterranean coast in Israel also demonstrate soil variations with position in landscape.[10] Soils on the tops and sides of the sand dunes have well-developed Bt horizons, with color hues of 2.5YR, overlying a C horizon of low clay content (Fig. 9-11). In the depressions, however, the soil for 3 m depth is close to 60 percent clay, and the soil color meets the criteria for a gleyed horizon. Other notable differences below the surface from ridgetop to depression are decrease in content of free iron oxide, a change from predominantly kaolinite to montmorillonite, and an increase in pH.

Soil variation of the Israel sand dunes is explained by eolian influx combined with slope processes and pedogenesis. Much of the clay in these soils results from eolian influx, some of which, after reaching the surface, is translocated downward in the soil. During pedogenesis, the clay minerals are thought to have altered from montmorillonite to kaolinite under favorable chemical and leaching conditions. The soil in the depressions is thought to have been mechanically transported into those positions and not to have formed in place, mainly because there are few weatherable mineral grains in the landscape that can alter to clay. The main process envisaged is downslope transfer of the eolian dust on loose organic litter by running water. It is also suggested that some clay transfer could take place within the soil by laterally moving soil water at the top of the strongly developed Bt horizon (about 20 cm depth). It is not felt that the evidence supports the transfer of sand-dune sediment from the slopes to the depression because so little sand is found in the depression. Leaching conditions in the depressions are slight and thereby favor retention of

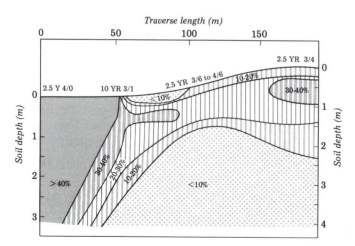

Fig. 9-11 Generalized distribution of clay-sized particles with depth in soils developed from sand-dune deposits, Israel. (Modified from Dan and others,[10] Fig. 9.) The difference in altitude from the ridge top to the depression is 5 m. Munsell color notations are for Bt horizons of soils formed above the depression and for soil material at a comparable depth within the depression.

montmorillonite as the main clay mineral. Pedogenesis is going on in these different parent materials during slope erosion and deposition in the depressions.

Soil catenas in more arid environments are characterized by lateral translocation of only the more soluble compounds—the salts. For cold deserts, Glazovskaya[20] notes the occurrence of the more soluble salts in soils at progressively lower portions of the landscape. Thus, $CaCO_3$ could exist in all the catena soils, but in the downslope direction, more soluble salts appear in the order of their solubility; next would be gypsum and Na_2SO_4, whereas chlorides of Ca, Mg, and Na would appear in the lowest parts of the landscape. Lateral salt movement could be with throughflow waters.[6] Iron released by weathering, and this can be very rapid in saline environments, would stay at the site of release and, along with aluminum, probably not be translocated laterally. Clay minerals would most likely be montmorillonite, but conditions appropriate for the formation of the fibrous clays also could exist.

Soils along the Dead Sea rift are representative of soil catenas related to alluvial fans in aridic areas.[9] The mean annual precipitation is about 3 to 4 cm/year. The fan material consists of coarse gravel at the mountain front and fines progressively in a downfan direction to sands, loams, and clay loams. The water table is deep (>5 m) in the gravelly materials and becomes progressively shallower toward the toe of the fan (1.5 to 2 m depth). Vegetation is sparse in the gravelly materials, increases in abundance in the finer-grained materials, and eventually decreases and is absent where the salt content is too high. Calcium carbonate is present in all of the soils, and gypsum and more soluble salts increase in abundance toward the lower edge of the fan. In the lower parts of the fan, salts have their highest contents at the surface, and gypsum content can exceed 50 percent. This indicates that the salts precipitate from water delivered to the upper parts of the soil by capillary rise from the shallow water table, a common feature in many deserts.[2]

SUMMARY

Catena studies are a good blend of pedology, geomorphology, geochemistry, and mineralogy, at least in the more simple cases. Many generalizations on trends can be made. Tardy,[51] for example, reviews the clay-mineral aspects of catenas with respect to climate and suggests that clay minerals of higher silica content usually occur in the lower parts of the landscapes. In his climatic transect, the clay minerals in the lower, more poorly drained parts of the landscape in one climatic regime could be the stable clay minerals in the well-drained site of a catena in a drier climate (Fig. 9-12). In the driest climates, Na accumulates and zeolites form in the lower topographic positions. Thus, he and others see a linkage of soils and landscapes along climatic transects. In all these studies, however, one

Humid tropical

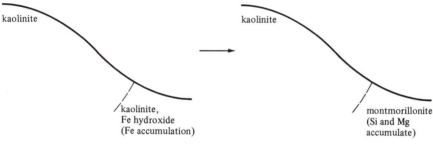

Humid tropical with dry season

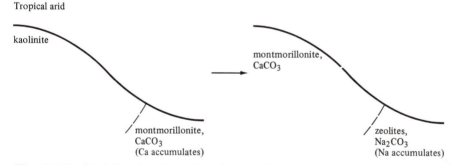

Tropical arid

Fig. 9-12 Variation in clay mineralogy with slope position in different climates. Within each climatic regime, the arrow points toward relations expected in the relatively drier climates. (Taken from Tardy and others,[51] Fig. 1.)

should be cautioned of the importance of adequate stratigraphic studies, for there are places where soils on portions of a slope are much older and better developed than those on the rest of the slope, and this will confound the interpretation of the catena properties.

Finally, one could expand catena studies to include the soilscape analysis of Hole and co-workers.[21,23,46] Soilscape is the soil portion of the landscape, and the analysis quantifies the soil pattern related to various landscapes. Examples of soilscape properties that can be quantified are the shapes and orientation of soil bodies, complexity of soil patterns as expressed as soil series/km^2, and boundary rankings between soil bodies. This latter puts greater weight on boundaries that separate mapping units at higher levels of soil classification. Variations in soilscapes are due to all the factors of soil formation, but if field areas are carefully selected, then most of the variation can be attributed to the topographic factor in different climatic settings.

REFERENCES

1. Al-Janabi, A.M., and Drew, J.V., 1967, Characterization and genesis of a Sharpsburg-Wymore soil sequence in southeastern Nebraska: Soil Sci. Soc. Amer. Proc., v. 31, p. 238–244.
2. Alperovitch, N., and Dan, J., 1972, Sodium-affected soils in the Jordan Valley: Geoderma, v. 8, p. 37–57.
3. Burke, R.M., and Birkeland, P.W., 1983, Toposequences and soil development as a relative dating tool on a two-fold chronosequence of eastern Sierra Nevada morainal slopes, California: Geol. Soc. Amer. Absts. with Prog., v. 15, no. 5, p. 327.
4. Burns, S.F., and Tonkin, P.J., 1982, Soil-geomorphic models and the spatial distribution and development of alpine soils, p. 25–43 *in* C.E. Thorn, ed., Space and time in geomorphology: Allen & Unwin, London.
5. Butler, B.E., 1959, Periodic phenomena in landscapes as a basis for soil studies: CSIRO Australia Soil Publ. No. 14.
6. Conacher, A.J., 1975, Throughflow as a mechanism responsible for excess soil salinization in non-irrigated, previously arable lands of the western Australian wheatland: A field study: Catena, v. 2, p. 31–68.
7. Conacher, A.J., and Dalrymple, J.B., 1977, The nine unit landsurface model: An approach to pedogeomorphic research: Geoderma, v. 18, p. 1–154.
8. Coultas, C.L., Clewell, A.F., and Taylor, E.M., Jr., 1979, An aberrant toposequence of soils through a titi swamp: Soil Sci. Soc. Amer. Jour., v. 43, p. 377–383.
9. Dan, J., 1981, Soils of the Arava Valley, p. 297–342 *in* J. Dan, R. Gerson, H. Koyumdjisky, and D.H. Yaalon, eds., Aridic soils of Israel: Agricultural Research Organization of Israel, Spec. Publ. No. 190.
10. Dan, J., Yaalon, D.H., and Koyumdjisky, H., 1968, Catenary soil relationships in Israel, 1. The Netanya catena on coastal dunes of the Sharon: Geoderma, v. 2, p. 95–120.
11. Daniels, R.B., and Gamble, E.E., 1967, The edge effect in some Ultisols in the North Carolina Coastal Plain: Geoderma, v. 1, p. 117–124.
12. Daniels, R.B., Gamble, E.E., and Nelson, L.A., 1967, Relation between A2 horizon characteristics and drainage in some fine loamy Ultisols: Soil Sci., v. 104, p. 364–369.
13. Daniels, R.B., Gamble, E.E., and Bartelli, L.J., 1968, Eluvial bodies in B horizons of some Ultisols: Soil Sci., v. 106, p. 200–206.

14. Daniels, R.B., Gamble, E.E., and Cady, J.G., 1971, The relation between geomorphology and soil morphology and genesis: Adv. in Agron., v. 23, p. 51–88.

15. Finney, H.R., Holowaychuk, N., and Heddleson, M.R., 1962, The influence of microclimate on the morphology of certain soils of the Allegheny Plateau of Ohio: Soil Sci. Soc. Amer. Proc., v. 26, p. 287–292.

16. Follmer, L.R., 1982, The geomorphology of the Sangamon surface: Its spatial and temporal attributes, p. 117–146 *in* C.E. Thorn, ed., Space and time in geomorphology: Allen & Unwin, London.

17. Frye, J.C., Shaffer, P.R., Willman, H.B., and Ekblaw, G.E., 1960a, Accretion gley and the gumbotil dilemma: Amer. Jour. Sci., v. 258, p. 185–190.

18. Frye, J.C., Willman, H.B., and Glass, H.D., 1960b, Gumbotil, accretion-gley, and the weathering profile: Ill. State Geol. Surv. Circ. 295, 39 p.

19. Gerrard, A.J., 1981, Soils and landforms: Allen & Unwin, London, 219 p.

20. Glazovskaya, M.A., 1968, Geochemical landscapes and types of geochemical soil sequences: Trans. 9th Internat. Cong. Soil Sci., v. 4, p. 303–312.

21. Habermann, G.M., and Hole, F.D., 1980, Soilscape analysis in terms of pedogeomorphic fabric: An exploratory study: Soil Sci. Soc. Amer. Jour., v. 44, p. 336–340.

22. Hamilton, R., 1964, Microscopic studies on laterite formation, p. 269–276 *in* A. Jongerius, ed., Soil micromorphology: Elsevier Publ. Co., New York, 540 p.

23. Hole, F.D., 1978, An approach to landscape analysis with emphasis on soils: Geoderma, v. 21, p. 1–23.

24. Huggett, R.J., 1975, Soil landscape systems: A model of soil genesis: Geoderma, v. 13, p. 1–22.

25. ———, 1976, Lateral translocation of soil plasma through a small valley basin in the Northaw Great Wood, Hertfordshire: Earth Surf. Proc., v. 1, p. 99–109.

26. Hussain, M.S., and Swindale, L.D., 1974, The physical and chemical properties of the gray hydromorphic soils of the Hawaiian Islands: Soil Sci. Soc. Amer. Proc., v. 38, p. 935–941.

27. Jenny, H., 1958, Role of the plant factor in the pedogenic functions: Ecology, v. 39, p. 5–16.

28. ———, 1980, The soil resource: Springer-Verlag, New York, 377 p.

29. Kantor, W., and Schwertmann, U., 1974, Mineralogy and genesis of clays in red-black soil toposequences on basic igneous rocks in Kenya: Jour. Soil Sci., v. 25, p. 67–78.

30. Kay, G.F., 1916, Gumbotil, a new term in Pleistocene geology: Science, v. 44, p. 637–638.

31. ———, 1931, Classification and duration of the Pleistocene period: Geol. Soc. Amer. Bull., v. 42, p. 425–466.

32. ———, and Pearce, J.N., 1920, The origin of gumbotil: Jour. Geol., v. 28, p. 89–125.

33. Leighton, M.M., and MacClintock, P., 1962, The weathered mantle of glacial tills beneath original surfaces in north-central United States: Jour. Geol., v. 70, p. 267–293.

34. Lotspeich, F.B., and Smith, H.W., 1953, Soils of the Palouse loess: I. The Palouse catena: Soil Sci., v. 76, p. 467–480.

35. Maignien, R., 1960, Review of research on laterites: UNESCO, Natural Resources Res. IV, 148 p.

36. Malo, D.D., Worcester, B.K., Cassel, D.K., and Matzdorf, K.D., 1974, Soil-landscape relationships in a closed drainage system: Soil Sci. Soc. Amer. Proc., v. 38, p. 813–818.

37. McKeague, J.A., 1965, Properties and genesis of three members of the uplands catena: Canad. Jour. Soil Sci., v. 45, p. 63–77.

38. Milne, G., 1935a, Some suggested units for classification and mapping, particularly for East African soils: Soil Res., Berlin, v. 4, p. 183–198.

39. ———, 1935b, Composite units for the mapping of complex soil associations: Trans. 3rd Internat. Cong. Soil Sci., v. 1, p. 345–347.

40. Mohr, E.C.J., van Baren, F.A., and van Schuylenborgh, J., 1972, Tropical soils: Mouton-Ichtiar Baru-Van Hoeve, The Hague, Netherlands, 481 p.

41. Muhs, D.R., 1982, The influence of topography on the spatial variability of soils in Mediterranean climates, p. 269–284 *in* C.E. Thorn, ed., Space and time in geomorphology: Allen & Unwin, London.

42. Nettleton, W.D., Flach, K.W., and Borst, G., 1968, A toposequence of soils on tonalite grus in the southern California Peninsular Range: U.S. Dept. Agri., Soil Cons. Serv., Soil Surv. Invest. Rep. no. 21, 41 p.

43. Nye, P.H., 1954, Some soil-forming processes in the humid tropics. I. A field study of a catena in the west African forest: Jour. Soil Sci., v. 5, p. 7–21.

44. Ollier, C.D., 1973, Catenas in different climates, p. 137–169 *in* E. Derbyshire, ed., Geomorphology and climate: John Wiley and Sons, New York.

45. Panabokke, C.R., 1959, A study of some soils in the dry zone of Ceylon: Soil Sci., v. 87, p. 67–74.

46. Pavlik, H.F., and Hole, F.D., 1977, Soilscape analysis of slightly contrasting terrains in southeastern Wisconsin: Soil Sci. Soc. Amer. Jour., v. 41, p. 407–413.

47. Radwanski, S.A., and Ollier, C.D., 1959, A study of an east African soil catena: Jour. Soil Sci., v. 10, p. 149–168.

48. Ruhe, R.V., 1956, Geomorphic surfaces and the nature of soils: Soil Sci., v. 82, p. 441–455.

49. ———, 1965, Paleopedology, p. 755–764 *in* H.E. Wright, Jr., and D.G. Frey, eds., The Quaternary of the United States: Princeton Univ. Press, Princeton, 922 p.

50. Ruhe, R.V., and Walker, P.H., 1968, Hillslope models and soil formation. I. Open systems: Trans. 9th Internat. Cong. Soil Sci., v. 4, p. 551–560.

51. Tardy, Y., Bocquier, G., Paquet, H., and Millot, G., 1973, Formation of clay from granite and its distribution in relation to climate and topography: Geoderma, v. 10, p. 271–284.

52. Tedrow, J.C.F., 1977, Soils of the polar landscapes: Rutgers Univ. Press, New Brunswick, N.J., 638 p.

53. Tonkin, P.J., Young A.W., McKie, D.A., and Campbell, A.S., 1977, Conceptual models of soil development and soil distribution in hill country, central South Island, New Zealand. Part I. The analysis of the changes in soil pattern: New Zealand Soc. Soil Sci., 25th Jubilee Conf., Summaries of presented papers, p. 25–27.

54. Trowbridge, A.C., 1961, Discussion: accretion-gley and the gumbotil dilemma: Amer. Jour. Sci., v. 259, p. 154–157.

55. Valentine, K.W.G., and Dalrymple, J.B., 1975, The identification, lateral variation, and chronology of two buried paleocatenas at Woodhall Spa and West Runton, England: Quaternary Res., v. 5, p. 551–590.
56. Vreeken, W.J., 1973, Soil variability in small loess watersheds: Clay and organic carbon content: Catena, v. 1, p. 181–196.
57. Walker, P.H., 1966, Postglacial environments in relation to landscape and soils on the Cary drift, Iowa: Iowa State Univ., Agri. and Home Econ. Exp. Sta. Res. Bull. 549, p. 835–875.
58. Walker, P.H., and Ruhe, R.V., 1968, Hillslope models and soil formation. II. Closed systems: Trans. 9th Internat. Cong. Soil Sci., v. 4, p. 561–568.
59. Watson, J.P., 1964, A soil catena on granite in southern Rhodesia. I. Field observations: Jour. Soil Sci., v. 15, p. 238–250. II. Analytical data: Jour. Soil Sci., v. 15, p. 251–257.
60. Whipkey, R.Z., and Kirby, M.J., 1978, Flow within the soil, p. 121–144 *in* M.J. Kirby, ed., Hillslope hydrology: John Wiley and Sons, Chichester.
61. Yaalon, D.H., 1975, Conceptual models in pedogenesis: Can the soil-forming functions be solved?: Geoderma, v. 14, p. 189–205.
62. Young, A., 1976, Tropical soils and soil survey: Cambridge University Press, Cambridge, 468 p.
63. Young, A.W., Tonkin, P.J., McKie, D.A., and Campbell, A.S., 1977, Conceptual models of soil development and soil distribution in hill country, central South Island, New Zealand. Part II. Chemical and mineralogical properties: New Zealand Soc. Soil Sci., 25th Jubilee Conf., Summaries of presented papers, p. 28–30.

10

Vegetation-soil relationships

The biotic factor in pedogenesis is difficult to assess because of the dependence of both vegetation and soil on climate and the interaction of soil and vegetation. Jenny[16,17,18] depicts the interrelationship of these three factors thus

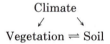

Here we are concerned with the lower part of the triangle, specifically, the influence of vegetation on the soil. Field sites can be found where vegetation is the most important variable producing differences in soil properties. In many places, these vegetational effects could be interpreted as microclimatological influences brought about by vegetation differences. As in the discussion of the other factors, all factors except vegetation will be kept constant. A constant climate is assumed by considering only the regional climate; in areas where this is constant, we can determine the influence of vegetation on the soils. Two aspects of the vegetation-soil relationship will be considered here: vegetation and soil morphology and vegetation and soil chemistry. In addition, the influence of former vegetation on a soil will be reviewed.

SOIL VARIATION AT THE FOREST-PRAIRIE BOUNDARY

An often-cited example of vegetation-soil relationships is the comparison of forested and grassland soils at the forest-prairie boundary of the midcontinent[1,16,29,33] and of Canada.[28] The study of White and Riecken[42] in the midcontinent is used here (Fig. 10-1). Alfisols are present in the deciduous forests, and they possess an A/E/B/C soil-horizon sequence.

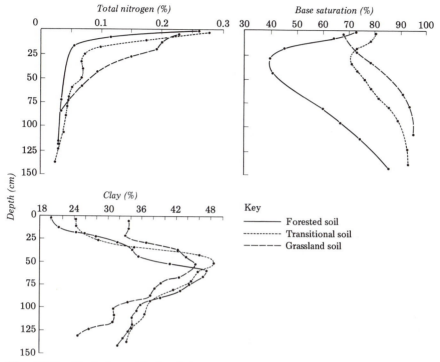

Fig. 10-1 Variation with depth in total nitrogen, percent base saturation, and clay content for soils formed under different vegetation covers. (Taken from White and Riecken,[42] Figs. 1, 3, and 5.)

Mollisols in the grassland possess an A/B/C soil-horizon sequence. Transitional soils commonly occur between these two kinds of soils in the field, and their properties are intermediate between the two end members.

Several diagnostic soil properties are closely related to differences in vegetation. The most obvious one is organic matter distribution. Although all the soils have an equally high content of organic matter at the surface, the distribution with depth varies with vegetation. Forested A horizons are relatively thin, and the organic matter content decreases rapidly with depth, whereas grassland A horizons are thick, and the organic matter content remains high for a considerable depth. These differences occur partly because of the manner in which organic matter enters the soil. In forested soils, the main input of organic matter is by litterfall to the soil surface, whereas in grassland soils the organic matter input is both by litterfall and by root decay at depth. The overall percentage base saturation tends to be higher in grassland than in forests. This difference could be a function of greater leaching in the forests along with greater annual biomass production and cycling of cations in grasslands. The greater leaching in the forests relative to the grasslands could be due, in part, to lower evapotranspiration rates and the presence of more chelating agents and more acid leaching waters under a forest canopy. In addition, the ratio of free Fe_2O_3 in the A and/or E horizons relative to that in the B horizons is lower in forested than in grassland soils. The origin of this difference probably lies in the differences in chelating ability of the organic compounds formed during decomposition in the two environments. Another striking morphological feature is clay distribution with depth. The surface layers of grassland soils have higher clay contents than do comparable layers in forested soils, indicating that the environment in the forests is such that clay particles can be translocated to a greater extent than in the grasslands.

Thus, differences in soils at the forest-prairie boundary are striking. To review, forested sites commonly show greater leaching of cations, correspondingly lower pHs, and greater clay translocation than do adjacent grassland sites. In addition, organic matter content is higher at greater depths in the grassland sites.

A detailed study of the chemistry of soils at the forest-prairie boundary in western Washington reveals some significant trends.[40] The grassland soils have more carbon, about twice as much nitrogen, and about one-half the C/N ratio relative to the forested soils. Some of the Fe and Al data do not group as one might expect. However, pyrophosphate-extractable Fe seems higher in the forest soils, whereas pyrophosphate-extractable Al is higher in the grassland soils. Amorphous Fe is higher in the forest soils and displays an accumulation bulge in two of the profiles; pH and clay mineralogy are not affected by the vegetation differences.

If the forest-prairie boundary shifts, then the soil properties under the new vegetation will be somewhat out of phase with the vegetation; this leads to soils with properties transitional between the end members discussed above[42] (Fig. 10-1). It is conceivable that organic matter distribution and base saturation with depth in the transitional soils would be the same if the change at the site was either forest to grassland or the reverse. The time necessary to reach the new steady state for these properties probably is in the order of 10^3 years or more. If the change were grassland to forest, iron and clay particles would start to move, and the time necessary to reach profiles characteristic of the forest is not known. If, on the other hand, the change were forest to grassland, the iron and clay profiles might persist, and these can be cautiously used to indicate former vegetational influences. In this latter case, the E horizon might persist if the grass roots are sufficiently shallow.

VARIATION IN SOIL WITH DISTANCE FROM TREES AND WITH TREE SPECIES

Within a constant regional climate, tree species will vary from site to site, and thus soil properties may vary from site to site. Some soil variations are quite subtle and can be observed only as changes in soil chemistry; these include pH or exchangeable cations. Other soil variations are quite striking and can be observed as changes in soil morphology. An extreme example of the latter variation is found in New Zealand, where local podzolization occurs under *Podocarpus sp.*[14] and Kauri pine *(Agathis australis.)*[8] The influence of the tree is such that the Spodosols occur only under each tree and not beyond the influence of the tree. Here we will present data on soil properties with distance outward from several tree species.

Zinke[47] has demonstrated the effect of a single shore pine tree, *Pinus contorta,* on surface soil properties. The tree was 45 years old and growing in a sand dune containing shell material. Some marked changes in soil properties are associated with distance from the tree and whether the sample is taken under or beyond the tree crown (Fig. 10-2). The soil is much more acid beneath the tree relative to the nonvegetated sand, the difference in pH being 1.5 or more. Nitrogen is almost an order of magnitude greater under the tree than it is beyond the influence of the tree. Because cation exchange capacity is derived mostly from the organic matter, it follows the nitrogen trends, being about twice as great beneath the tree (approximately 28 me/100 g soil) as beyond the tree (approximately 16 me/100 g soil).

Zinke[47] also studied surface-soil variation with radial distance out from several different tree species and found that all the species studied had similar trends (Fig. 10-3). Differences between species, although evident, may not reflect only a species effect because samples were taken from sites throughout California, and other soil-forming factors may not have been

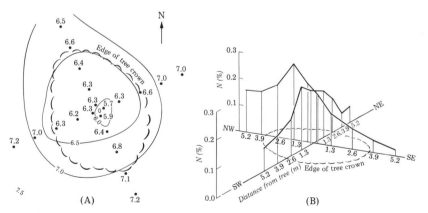

Fig. 10-2 Variation in surface-soil properties with distance from a pine tree *(Pinus contorta)*. (A) Recorded pH values and isolines of approximately equivalent pH values. The NW-SE diameter is approximately 8 m. (B) Nitrogen content in percent by weight. Samples were taken from the 0 to 6.40-cm-depth layer. High values extend NW beyond the tree crown edge, probably because of the prevailing winds. (Taken from Zinke,[47] Figs. 1 and 2, by permission of the Duke University Press, Durham, North Carolina.)

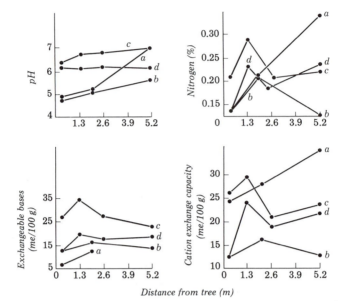

Fig. 10-3 Variation in surface soil properties with distance from old forest trees of different species in California. Samples were taken from the 0 to 6.4-cm-depth layer. The outer sampling point always extends beyond the influence of the crown canopy, whose edge is 2.6 to 5.2 m from the trunk. The tree species are (a) and (b) ponderosa pine *(Pinus ponderosa)*, (c) incense cedar *(Libocedrus decurrens)*, and (d) Douglas fir *(Pseudotsuga menziesii)*. The high nitrogen value for (a) is due to sampling near a nitrogen-fixing plant. (Taken from Zinke,[47] Fig. 4, by permission of the Duke University Press, Durham, North Carolina.)

the same. The general trend, however, is decreasing pH toward the trees. Nitrogen content, exchangeable bases, and cation exchange capacity all are low near the tree stems, increase to a maximum some distance from the stem, but within the area covered by the crown, and generally decline outward.

The above trends are interpreted as being due primarily to the influence on the soil of bark and leaf fall near the tree trunk, relative to no tree litter in the opening between the trees.[47] Bark litter predominates in a ring surrounding the tree closest to the stem, and leaf litter predominates in a ring out to about the edge of the crown. The trends in soil properties, therefore, are a function of the predominance either of bark or leaf litter. Of the two kinds of litter, bark litter is the more acid and has a lower cation and nitrogen content. Thus the soils directly reflect the composition of the litter. Zinke[47] also suggests that the water that reaches the soil, whether by stemflow, drip from foliage, or without vegetation interception, may exert some control on the soil-property trends because these waters differ in their chemical composition. No data, however, were reported on the composition of these waters.

Variation in soil properties outward from a beech tree, *Fagus gran-difolia,* in the midcontinent provides data on the influence of stemflow water on pedogenesis.[11,12] The parent material is till containing $CaCO_3$, and the soil formed is a Udalf. Trends with distance from the stem are quite apparent and vary more for chemical properties than for physical properties (Fig. 10-4). Close to the tree the B horizons are more mottled, and the E horizons thicker; there is less fine clay in the B horizons relative to sites 200 cm from the tree. The main trends in chemical properties in going away from the tree are decreasing amounts of total nitrogen and organic carbon, increasing percentage base saturation, increasing pH and decreasing exchangeable H^+, and higher free Fe_2O_3 along with greater differentiation of Fe_2O_3 with depth.

The above trends reflect the amount and composition of stemflow water. Stemflow water is that water derived from rainfall that moves along the branches to the tree stem and then down the stem to the ground. Thus, a larger quantity of water can enter the soil near a tree stem than can enter a soil located beyond tree-stem influences. Gersper and Holowaychuk[12] estimate that the soil at the tree stem can, by this means, receive as much as five times the quantity of water than can a soil in the open that receives its water from rainfall alone. Furthermore, they cite literature indicating that water in contact with vegetation has a much higher content of chemical elements than has rainfall, and that stemflow water has a higher content of elements than water that drips from the crown. The estimated amount of material added to the soil each year is quite large (Table 10-1). A major source of the elements in stemflow water is thought to be dust and other matter adsorbed on tree surfaces, as well as insects and bark washed from the tree. Gersper and Holowaychuk[12] feel that the main factors responsible for soil trends are variations in influx of water and chemical elements delivered to the soil next to the tree stem relative to influx some distance from the tree stem. Designated "biohydrologic" soil-forming factor, it is important in those soils in which the interaction of precipitation with vegetation affects soil formation.

The relative importance of stemflow or of litterfall on soil variation with distance from a specific tree seems to vary with the nature of the bark.[13] Trees with smooth bark have high rates of stemflow, and thus soil-property variations are largely controlled by stemflow. However, trees with rough bark (such as those reported by Zinke, above) have low rates of stemflow, and soil-property variation with distance from the tree depends on the amount and composition of the litterfall.

Many workers have recognized that some soil properties vary with tree species. Perhaps the data with the best environmental control are those from plantations in which various tree species have been growing for several decades.[7,24–27] Organic carbon and total nitrogen, pH, and the exchangeable cations commonly vary with tree species. Ovington's[24–27] work shows, in a very general way, that soils under conifers have greater

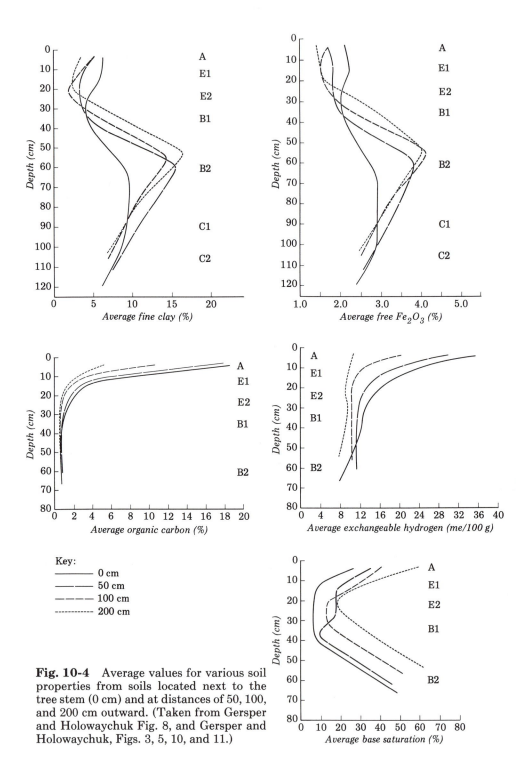

Fig. 10-4 Average values for various soil properties from soils located next to the tree stem (0 cm) and at distances of 50, 100, and 200 cm outward. (Taken from Gersper and Holowaychuk Fig. 8, and Gersper and Holowaychuk, Figs. 3, 5, 10, and 11.)

**Table 10-1
Estimated Annual Quantities of
Elements Added to a Udalf
Through Stemflow near the
Stem of an American Beech
Tree** *(Fagus grandifolia),* **per
Hectare**

Element	Quantity added (kg/ha)
C	10,385
Ca	1,033
K	842
Na	202
Mg	191
P	112
Fe	22

(Taken from Gersper and Holowaychuk,[12] Table 2.)

amounts of organic carbon and total nitrogen; lower pH values; greater amounts of sodium, potassium, and phosphorus; and lower amounts of calcium and magnesium, relative to the soils under deciduous trees. These soil properties are among those that can change most rapidly with changing environmental conditions. Thus, if a new tree species occupies the soil site formerly occupied by another tree species, the soil properties probably would alter rather quickly (10^2–10^3 years) to values determined by the new tree species. In short for these properties, the influence of former tree species on the soil site may not linger long after a change in species.

For older soils, there can be more marked differences in soils with tree species. Such a comparison was made for B subhorizons in Spodosols and spodosolic soils developed in till in southeastern Quebec and Maine.[9] Annual rainfall is about 100 cm; soils developed under conifers (fir and spruce) are compared with those developed under deciduous vegetation (maple and beech). In general, hues are redder under conifers than under deciduous trees. Organic carbon and pH are somewhat similar under both vegetation types, but the lowest single pH values are associated with conifers. Most dithionite- and pyrophosphate-extractable Fe and Al in the soils are significantly higher under the conifers (Table 10-2). Within the fulvic-acid fraction, the contents of C, Fe, and Al are generally higher under conifers, and these values make up over 50 percent of the total carbon, and over 90 percent of the pyrophosphate-extractable Fe and Al. These data suggest that the podzolization process is more intense under conifers than under deciduous trees.

Table 10-2
Variation in Mean Values (percent) of Soil Extracts* in B
Subhorizons with Vegetation Type

		SOIL				FULVIC ACIDS		
		Fe_d	Al_d	Fe_p	Al_p	C	Fe	Al
Conifers	Bs1	5.0***	1.1**	2.5**	0.83**	3.2***	2.3**	0.76**
	Bs2	3.6	1.8***	1.3	1.3**	2.9	1.1	1.2**
Deciduous trees	Bs1	2.7***	0.39**	1.7**	0.39**	1.5***	1.6**	0.35**
	Bs2	3.0	0.76***	1.6	0.71**	2.0	1.5	0.65**

*d, dithionite extract; p, pyrophosphate extract.
**Differences between the same horizon under different vegetation are significant at the 0.05 level.
***Differences between the same horizon under different vegetation are significant at the 0.01 level.
(Taken from DeKimpe and Martel,[9] Tables 2 and 3.)

DECIPHERING PAST VEGETATIONAL CHANGE BY SOIL PROPERTIES

One goal of the study of Quaternary soils is to assess past vegetational change by the examination of soil properties. One way of doing this is to establish the relationship between present vegetation and soils and then to examine carefully both surface and buried soils to determine if their properties might indicate formation under a past vegetation different from the present vegetation. One would hope to go one step further in the interpretation and correlate past changes in vegetation with past changes in climate. Sorenson and others[35] have attempted this for east-central Texas, based partly on soil morphology, but as they point out, their conclusions are tentative and require analysis of additional soil chemical and mineralogical properties.

One of the areas in which distinctly different soil patterns are closely linked with vegetation is at the forest-grassland boundary. As discussed earlier, soil properties in many places change markedly across the vegetational boundary. Under ideal conditions, Spodosols or Alfisols with an E horizon form in the forest, and Mollisols or Inceptisols form in the grasslands. Whether or not a change in past vegetation could be determined by soil properties would depend on the time elapsed since the vegetational change and on the ability of a specific soil property to persist relatively unchanged under the new conditions.

The time required for adjustment to a new environment depends upon the specific surface-soil property. Properties like pH and exchangeable bases change quickly and thus carry little if any legacy of past vegetation. The amount and distribution of organic matter with depth probably would persist for longer periods of time, but perhaps not as long as it would take to reach steady-state values for that environment beginning with an unvegetated, unweathered surface (10^{2+} to 10^{3+} years). For example, in those ecosystems where the rate of turnover of organic matter is rapid, the change to new steady-state values would be rapid. An example of this might be the grassland areas of the humid midcontinent. If, however, the rate of turnover of organic matter is slow, such as in the forested and tundra ecosystems of the northern latitudes or of high altitudes, changes with new vegetation would be quite slow, and properties related to the former vegetation might persist for thousands of years. Furthermore, if the change is from forest to grassland, the E horizon may persist for long periods of time if the grassland A horizon is comparable in thickness to the forest A horizon or if turnover rates of organic matter are slow in the grassland ecosystem. Jungerius[20] discusses some of these changes in vegetation and soils in southern Alberta and points out that in places the platy structure of the E horizon of the former forested soil can persist in present-day Mollisols and thus provide a clue to past vegetational patterns. In contrast, a change from grassland to forest can be accompanied by the formation of an incipient E horizon at the base of the grassland A horizon.[31]

Soil chemical extracts and clay mineralogy might also be used for paleovegetation reconstruction based on soil properties. For example, trends in Fe and Al are predictable in some podzolic environments, and if these were expressed strongly enough, they would persist in a subsequent grassland environment or perhaps even in a forest environment with different tree species. I do not know if the Fe_o/Fe_d ratio would persist unchanged, however. Phosphorus extracts exhibit good trends with leaching environment, and perhaps also with vegetation, so perhaps they could be used. Clay minerals are good vegetation indicators in some places. In the Colorado Front Range, for example, vermiculite tends to be associated with conifer forests.[23] Where it is found in tundra soils a short distance above present-day treeline, it could point to a former higher treeline.[32] Careful analysis of beidellite occurrence might also be helpful in paleovegetation studies, as it could form in a podzolic E horizon and persist with a subsequent change in vegetation.

Chances for deciphering past vegetational trends farther back in time ($> 10^4$ years) probably are best for buried soils because these can be compared with the present-day surface soils in the same region. Soil morphological features seem to provide the best data for these interpretations. Ruhe and Cady[30] compared the morphology of the buried Sangamon and late-Sangamon soils in Iowa with that of the surface Mollisols and concluded that the former, because they contain an E horizon, had formed under a forest cover. Jungerius[20] compared surface- and buried-soil morphology to estimate treeline fluctuations in Alberta since the Wisconsin glaciation. Sorenson and others[34] used the distribution and morphology of surface and buried soils west of Hudson Bay to demonstrate fluctuations in the forest-tundra boundary over the past 6000 years (Fig. 10-5). Spodosols are formed in the forests, whereas Inceptisols with A/C profiles are formed on well-drained sites in the tundra, a distribution similar to that seen in Alaska.[38] The latitudinal distribution of these contrasting soil morphologies, therefore, provides data on the former positions of the forest-tundra boundary.

Although the interpretation that non-forested soils with E horizons north of present treeline were forested in the past probably is correct for many places, some E horizons in well-drained soils do not seem to require a forest cover for their formation. For example, some surface soils in the tundra zone just north of the treeline show evidence of weak podzolization.[37,38] In these cases, either the forest-tundra border was farther north in the recent past or podzolization can take place in well-drained sites in the southern part of the tundra zone in the absence of trees.

Other soils with E horizons in northern regions seem to have formed under non-forested conditions. For example, James[15] reports on soils with an E horizon 480 km north of the present treeline along the west coast of Hudson Bay that probably formed under the present-day, lichen-heath vegetation. Spodosol-like soils with E horizons are present in Greenland,[39] again far north of the treeline. In addition, Spodosol-like soils are found on the north slope of the Brooks Range under dwarf heath vegetation.[5,6] Finally, in a recent detailed study, also in northern Alaska, Ugolini and others[41] used an array of morphological, chemical, and isotopic data to demonstrate that spodosolic soils developed under lichens and mixed-heath vegetation cannot be distinguished from those developed under

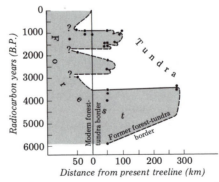

Fig. 10-5 Tentative Holocene migrations of forest-tundra boundary, southwest Keewatin, Canada. (Taken from Sorenson and others.[34]) Filled circles denote radiocarbon dates.

spruce trees. Thus, one should be cautious in relating all non-forested, Spodosol-like soils to formation under a forest vegetation.

It might be possible to relate variations in soil morphology to fluctuation in the forest-tundra boundary in the high mountains. Bliss and Woodwell[3] report on the presence of Spodosols both under fir-spruce *(Abies-Picea)* krummholz and at a nearby site under sedge-heath vegetation. Spodosols also occur above treeline in the European Alps[4,21] and in Scotland.[36] These alpine Spodosols apparently formed under the present-day vegetation. Work by myself and students in various parts of the Rocky Mountains, however, has failed to demonstrate the presence of E horizons in surface soils above the present treeline; a suggested interpretation is that the treeline has not gone to altitudes higher than the present in the recent past. Thus, although soils above treeline generally lack E horizons, they have been reported in certain environments, and one has to be cautious in using only the presence or absence of an E horizon in the reconstruction of former treelines.

BIOGENIC OPAL IN SOILS AS AN INDICATOR OF PAST VEGETATION

Many plants secrete opal that conforms to the shape of the cell from which it forms; upon decomposition, the opal remains in the soil.[22,44] These opal forms are called phytoliths. Wilding and others[45] provide a useful review of the application of these to soils studies. Phytolith morphologies vary with plant species, and more work needs to be done on the taxonomy of these forms. They appear to persist for long periods of time, however, as shown by radiocarbon dates of about 13,000 years for phytoliths in Ohio.[43] Dissolution of these takes place at a rate that depends on vegetation type.[2]

Forested and grassland soils differ in their phytolith content, and this difference in content can be used to reconstruct past vegetation. Grasslands produce more phytoliths per unit area than do forests, and these differences show up in the phytolith contents of the surface soils. To decipher vegetational change, therefore, one has to sample a transect across the two vegetation types. If the change in phytolith content coincides with the present forest-grassland boundary, the interpretation is that the boundary has been stable for some time.[46] If, however, both the forested and grassland soils have similar phytolith contents, the boundary apparently has shifted. In some areas, the vegetational change seems to have been from grassland to forests,[19] whereas in others, the evidence favors a shift from forests to grassland.[44]

Phytoliths are also useful in deciphering the environment of formation for buried soils. Dormaar and Lutwick[10] used phytoliths in combination with the infrared spectra of humic acids to identify buried soils as such and to reconstruct their vegetational history. Because it is difficult to reconstruct the environment of formation for buried soils from morphology alone, phytoliths, which can persist long after the soil organic matter has been removed by processes following burial, may provide important clues to past vegetation.

REFERENCES

1. Bailey, L.W., Odell, R.T., and Boggess, W.R., 1964, Properties of selected soils developed near the forest-prairie border in east-central Illinois: Soil Sci. Soc. Amer. Proc., v. 28, p. 257–263.
2. Bartoli, F., and Wilding, L.P., 1980, Dissolution of biogenic opal as a function of its physical and chemical properties: Soil Sci. Soc. Amer. Jour., v. 44, p. 873–878.
3. Bliss, L.C., and Woodwell, G.M., 1965, An alpine podzol on Mount Katahdin, Maine: Soil Sci., v. 100, p. 274–279.
4. Bouma, J., Hoeks, J., van der Plas, L., and van Scherrenburg, B., 1969, Genesis and morphology of some alpine podzol profiles: Jour. Soil Sci., v. 20, p. 384–398.
5. Brown, J., 1966, Soils of the Okpilak River region, Alaska: Cold Regions Res. and Engr. Lab. (Hanover, N.H.), Res. Report 118, 49 p.
6. Brown, J., and Tedrow, J.C.F., 1964, Soils of the northern Brooks Range, Alaska: 4. Well-drained soils of the glaciated valleys: Soil Sci., v. 97, p. 187–195.
7. Challinor, D., 1968, Alteration of surface soil characteristics by four tree species: Ecol., v. 49, p. 286–290.
8. Crocker, R.L., 1952, Soil genesis and the pedogenic factors: Quat. Rev. Biol., v. 27, p. 139–168.
9. De Kimpe, C.R., and Martel, Y.A., 1976, Effects of vegetation on the distribution of carbon, iron, and aluminum in the B horizons of northern Appalachian Spodosols: Soil Sci. Soc. Amer. Jour., v. 40, p. 77–80.
10. Dormaar, J.F., and Lutwick, L.E., 1969, Infrared spectra of humic acids and opal phytoliths as indicators of palaeosols: Can. Jour. Soil Sci., v. 49, p. 29–37.
11. Gersper, P.L., and Holowaychuk, N., 1970a, Effects of stemflow water on a Miami soil under a beech tree: I. Morphological and physical properties: Soil Sci. Soc. Amer. Proc., v. 34, p. 779–786.
12. Gersper, P.L., and Holowaychuk, N., 1970b, Effects of stemflow water on a Miami soil under a beech tree: II. Chemical properties: Soil Sci. Soc. Amer. Proc., v. 34, p. 786–794.
13. Gersper, P.L., and Holowaychuk, N., 1971, Some effects of stem flow from forest canopy trees on chemical properties of soils: Ecol., v. 52, p. 691–702.
14. Jackson, M.L., and Sherman, G.D., 1953, Chemical weathering of minerals in soils: Advances in Agron., v. 5, p. 219–318.
15. James, P.A., 1970, The soils of the Rankin Inlet area, Keewatin, N.W.T., Canada: Arctic and Alpine Res., v. 2, p. 293–302.
16. Jenny, H., 1941, Factors of soil formation: McGraw-Hill, New York, 281 p.
17. ———, 1958, Role of the plant factor in the pedogenic functions: Ecol., v. 39, p. 5–16.

18. ——, 1980, The soil resource: Springer-Verlag, New York, 377 p.

19. Jones, R.L., and Beavers, A.H., 1964, Variation of opal phytolith content among some great soil groups of Illinois: Soil Sci. Soc. Amer. Proc., v. 28, p. 711–712.

20. Jungerius, P.D., 1969, Soil evidence of postglacial tree line fluctuations in the Cypress Hills area, Alberta, Canada: Arctic and Alpine Res., v. 1, p. 235–245.

21. Kubiëna, W.L., 1953, The soils of Europe: Thomas Murby and Co., London, 314 p.

22. Lutwick, L.E., 1969, Identification of phytoliths in soils, p. 77–82 *in* S. Pawluk, ed., Pedology and Quaternary Research: Univ. Alberta Printing Dept., Edmonton, 218 p.

23. Netoff, D.I., 1977, Soil clay mineralogy of Quaternary deposits in two Front Range-Piedmont transects, Colorado: Unpubl. Ph.D. thesis, Univ. of Colorado, Boulder, 169 p.

24. Ovington, J.D., 1953, Studies of the development of woodland conditions under different trees. I. Soils pH: Jour. Ecol., v. 41, p. 13–34.

25. ——, 1956, Studies of the development of woodland conditions under different trees. IV. The ignition loss, water, carbon and nitrogen content of the mineral soil: Jour. Ecol., v. 44, p. 171–179.

26. ——, 1958a, Studies of the development of woodland conditions under different trees. VI. Soil sodium, potassium and phosphorus: Jour. Ecol., v. 46, p. 127–142.

27. ——, 1958b, Studies of the development of woodland conditions under different trees. VII. Soil calcium and magnesium: Jour. Ecol., v. 46, p. 391–406.

28. Pettapiece, W.W., 1969, The forest grassland transition, p. 103–113 *in* S. Pawluk, ed., Pedology and Quaternary Research: Univ. Alberta Printing Dept., Edmonton, 218 p.

29. Ruhe, R.V., 1969, Soils, paleosols, and environment, p. 37–52 *in* W. Dort, Jr., and J.K. Jones, Jr., eds., Pleistocene and Recent environments of the central Great Plains: Univ. Press of Kansas, Lawrence, 433 p.

30. Ruhe, R.V., and Cady, J.C., 1969, The relation of Pleistocene geology and soils between Bentley and Adair in southwestern Iowa: U.S. Dept. Agri. Tech. Bull. 1349, p. 1–92.

31. Sawyer, C.D., and Pawluk, S., 1963, Characteristics of organic matter in degrading chernozemic surface soils: Canada Jour. Soil Sci., v. 43, p. 275–286.

32. Shroba, R.R., and Birkeland, P.W., 1983, Trends in late-Quaternary soil development in the Rocky Mountains and Sierra Nevada of the western United States, p. 145–156 *in* S.C. Porter, ed., Late-Quaternary environments of the United States. Vol. 1. The late Pleistocene: University of Minnesota Press, Minneapolis.

33. Smith, G.D., Allaway, W.H., and Riecken, F.F., 1950, Prairie soils of the upper Mississippi Valley: Advances in Agron., v. 2, p. 157–205.

34. Sorenson, C.J., Knox, J.C., Larsen, J.A., and Bryson, R.A., 1971, Paleosols and the forest border in Keewatin, N.W.T.: Quaternary Res., v. 1, p. 468–473.

35. Sorenson, C.J., Mandel, R.D., and Wallis, J.C., 1976, Changes in bioclimate inferred from palaeosols and palaeohydrologic evidence in east-central Texas: Jour. Biogeogr., v. 3, p. 141–149.

36. Stevens, J.H., and Wilson, M.J., 1970, Alpine podzol soils on the Ben Lawers massif, Perthshire: Jour. Soil Sci., v. 21, p. 85–95.
37. Tedrow, J.C.F., 1968, Pedogenic gradients of the polar regions: Jour. Soil Sci., v. 19, p. 197–204.
38. Tedrow, J.C.F., Drew, J.V., Hill, D.E., and Douglas, L.A., 1958, Major genetic soils of the arctic slope of Alaska: Jour. Soil Sci., v. 9, p. 33–45.
39. Ugolini, F.C., 1966, Soils of the Mesters Vig District, northeast Greenland. 1. The arctic brown and related soils: Meddelelser om Grønland, Bd. 176, no. 1, p. 1–22.
40. Ugolini, F.C., and Schlichte, A.K., 1973, The effect of Holocene environmental changes on selected western Washington soils: Soil Sci., v. 116, p. 218–227.
41. Ugolini, F.C., Reanier, R.E., Rau, G.H., and Hedges, J.I., 1981, Pedological, isotopic, and geochemical investigations of the soils at the boreal forest and alpine tundra transition in northern Alaska: Soil Sci., v. 131, p. 359–374.
42. White, E.M., and Riecken, F.F., 1955, Brunizem-gray brown podsolic soil biosequences: Soil Sci. Soc. Amer. Proc., v. 19, p. 504–509.
43. Wilding, L.P., 1967, Radiocarbon dating of biogenic opal: Science, v. 156, p. 66–67.
44. Wilding, L.P., and Drees, L.R., 1969, Biogenic opal in soils as an index of vegetative history in the Prairie Peninsula, p. 96–103 *in* R.E. Bergstrom, ed., The Quaternary of Illinois: Univ. Illinois Coll. Agri. Spec. Publ. no. 14, 179 p.
45. Wilding, L.P., Smeck, N.E., and Drees, L.R., 1977, Silica in soils: Quartz, cristobalite, tridymite, and opal, p. 471–552 *in* J.B. Dixon and S.B. Weed, eds., Minerals in soil environments: Soil Sci. Soc. Amer., Madison, Wisconsin.
46. Witty, J.E., and Knox, E.G., 1964, Grass opal in some chestnut and forested soils in north central Oregon: Soil Sci. Soc. Amer. Proc., v. 28, p. 685–688.
47. Zinke, P.J., 1962, The pattern of individual forest trees on soil properties: Ecol., v. 43, p. 130–133.

11

Climate-soil relationships

The climatic factor is considered by many to be the most important factor in determining the properties of many soils. This climate-soil relationship can be seen by comparing the U.S. soil map (Fig. 2-4) with a map of precipitation and temperature (Appendix 2); most soil orders and suborders are restricted to certain climatic regions. In this chapter, we will look into some of these large-scale relationships, mention specific soil properties that are related to climate, and discuss those properties that might be most useful in reconstructing past climates.

Moisture and temperature are the two aspects of climate most important in controlling soil properties. Moisture is important because water is involved in most of the physical, chemical, and biochemical processes that go on in a soil, and the amount of moisture delivered to the soil surface influences the weathering and leaching conditions with depth in the soil. Temperature influences the rate of chemical and biochemical processes. Jenny[58] demonstrates how to derive separate functions for precipitation and for temperature in some areas, provided other aspects of the climatic factor and the other soil-forming factors can be considered constant.

A point discussed in parts of this chapter is that when one compares some soil properties in quite different climates, pedologic differences that one might predict from the climatic data are not always apparent. This causes some problems in explaining, but it is good for stratigraphic studies in which soils are used in correlation.

CLIMATIC PARAMETERS

A numerical value for the climate can be used to demonstrate, quantitatively, the functional relationships between climate and the various soil properties. In places, the mean annual values of either precipitation or temperature can be used as an approximation of the climate. The use of mean annual values, however, fails to take into account the monthly distribution of precipitation and temperature (Fig. 11-1). It is important to know, for example, whether or not precipitation is seasonal, and if it is, whether the precipitation maximum coincides with the annual temperature maximum or minimum, because these climatic variables strongly influence soil leaching and soil-water chemistry, both of which are important in determining key soil properties. For example, some areas have strong seasonal contrasts with relatively wet winters and dry summers. Under such a fluctuating moisture regime, it could be that kaolinite is the clay mineral stable with the chemistry of the winter soil waters, whereas

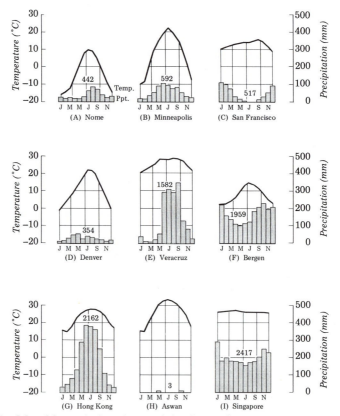

Fig. 11-1 Monthly variations in precipitation and temperature for various climatic stations around the world. Precipitation given on each graph (mm) is total mean annual. (Reproduced by kind permission of John Bartholomew & Son Ltd. from *The Times Atlas of the World*,[107] pp. XXVIII–XXIX.) Climate and altitude of each station as follows. (A) Tundra, 7 m; (B) continental, 285 m; (C) Mediterranean, 47 m; (D) steppe, 1613 m; (E) tropical savanna, 16 m; (F) marine, 43 m; (G) humid subtropical, 33 m; (H) desert, 111 m; (I) tropical rain forest, 5 m. Upper curve, temperature; histogram, precipitation.

smectite is stable with the summer soil waters. If, in studying clay-mineral stability, we collect soil waters that are representative only of the mean annual conditions, we might err in our interpretation. Few such detailed analyses have been made, but the point is that many of the details of the soil-climate connection still have to be worked out.

Water balance can be used to describe some of the climatic characteristics of a region that are important to soil formation.[2,3,4] This balance represents the gains and losses of soil moisture over a certain interval of time (week, month, year). Gains are from precipitation (P); these data come from climatological stations. Losses, however, are from evapotranspiration; values for potential evapotranspiration (ET_p) are calculated by the method of Thornthwaite and Mather.[108] If Manhattan, Kansas is used as an example (Table 11-1), the water balance shows the approximate monthly variation in soil moisture. For those months with positive values for ($P - ET_p$), water is stored in the soil at moisture values above permanent wilting point; this water is available for plant growth, weathering reactions, and if the water is moving, it can translocate material within the soil. In contrast, months with negative values of ($P - ET_p$) are marked by the removal of water from the soil. Water removal can go on until the soil reaches water contents approaching permanent wilting

Table 11-1
Water Balance at Manhattan, Kansas, for a Soil with 10.2-cm Available Water-Holding Capacity

	J	F	M	A	M	J	J	A	S	O	N	D	ANNUAL
Potential evapotranspiration (ET_p)	0.0	0.0	1.9	5.1	9.3	13.9	16.4	15.0	10.1	5.2	1.4	0.0	78.3 cm
Precipitation (P)	2.0	3.0	3.8	7.1	11.1	11.7	11.5	9.5	8.6	5.8	3.8	2.2	80.1 cm
Soil-moisture gains, $P - ET_p = (+)$	2.0	3.0	1.9	2.0	1.8					0.6	2.4	2.2	15.9 cm*
Soil-moisture losses, $P - ET_p = (-)$						2.2	4.9	5.5	1.5				14.1 cm
Soil-moisture storage at end of month (A and B horizons)	7.2	10.2	10.2	10.2	10.2	8.0	3.1	0.0	0.0	0.6	3.0	5.2	
Possible deep percolation below B horizon**			1.9	2.0	1.8								5.7 cm

*This annual value is the leaching index of Arkley.[3]
**In this case it is assumed that the water-holding capacity of the A and B horizons is 10.2 cm. Therefore any surplus of water over 10.2 cm can move to greater depths in the soil.

point. If the water-holding capacity of the soil is known, the amount of water percolating to depths beneath the B horizon can be calculated.

Plots of water balances for several different climatic regions indicate significant soil-moisture differences of pedologic importance (Fig. 11-2). For example, both the California coast (Half Moon Bay) and the California Great Valley (Sacramento) are characterized by soil-moisture buildup for about the same months, but a greater amount of water leaches through the soil at the coastal site than at the inland site. Fallon, Nevada, located in the rain shadow of the Sierra Nevada, is characterized by a short period of soil-moisture storage and slight leaching; thus, soil profiles there are shallow and slightly leached and contain Bk horizons. However, Manhattan, Kansas is in an area of summer precipitation and thus is characterized by fairly high soil-moisture contents throughout much of the year, including some of the warm summer months. These data point out the basic differences between the climate of the midcontinent and that of California and the Basin and Range Province; that is, in the midcontinent, the soils are moist during much of the warm season, whereas the western areas are winter wet and summer dry. From this it can be predicted that the rates of biological and chemical processes will vary between these regions. Compared with western sites, sites in the midcontinent should produce more above-ground organic matter annually, and they should have higher rates of organic matter decomposition, mineral weathering, and clay-mineral transformation.

The data on water balance and water-holding capacity of the soil can be combined to approximate water movement with depth in soils.[3] Knowing the frequency of wettings per year, and assuming that downward movement of water takes place only when field capacity for that part of the soil has been reached and that moisture can be removed from the soil by evapotranspiration down to permanent wilting point, one can construct curves depicting water movement (Fig. 11-3). These curves then can be compared with soil data, such as clay-mineral variation with depth, or the top of the Bk horizon, in order to determine which soil features can be attributed to present-day water movement and which may have formed under some past water-movement regime.

Although a single number cannot be derived from the water-balance data to characterize adequately a particular climate, water balances do provide data from which one can rank soils from various regions by the amount of water leaching through the soil. Arkley,[3] for example, has derived a number called the leaching index, which is the mean seasonal excess of $(P - ET_p)$ for those months of the year in which $P > ETp$ (Table 11-1). Soils located in areas characterized by a high leaching index commonly have properties associated with large amounts of water percolating through the soil, such as 1:1 clays, whereas soils in regions with a low leaching index have properties associated with slight leaching, such as 2:1 clay minerals or a Bk horizon close to the surface. McFadden[72] and

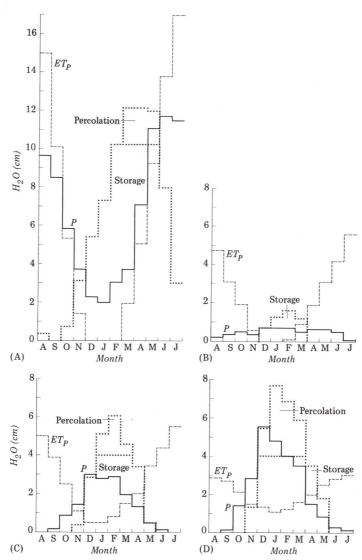

Fig. 11-2 Monthly values of precipitation (*P*), potential evapotranspiration ((ET_p), soil-moisture storage in the A and B horizons (10.2-cm available water-holding capacity), and soil-moisture percolation below the Bw or Bt horizon for various climatic stations. Mean temperature as follows:

	Mean temperature (°C)	
Station	Jan.	July
(A) Manhattan, Kan.	−1.6	26.6
(B) Fallon, Nev.	−1.2	22.9
(C) Sacramento, Calif.	7.5	23.2
(D) Half Moon Bay, Calif.	9.9	17.0

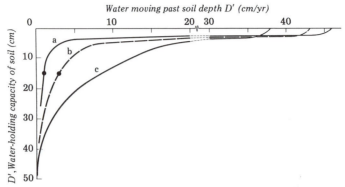

Fig. 11-3 Amount of water moving past various depths of the soil, calculated from water-balance data. Depth is in water-holding-capacity units (D), but actual soil depths can be calculated if data on water-holding capacities for specific soil profiles are known. (Some data from Arkley,[3] Fig. 4, © 1963, The Williams & Wilkins Co., Baltimore.) Filled circles, depth to the top of the $CaCO_3$ accumulation layer for nearby soils. Key for curves: a, Clovis, N.M.; b, Boise City, Okla.; c, Santa Monica, Calif.

Reheis[86,87] use the leaching index as the climatic parameter in chronosequence studies in several climatic regimes.

Because most data in the literature on climate-soil relationships use mean annual values of precipitation and temperature, these will generally be used here. Water-balance data might be more meaningful, however.

REGIONAL SOIL TRENDS RELATED TO CLIMATE

Many soil properties show distinct trends with regional climate in going from the equator to poles, as is shown by the diagram of Strakhov[104] (Fig. 11-4) and the biomass data of Rodin and Bazilevič.[91] These variations in the soils originate in such processes as organic matter influx and decomposition, presence or absence of chelating agents, soil-water chemistry, and the depth and rate of leaching of water through the soil. These processes, in turn, are controlled by the climate. The tropical forest regions are characterized by intense, deep weathering, with iron and aluminum oxides and hydroxides predominant close to the surface. With depth, clay minerals of the 1:1 and finally the 2:1 varieties are found. Organic matter in these soils is relatively low because, even though the amount of organic matter annually added to the soil is high, the decomposition rate of organic matter also is high. The above trends diminish to the north in the savanna region. The deserts are characterized by low organic matter input relative to the rate of decomposition, and low organic matter content in the soil results. Slight leaching produces 2:1 clays, pedogenic carbonate at depth, and gypsum at depth under extreme aridity. Increased precipitation and decreased evapotranspiration character-

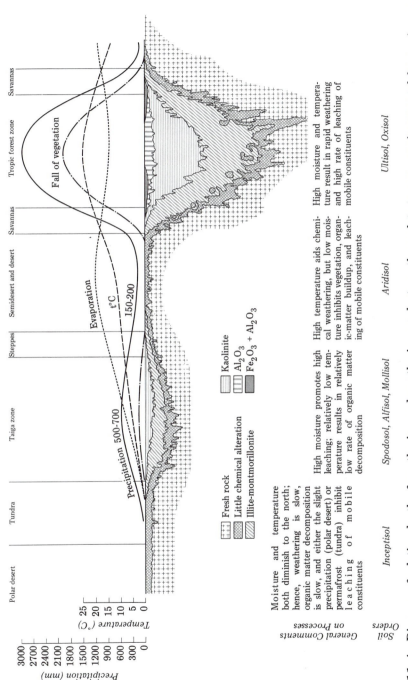

Fig. 11-4 Diagram of relative depth of weathering and weathering products as they relate to some environmental factors in a transect from the equator into the north polar region. (Taken mostly from Strakhov,[104] Fig. 2.)

ize the steppes compared to conditions in the deserts; the result is a fairly thick cover of vegetation and thick A horizons rich in organic matter, with moderately leached subsurface horizons. We find $CaCO_3$ in soils only in the more arid part of the steppes. North of the steppes is the taiga, a region of fairly high soil leaching. The low temperatures of the taiga result in fairly low rates of organic matter decomposition and prominent O and/ or A horizons form. Where conditions favor iron and aluminum movement, an E horizon can form. The tundra lies north of the taiga, and the combination of fairly low precipitation, low temperature, and permafrost close to the surface in places result in a moist soil with relatively slow rates of organic matter influx and decomposition. These conditions produce a soil in which an A horizon with a fairly high organic content commonly overlies a gleyed horizon. Continuing northward, the tundra gives way to the polar desert, a region of low precipitation and temperature.[106] Because vascular plants are nearly absent in this desert, organic matter content in the soil is fairly low and provided mainly by lichens, algae, and diatoms. Although these polar desert soils generally are permeable, the absence of appreciable moisture and water movement in the soil causes them to be saline and alkaline, with very little weathering. In many respects, these soils are similar to those of some hot deserts.[28]

The above variation in soil properties and processes forms what Tedrow[106] calls a pedogenic gradient. His examples come from the northern latitudes, as does that of Moore,[75] but the term is appropriate for any pedologic variation associated with climatic gradients. For example, McFadden's[72] work is an example of a pedogenic gradient along a climatic gradient from arid to xeric, and Gile's[41,42] is one from arid to semiarid.

Very little has been done on the relationship between climatic parameters and large-scale soil-classification units, probably because the relationship is an extremely complex one. Arkley[4] has attempted such a correlation for the western United States, in which soils shown on a regional soil map were related to water-balance climatic parameters obtained from climatic stations located within the mapped soil units (Fig. 11-5). The parameters chosen were calculated actual evapotranspiration, based on a soil-moisture storage capacity of 15.2 cm, leaching index, and mean annual temperature. These parameters were considered to relate best to the major processes responsible for the various soil-classification units. The plot of data indicates that, although there is considerable overlap of some soil orders and although some orders range widely in values for the climatic parameters, most soil orders seem to fall within well-defined values for the climatic parameters. In some places, the overlap might occur because the climatic parameters chosen were not those most highly correlated with the particular orders; in other places, the overlap might occur because both orders are stable in that climatic regime, and one soil order may grade into another with time. These plots also bring up a major problem in working with soil morphology for paleoclimatic reconstruction.

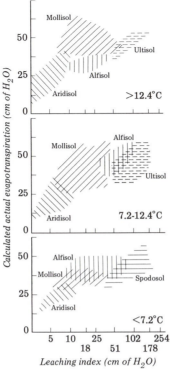

Fig. 11-5 Relationship among soil orders, leaching index, calculated actual evapotranspiration (soil-moisture-storage capacity 15.2 cm), and mean annual temperature for the western United States. (Data from Arkley,[4] Fig. 5, © 1967, The Williams & Wilkins Co., Baltimore.) Soil orders are plotted instead of the great soil groups of the old U.S. classification. Horizontal axis scale is linear between recorded values.

That is that each order is represented by a large variation in climatic parameters, and thus soil classification units do not give very precise data on paleoclimate, except in those transitional regions that separate soils of markedly different morphologies. For more precise information it might be best to construct figures for soil suborders rather than orders.

VARIATION IN SPECIFIC SOIL PROPERTIES WITH CLIMATE

The main soil morphological and mineralogical properties that correlate with climate are organic matter content, clay content, kind of clay and iron minerals, color, various chemical extracts, the presence or absence of $CaCO_3$ and more soluble salts, and depth to the top of salt-bearing horizons. Some trends will be given here and others can be obtained from soil chronosequences (Ch. 8), because many of the latter are in different climatic regimes. For example, one can obtain data on soils of a particular age, say the 100,000-year-old soils, and relate the data with some parameter of climate in an X-Y plot.

Organic-Matter Content

Jenny[54,57,58] has studied many climatic transects to determine the trends in organic matter constituents in the soil (organic carbon and total nitrogen). In general, he finds that soil nitrogen increases logarithmically with increasing moisture and decreases exponentially with rising temperature. These relationships hold for such diverse climatic regions as the Great Plains, India, and California (Figs. 11-6 and 11-7). In general, this means that, at fairly low values of either precipitation or temperature, each increment of change in either climatic parameter has a much greater effect on the amount of organic carbon or of nitrogen in the soil than has an identical increment of change at higher values of precipitation and temperature. The figures also indicate the obvious general organic matter content trends with climatic region; that is, low in deserts, intermediate in temperate regions and high in some tropical regions. Total organic carbon and nitrogen contents per unit surface area in soils of the temperate and tropical regions also follow these trends (Table 11-2).

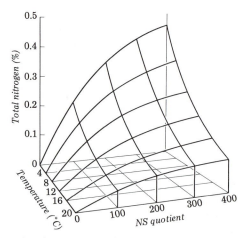

Fig. 11-6 Relationship between total nitrogen content of the surface soil and climate for grassland soils of the Great Plains. (Taken from Jenny,[54] Fig. 92.) Moisture is given as the NS quotient

$$m = \frac{\text{Precipitation (mm)}}{\text{Absolute saturation deficit of air (mm Hg)}}$$

The equation that describes the surface is

$$N = 0.55e^{-0.08\,T}(1 - e^{-0.005\,m})$$

where N is total nitrogen; T, mean annual air temperature; and m is the NS quotient.

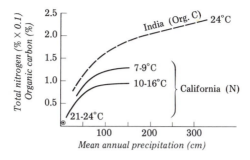

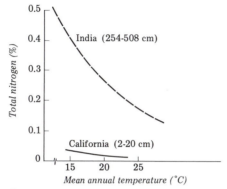

Fig. 11-7 Idealized trends of organic carbon and nitrogen with mean annual precipitation and temperature: India and California. (Taken from Jenny,[57] Missouri Agri. Exp. Sta. Res. Bull. 765, 1961, and Harradine and Jenny,[47] Figs. 5 and 6, © 1958, The William & Wilkins Co., Baltimore.) Temperature values are mean annual (upper graph); the numbers in parentheses (lower graph) are mean annual precipitation values.

The above climatic trends are related to yearly gains and losses of organic matter. Jenny[56,57] compared the dynamic nature of several tropical and temperate ecosystems to find some basic differences (Table 11-2). The tropical soils he studied contain more organic carbon and total nitrogen than the temperate soils, but the latter have a higher proportion of organic constituents in the forest floor. This is the case in spite of data indicating that litterfall is much greater in tropical forests. The annual rates of decomposition of the litterfall and forest floor material explain this apparent discrepancy, for the rates are much higher in the tropical region. In California, the rates of decomposition also have been shown to decrease with elevation and thus with climate[59] (Table 11-2).

These studies can serve as models for further study of ecosystems in other climates. Thus, to understand the relationship between climate and the organic matter constituents in a soil to the extent that Jenny has, we must know the gains and losses involved, the decomposition constants,

Table 11-2
Annual Gains, Losses, and Decomposition Constants for Several Forest Ecosystems

| | TROPICAL FORESTS OF COLUMBIA | | CALIFORNIA | | |
| | | | Temperate forest (1640 m) | | Cold forest (3280 m) |
	Sealevel	1540 m	Oak	Pine	Pine
Litterfall (g/m²)					
Weight*	730	935	149	305	101
Total nitrogen	10.4	15.7	1.27	1.54	—
Organic carbon	391	510	74.6	164	—
*Forest floor** (g/m²)*					
Weight*	432	1,455	2,517	12,635	11,081
Total nitrogen	8.8	35.4	31.1	117	—
Organic carbon	225	795	1,224	6,463	—
Decomposition constant of litterfall and forest floor (annual percent loss)					
Weight*	62.8	39.1	5.6	2.4	0.90
Total nitrogen	54.2	30.7	3.9	1.3	—
Organic carbon	63.5	39.1	5.7	2.5	—
*Total weights in soil profile*** (g/m²)*					
Total nitrogen	2,502	3,521	633	650	—
Organic carbon	36,681	45,196	11,606	17,317	—

*Volatile weight only, because the forest floor was contaminated with sand grains.
**Forest floor is the fresh and partially decomposed plant debris overlying the mineral soil.
***Includes forest floor and amounts in mineral soil.
(Taken from Jenny,[56] Table 1, © 1950, The Williams & Wilkins Co., Baltimore, and Jenny,[57] Table 2, Missouri Agri. Exp. Sta. Res. Bull. 765.)

and the total weight of the organic matter constituents, as well as the climatic parameters.

Clay Content

It is commonly reported that clay content in soils varies with climate, but precise data on this relationship are not often available because, in many places, factors other than climate have influenced clay production, and their influence is hard to quantify.

Two of the early quantitative relationships between soil clay content and climate are the moisture and temperature functions of Jenny,[53] in which he found a linear relationship with moisture and an exponential relationship with temperature (Fig. 11-8). Thus, if all other factors are constant, one would expect low rates of clay production in cold-dry, cold-wet, and hot-dry environments, increasing rates with increasing moisture

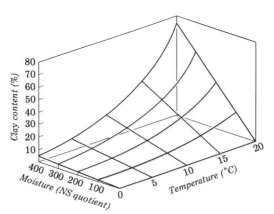

Fig. 11-8 Idealized relationship between percent clay in the top 92 to 102 cm of soil derived from granites and gneisses and mean annual moisture and temperature. Moisture is given as the NS quotient (see Fig. 11–6 capiton). (Taken from Jenny,[53] Fig. 8, © 1935, The Williams & Wilkins Co., Baltimore.)

contents, and highest rates in hot-humid environments. These data were obtained before we knew much about Quaternary stratigraphy and the relationships between soil morphology and age, so, although Jenny's generalized trends may still hold, the absolute amounts of clay for any climate might not coincide too precisely with the data of Fig. 11-8.

To determine the precise relationship between clay production and climate, it would be best to have data on total clay present per unit area of the ground surface (g/cm^2/soil column). Percentage values, although based on the less than 2 mm fraction, do not reflect the true amounts of clay in the profile because variations in gravel content and bulk density have a marked effect on absolute clay content (see Fig. 1-6). Until such data are collected, we will not be able to quantify precisely clay production as a function of any of the factors. An alternative would be to calculate the clay accumulation index of Levine and Ciolkosz.[64] This is defined as the difference in clay content between the Bt horizon or subhorizons and the C horizon (preferably the Cu horizon), multiplied by the thickness of the horizon or subhorizons, and summed for the profile. The data could be adjusted for gravel content. An equally serious problem here is the differentiation of clays of different origin. For example, some clays might be formed by weathering of primary grains, and others might be part of the solid atmospheric influx into the soil. Both processes produce pedogenic clays, yet the rates of clay accumulation by each process could be quite different.

Clay production can be qualitatively related to climate by comparing stratigraphically dated deposits and their respective soils in many different environments. One useful age datum is soil formed on late-Wisconsin tills during the late Pleistocene and Holocene. Such tills in many parts of the Cordilleran Region, in areas characterized by igneous and metamor-phic rocks of granitic composition, commonly have only A/Bw or A/Cox soil profiles; Bt horizons usually are absent. This relationship holds for soils in the Sierra Nevada,[25] the Cascade Range of Washington[85] and Ore-gon,[95] and the Rocky Mountains.[98] Soils of similar age formed from parent materials of at least partly volcanic rock in the high-precipitation areas of the Puget Sound Lowland, Washington or on Mt. Rainier, Washington also lack a Bt horizon.[29,30,63] These latter data come somewhat as a surprise because one would predict a greater rate of clay production in high-rain-fall areas with volcanic parent materials, since these materials are less resistant to alteration relative to parent materials derived from granitic rock. The conclusion to be drawn from these data is that soils formed from tills of somewhat similar texture look alike after 10^4 years of soil forma-tion, in spite of large variations in climate and parent materials.

Soils with Bt horizons occur in Illinoian tills at some of the localities mentioned above. For the Rocky Mountains, the Bt horizons are com-monly 35 to 75 cm thick and have 5 to 13 percent more absolute clay rel-ative to the parent material.[98] In contrast, the best developed Bt horizon of the same age in the Sierra Nevada is 65 cm thick and has 7 percent more absolute clay (i.e., 14 vs. 7 percent).[25] For an alpine till in north-eastern Oregon, the respective values are 42 cm and, relative to the Cu horizon, a 5 percent increase in the Bt1 and a 3 percent increase in the Bt2 horizon.[24] Thus, it is not unusual for soils of this age in tills in the western United States to double their clay content from the Cu to the Bt horizon and for the Bt to be about 50 cm thick. More samples within the region are needed to determine if trends with climate occur, because gross trends are not apparent with the limited data.

One problem with these soils is that because of their age they probably have undergone climatic changes; for example, they were forming when the late-Wisconsin glaciers advanced to positions quite near the older soils. What was the influence of that climate on the soils, and was the influence the same everywhere? Recall that the intermontane basins of Wyoming were about 10 to 13 °C colder then,[73] and amino-acid analysis suggests that parts of the northeastern Great Basin could have been equally as cold.[71] The point is that soils from quite different areas and environments across the Cordilleran Region are grossly similar. Compar-ison could be made with Bt horizons formed in tills in the humid climate of Pennsylvania (Fig. 8-10), since the latter have a clay accumulation index that ranges from 1299 to 1387,[64] but we need more data on factors other than climate that might cause the variation. One important factor is parent material, because tills in the western states are formed from crystalline rocks, whereas those in Pennsylvania are formed from sedi-mentary rocks.

Still older soils are hard to locate and difficult to date. However, one can compare the Bt horizons of about 0.5-my-old or older soils in the

Rocky Mountains[97] and the Sierra Nevada.[15] Such soils in the former area have a 92-cm-thick Bt, a maximum clay content of 28 percent, and an absolute increase relative to the Cu of at least 18 percent. In contrast, those soils in the latter area have a 145-cm-thick Bt, a maximum of 40 percent clay, and probably at least a 20 percent absolute increase in clay. One would be tempted to ascribe the differences to the warmer climate in California, but too many other unknown factors could be responsible.

The midcontinent is an area of seemingly rapid clay production. Soils formed on late-Wisconsin and older tills have more strongly developed Bt horizons than their Cordilleran counterparts.[36] Although part of this variation could result from parent materials with relatively high clay content in the midcontinent, at least some of the greater clay production of the midcontinent could result from the climate, which is characterized by significant rainfall during the warm summers. Soils suitable for a comparison of soil development on parent materials of somewhat similar textures include those on outwash deposits of the eastern Sierra Nevada (Fig. 11-9) and of Ohio (Fig. 8-10). Again, the midcontinent shows up as a region of rapid clay production, relative to the Sierra Nevada.

In a climatic transect from wetter and colder to drier and warmer in the western Great Basin, one sees systematic changes in the Bt horizons. An example of this is the transect from the eastern Sierra Nevada to the area east of Reno, Nevada, where the more aridic Bt horizons are thinner

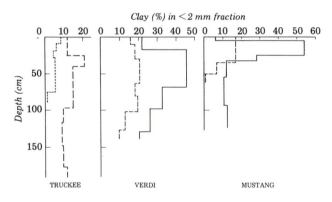

Fig. 11-9 Clay distribution in soils formed from stream terrace deposits and a lacustrine sand (younger soil at Mustang) on the Truckee River, California and Nevada. (Taken from Birkeland,[10] Fig. 3.) Truckee lies at the foot of the eastern Sierra Nevada; climatic parameters are 80-cm mean annual precipitation, 6°C mean annual temperature, and a leaching index of 65.8 cm. Comparable climatic parameters for Reno, Nevada (between Verdi and Mustang) are 18 cm, 9.5°C, and 7.4 cm. Summers are dry at all locations.

and where they have a greater buildup of clay relative to that of the presumed Cu horizon (Fig. 11-9). The origin of clays along this transect have not been investigated, but those in the wetter area could be formed from parent-material weathering, whereas those at the drier end of the transect could have a notable eolian component.

The best examples of the rapidity at which Bt horizons form in the arid regions are those in the Las Cruces area (Fig. 8-10), noted for its influx of eolian materials. Gile and others[44] have measured the solid influx and found that its yearly variation is from 9 to 26×10^{-4} g/cm^2, and of this 22 to 39 percent is clay. They have calculated the amount of clay for some profiles in their Table 28, and the above influx rates seem ample to explain the totals. I expect that most arid areas experience eolian influx of clay and that the rate would vary with area, as does the carbonate influx (Fig. 8-14).

A study with perhaps better control on most of the soil-forming factors is that of McFadden,[72] in Southern California. Chronosequences were studied in three climatic regimes—arid, semiarid, and xeric. The buildup of pedogenic clay appears to decline after an initial buildup, but there is a definite trend of a more rapid buildup in wetter climates (Fig. 11-10).

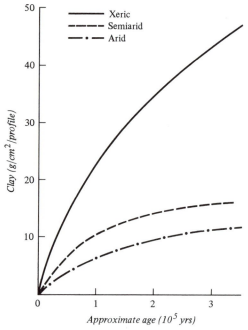

Fig. 11-10 The amount of pedogenic clay as a function of time in three climatic regimes, Southern California. The mean annual precipitations (cm) and temperatures (°C) for the three regimes are, respectively, arid—<12, <22; semiarid—12–25; 15–22; and xeric—25–75, 12–22. (Taken from McFadden,[72] Fig. 27.)

He uses several lines of chemical and mineralogical evidence to suggest that the clay in the more arid areas is derived mainly by eolian processes and that weathering processes become more important in producing clay in the wetter areas.

In contrast to the above, pedogenic clay accumulation is very slow in the hot desert of southern Israel and Sinai.[32] Mean annual precipitation is less than 5 cm, and soils several hundred thousand years old have B horizons that are still classified as cambic.

Finally, arctic and polar regions are characterized by virtually no clay accumulation. Many of the soils are formed in sandy tills made up of igneous and metamorphic rocks of granitic composition, so the parent materials are somewhat similar to those of the Cordilleran Region. No Bt horizons have been reported in soils formed in late-Wisconsin, Illinoian, and still older tills on Baffin Island.[12,16,35] In Antarctica, Bt horizons are not present, even in soils several million years old.[82] Bw horizons form in both environments, and those in the arctic form more rapidly.[17] Low values for mean annual precipitation and temperature keep pedogenic processes at a very low level in these extreme environments.

In summary, although many clay data are available for soils, we still do not have a very precise idea how clay content varies with climate, when all other factors are kept constant. As Quaternary stratigraphic studies proceed in many diverse climates, soils of different ages can be collected, and their clay production analyzed. We must improve our techniques, however, to obtain the best data. Bulk densities should be taken so that the rates can be based not on percents, but rather on weight of clay formed or accumulated per unit surface area for the entire soil. In addition, detailed studies of the non-clay fraction to determine the relative degree of weathering in the soil, as well as the absolute amounts of weathering of specific minerals, can be undertaken; these should show some relationship to the amounts of clay produced in more moist environments. Once this is done, we can attempt a newer version of Figure 11-8; my suspicion is that it would look much different than the original, for at least we know now that some desert regions are areas with high amounts of pedogenic clay. Furthermore, few soil-stratigraphic studies have been carried out in tropical areas.

Clay Mineralogy

Clay minerals formed in the soil vary with the water chemistry and the rate of leaching, and thus, with the climate. Many transects relating clay mineralogy to climate have been described, but only a few will be mentioned here. In any climatic transect, however, the clay mineralogy seems best correlated with precipitation or leaching index, and, in going toward areas of greater precipitation, the clay mineralogy trend usually is toward species containing progressively less silica. The reason for this trend is that soil leaching increases with increasing precipitation, and therefore more silica is lost from the soil during the course of weathering (Fig. 11-11). An interesting transect from Norway to Africa was made by Tardy,[105] in which he attempted to relate the chemistry of stream and spring waters to clay mineralogy with fairly good results. Data are given here for well-drained soils so that the mineralogy reflects regional climate, not topographic influence (Fig. 9-12). The discussion here will focus mostly on soil clay minerals formed from crystalline-rock parent materials. Clay minerals formed from sedimentary rocks or deposits derived largely from sedimentary rocks will not be discussed, because some of their clay minerals are inherited.

Clay-mineral associations with climate first will be made on a broad scale. Allen[1] does this by tabulating the mineralogy families in soil orders (Table 11-3), Marshall[69] also presents data on these, and Gradusov[45] has produced very useful world maps for some of the clay minerals. A mineralogy family has over one-half of the indicated mineral present, or in some cases, that mineral is dominant; mixed means a mixture of many minerals. Because orders commonly are associated with different climates, some broad trends are apparent. The best relationships are kaolinite with Oxisols and Ultisols. Smectites are dominant in Vertisols, but the latter are found in several climatic regimes. Although smectites are common in the Aridisols, they are equally as common in many other orders, especially the Mollisols. Mixed mineralogies are found in many orders and appear to be dominant in the Alfisols and Inceptisols. Spodosols represent a rather special pedogenic environment, and the clay mineralogy corresponds to this, as reviewed by Ross.[92] The clay mineralogy of the C horizon, for the soils he reviews, consists mainly of chlorite and mica. Vermiculite is common in the B horizon, in addition to the above two minerals. The A or E horizon is characterized by an absence of chlorite and abundant Al-rich 2:1 clay minerals, specifically, beidellite and dioctahedral vermiculite. One problem with broad relationships at this level is that many orders include such a wide variation in the soil-forming factors that

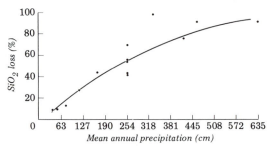

Fig. 11-11 Overall trend of silica loss with precipitation in a 10,000- to 17,000-year-old volcanic ash, Hawaii. (Taken from Hay and Jones,[48] Fig. 2, published by the Geological Society of America.) This is the same ash for which mineral weathering data are presented in Fig. 11–15.

Table 11-3
Clay-Mineral Families for Soil Orders in the United States, Puerto Rico, and the Virgin Islands

Order	CLAY MINERAL FAMILES (%)					
	Kaolinitic	*Illite*	*Smectitic*	*Vermiculitic*	*Mixed*	*n***
Alfisols	5	7	35	1	52	291
Aridisols	1	2	68		29	131
Entisols	1	2	62		35	116
Inceptisols	8*	8	28		56	135
Mollisols	2	1	59	1	37	499
Oxisols	100					6
Spodosols					100	2
Ultisols	32	4			64	81
Vertisols	3		84		13	68

*One of the 8 percent is halloysitic class.
**Number of families represented in each order.
(Taken from Allen,[1] Table 22-1.)

a good correlation is not possible. Correlation could be improved, perhaps, by isolating some of the factors.

Even though tropical regions typically are used to characterize the end products of intense weathering (aluminum and iron compounds), there is a definite mineralogical trend with climate.[66] Commonly, montmorillonite forms at low amounts of precipitation, kaolinite at higher amounts, and oxides and hydroxides of iron and aluminum at still higher amounts.[49,69,74,112] In Hawaii, Sherman[96] reports that montmorillonite predominates below about 100-cm precipitation, kaolinite between about 100 and 200, and the iron and aluminum compounds above 200 (Fig. 11-12).

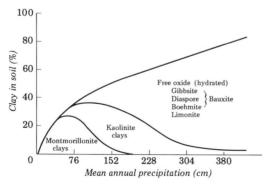

Fig. 11-12 Clay mineralogy as a function of precipitation under a continuously wet climate, Hawaii. (Taken from Sherman,[96] Fig. 3 in *Problems of Clay and Laterite Genesis,* published by AIME.) Sherman also presents data for an alternating wet and dry climate with a somewhat similar relationship of precipitation and soil-clay minerals.

In an ash layer in Hawaii dated at 10,000 to 17,000 years, Hay and Jones[48] found montmorillonite at 25- to 65-cm precipitation, and gibbsite at greater than 370-cm precipitation. In soils formed on mafic lavas and tuffs in the Caribbean, Beaven and Dumbleton[7] reported montmorillonite predominant at less than 150-cm precipitation, kaolinite predominant at greater than 200-cm precipitation, and mixtures of the two at intermediate values of precipitation.

Barshad[6] reported on the clay mineralogy of several hundred surface-soil samples (0 to 15 cm depth) in California and found a close correspondence of clay mineral with precipitation (Fig. 11-13). Generally, montmorillonite is only found at less than about 100-cm precipitation, and gibbsite at greater than about 100. Kaolinite and/or halloysite are present over a wide range of precipitations and are predominant above about 50-cm precipitation. Illite and vermiculite are also present, the former only in felsic igneous rocks, the latter in both felsic and mafic igneous rocks. The abundance of the various minerals varies somewhat for the same

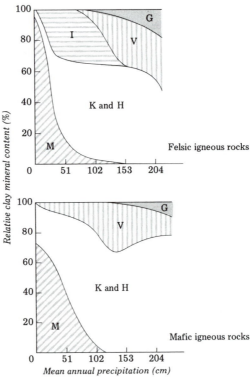

Fig. 11-13 Relative clay-mineral content as a function of precipitation for surface-soil samples, California. (Taken from Barshad,[6] Figs. 1 and 2.) Mean annual temperatures range from 10 to 15.6°C. Key to symbols: M, montmorillonite; K, kaolinite; H, halloysite; I, illite; V, vermiculite; G, gibbsite.

amount of precipitation between soils formed from felsic igneous rocks and those formed from mafic igneous rocks. The main reason for this shift in clay-mineral abundance with parent material is the cation content of the parent material. Thus, for example, montmorillonite can form under higher precipitation from mafic igneous rocks than from felsic igneous rocks because the former have a higher cation content, and high cation content in the soil solution favors formation of montmorillonite. In another area characterized by a Mediterranean climate, Israel, Singer[99] also finds that smectite makes up less than 50 percent of the clay-mineral fraction of basaltic soils where the mean annual precipitation exceeds 50 cm.

Clay-mineral studies of soils from a large number of till and outwash deposits of many ages along the east side of the Sierra Nevada in California and Nevada substantiate the conclusions of Barshad.[11,13] In general, in the northern part of the range in deposits with a mixed lithology of granitic, andesitic, and basaltic rocks, the change from high halloysite content to high montmorillonite content takes place at the pine-sagebrush vegetation boundary at about 40-cm annual precipitation (Fig. 11-14). However, near the southern end of the study area, soils formed from granitic till presently located below the pine-sagebrush boundary, in what is probably a drier climate, do not contain montmorillonite. Again, this variation with parent material can be predicted from the data of Barshad, although the exact climatic values at the transition from one predominant

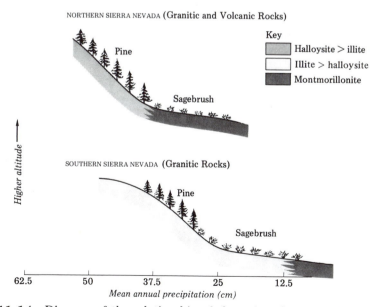

Fig. 11-14 Diagram of the relationship of clay mineralogy and climate with different parent materials along the eastern Sierra Nevada. (Data from Birkeland and Janda,[13] Table 3.)

clay mineral to another may vary with the precise environmental conditions, including any paleoclimatic influences.

The fibrous clay minerals palygorskite and sepiolite are associated with aridity,[5] since they require a fairly high pH and an Mg source,[102] such as are present in Bk and K horizons. In Morocco, palygorskite was reported to be stable at less than 30-cm precipitation.[81] As mentioned earlier (Ch. 8), sepiolite is formed from palygorskite, and the transformation is highly time-dependent.

To sum up, clay minerals do have a fairly good correlation with climate. Correlation is often best and most specific at either end of the climatic spectrum: that is, the 1:1 clay minerals in warm, humid environments; biedellite in cool, humid conditions, characterized by podzolization processes; and montmorillonite, palygorskite, and sepiolite in the more aridic environments. Clay-mineral associations in other environments may be more difficult to predict.

Mineral Weathering

The variation in rate of mineral etching with time for different regions (Fig. 8-7) is attributed, in part, to differences in climate. If weathering in soils alone is compared, one obtains the following ranking, from regions of greatest amount of weathering to least amount: St. Vincent = central Europe = Michigan > Missouri, Iowa > Colorado Piedmont and Rocky Mountains > eastern Sierra Nevada and the San Joaquin Valley. This suggests that hot-humid and cool-humid climates are those most conducive to mineral and rock alteration. It would not necessarily follow that clay-production rates follow these trends, since clays are produced only when the correct constituents in the right proportions are present and precipitate in the soil.

Work on a volcanic ash unit in a climatic transect in Hawaii clearly demonstrates the predominant influence of rainfall on mineral weathering.[48] The ash, originally about 95 percent vitric ash, was laid down from 17,000 to 10,000 years ago; it has been weathering under a precipitation that ranges from about 25 to 635 cm/yr. At less than 115-cm annual precipitation, only glass has altered; between 115- and 255-cm annual precipitation, plagioclase is the only mineral to show slight alteration; above 225-cm annual precipitation, all minerals are altered to various degrees; and at 570- to 635-cm annual precipitation, all except olivine are completely altered (Fig. 11-15). These are really exceptional rates of weathering, which again demonstrate that rapid weathering takes place in humid tropical regions.

Soil Redness and Iron-Bearing Minerals

The color of a soil, in particular its degree of redness, is generally related to climate. One has to be very cautious, in this regard, because at least part of the redness of old soils is a function of time of soil formation

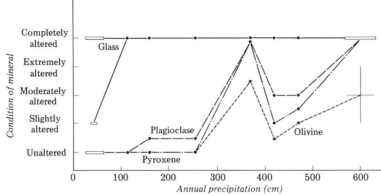

Fig. 11-15 Mineral weathering in volcanic ash as a function of annual precipitation, Hawaii. (Data from Hay and Jones,[48] Table 4.) The plagioclase is labradorite, the pyroxene, clinopyroxene. Bars indicate ranges in either precipitation or mineral condition. Weathering of olivine varies because only when the glass coating on the grains have dissolved does the mineral begin to weather. Mean annual temperatures range from 23 °C near sea level to 16 °C at 1202-m altitude. The more rapid weathering at 370-cm precipitation relative to that at 255 and 420 is unexplained; it could be due to a higher mean annual temperature, a finer grain size, or higher precipitation in the late Quaternary. (R.L. Hay, written commun., 1972.)

(Table 8-2). Part of the differences in soil color with climatic region, however, is a function of climate. Temperature probably is more important than precipitation in producing redness, because red soils are common in the humid southeastern United States, the humid tropics, and hot deserts of the southwestern United States and Baja California, Mexico; red soils are not common in areas of low temperatures and varying amounts of precipitation (e.g., along the northern borders of the United States and in polar regions).

The degree of redness of soils has been shown to be a function of iron content and mineralogy.[8,33] Recently, techniques have been developed to better identify the pedogenic iron minerals. These minerals bear a relationship to color in that hematite is associated with hues of 5R-2.5YR, goethite with 7.5YR-10YR, and ferrihydrite with 5YR-7.5YR (Table 4-1). Schwertmann and co-workers have done much of the work; one was an east-west transect in soils formed in glaciofluvial gravels in the foreland north of the European Alps.[94] The data suggest that the more hematitic soils are redder and that these occur in the warmer, drier climatic regimes (Table 11-4). The C horizons are 10YR hue and contain goethite. They also recognized a parent-material effect, in that silty till of the same age as the glaciofluvial gravels has no hematite and has 10YR hue (Table 11-4). With these data and those from soils in Spain, they obtained the following relationship between soil redness (*Y*), as defined by Hurst,[51] and

Table 11-4
Variation in Soil Color and Iron Extracts and Minerals for Bt
Horizons of Alfisols Formed from Late-Wisconsin Glaciofluvial
Gravels, European Alps

		COLOR		CLAY FRICTION(%)***			Ratio †
MAT^* (°C)	MAP^{**} (cm)	Soil	Clay	Fe_o	Fe_d	Fe_o/Fe_d	Hm: Hm + Gt
10.3	57.8	5YR 3/4	5YR 4/6	0.80	4.94	0.16	0.29
8.0	75.0	5YR 4/8	5YR 3/6	0.38	3.83	0.10	0.09
6.5	137.9	7.5YR 4/4	5YR 4/4	0.40	4.64	0.09	0.08
7.5	106.7	7.5YR 4/4	7.5YR 4/6	0.81	5.27	0.15	0.05
7.6‡	118.7	2.5Y 4/4	10YR 4/4	0.40	3.43	0.12	0.0

*Mean annual temperature.
**Mean annual precipitation.
***Fe_o is the oxalate extract, Fe_d the dithionite extract.
†Hm is hematite, Gt goethite.
‡Parent material is silty till.
(Data from Schwertmann and others,[94] Table 1.)

hematite content (X):

$$Y = 0.81 + 8.4X - 0.75X^2 (r = 0.97; n = 21)$$

In another transect with Oxisols and Ultisols, in southern Brazil, hematite is favored over goethite at mean annual temperatures >17 °C, mean annual precipitation minus mean annual evapotranspiration values >90 to 100 cm, and a low content of organic matter (<3 percent).[60]

Calcium Carbonate, and More Soluble Salts

Several features of $CaCO_3$ accumulation in soils are related to the climate, more specifically to that fraction of the precipitation that leaches downward in the soil. Here we will briefly discuss the amount of precipitation at which $CaCO_3$ begins to appear in soils, the relationship between the amount of precipitation and the depth to the top of the $CaCO_3$-bearing horizon, and regional rates of development.

In any climatic transect running into a sufficiently dry region, $CaCO_3$ usually appears in the more arid soils. The climatic value at which it first appears is obviously a function of the distribution of moisture and temperature throughout the year. In the midcontinent, for example, an area of significant summer precipitation, the boundary between the calcic and non-calcic soils (Fig. 11-16) is at about 50-cm mean annual precipitation and 5 to 6 °C mean annual temperature in northern Minnesota and at about 60-cm mean annual precipitation and 22 °C in southern Texas. However, in going from the northern Sierra Nevada to the Basin and Range Province, $CaCO_3$ first appears in the soils near Reno, Nevada with a mean annual precipitation of 18 cm and a mean annual temperature of 9.5 °C. The calcic–non-calcic soil boundary occurs at a lower precipitation

Fig. 11-16 Approximate position of the calcic–non-calcic soil boundary for the midcontinent. (Taken from Jenny,[54] Fig. 99.)

at Reno relative to the midcontinent partly because soil leaching per unit of rainfall is more effective in a winter-wet, summer-dry climate such as Reno's. Another factor that might explain the moister climate at the midcontinent calcic–non-calcic soil boundary is texture, since midcontinent parent materials commonly are finer textured than are the parent materials near Reno. Still another factor to be considered in explaining this boundary is the calcium content of the parent materials, because Jenny[55] has shown that $CaCO_3$ horizons can persist to higher values of precipitation in soils with calcium-rich parent materials relative to parent materials low in calcium.

Although Figure 11-16 accurately depicts the calcic–non-calcic soil boundary for the midcontinent, it is not accurate for the western United States because here the mountains and intermontane basins have markedly different climates and soils. Nearly every mountain-basin transect is characterized by a relatively moist-cool climate in the mountains and a relatively dry-warm climate in the basins. Under appropriate conditions, therefore, soils on the lower slopes of the mountains and in the basins will contain $CaCO_3$. These relationships are repeated over and over again in the western United States. Richmond[88] has made a particularly detailed study of these relationships in the La Sal Mountains of Utah and finds that the calcic–non-calcic soil boundary is related not only to present altitude and climate (Fig. 11-17), but also to past conditions of pedogenesis. He also found that with lower altitude the Bk horizons of all soils of a

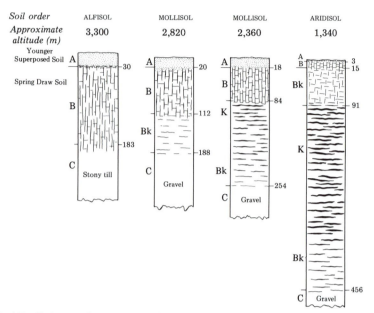

Fig. 11-17 Relationship of soil order and morphology with altitude, the La Sal Mountains, Utah. (Taken from Richmond, Fig. 18.) The soil shown is of pre-Wisconsin age. The numbers beside the profiles are soil-horizon boundary depths in centimeters.

particular age get thicker, their pH shows a slight increase, and mixtures of B and Bk horizons are more common (Fig. 12-6).

Although the amount of $CaCO_3$ in the soil is primarily a function of the duration of pedogenesis, depth to the top of the Bk or K horizon is closely related to the amount of precipitation. This is qualitatively shown by the La Sal Mountains data (Fig. 11-17). Depth relationships have been quantitatively assessed in two regions of the United States. Jenny[54] worked on a transect from the relatively dry climate of the Colorado Great Plains eastward to the more humid regions of Missouri and found the following relationship to be fairly representative

$$D = 2.5(P - 12)$$

where D is the depth to the top of the Bk or K horizon, and P is the mean annual precipitation. Arkley[3] analyzed the same relations for a number of California and Nevada soils and got a different relationship

$$D = 1.63(P - 0.45)$$

The difference between the two relationships is ascribed to climate. In California and Nevada, with a predominant winter rainfall, a unit increment of rainfall is much more effective in leaching $CaCO_3$ to a particular depth than it is in the midcontinent, where significant rainfall comes during the season of high evapotranspiration. One should be aware of local factors that will confound these latter relationships. For example, subsequent accumulation of surficial material might result in the top of the carbonate horizon being deeper than predicted, or erosion of surficial material may result in the top being too shallow.

The rate of carbonate buildup varies with region (Fig. 8-14), but only part of the variation is related to the climate. Part of the variation is a function of the carbonate content of dust and the rate of dust influx. For example, areas with limestone bedrock in the upwind dust source area have high influx rates of carbonate (Roswell-Carlsbad area), whereas those with limited limestone outcrop upwind have low influx rates (Vidal Junction). Part of the impact of the climatic factor could be the chemical composition of the precipitation, for it has been shown to vary with region. Many of the calcic soils in the western United States lie in areas of relatively high concentration of Ca^{2+} in the rainwaters, and this is important to regional rates of carbonate buildup. Recall that over one-half of the Ca^{2+} involved in the carbonate buildup at Las Cruces is ascribed to that dissolved in rainwater (Ch. 5).

There is a complex relationship between the amount of precipitation and potential Ca^{2+} influx into the soil, as pointed out by Machette and others.[68] Some areas are characterized by low rainfall relative to Ca^{2+} influx; the result could be that not all the Ca^{2+} may be fully leached into the soil, especially if a substantial portion of the latter is carbonate dust. Wind, for example, could move the particles on before they dissolve and Ca^{2+} moves into the profile. At the other extreme are those areas in which rainfall is high relative to potential Ca^{2+} influx. In this case, not all the Ca^{2+} will accumulate in the soil, for some can be leached from the soil during wetter years. It is difficult to predict the relative balance between rainfall and Ca^{2+} influx for specific areas, but the authors give possible examples for the southwestern United States.

In still drier climates, gypsum and more soluble salts are present in well-drained soils lying above the influence of groundwater. The origin of the salts is mainly atmospheric, as solid wind-blown particles and dissolved in the rainwater. The salts occur at the depth to which soil waters penetrate during the wetter years,[79] and because of their high solubility, are very sensitive to climatic changes that might alter this depth of penetration. Page[80] studied gypsum soils in Tunisia and found them to be more common in areas with less than about 17.5-cm mean annual precipitation and a leaching index (defined in Table 11-1) of less than about 12 mm. In the northeastern Bighorn Basin, Wyoming, gypsum is accumulating in a climate with about 16.5-cm mean annual precipitation and a leaching index of 26 mm.[87] In contrast, gypsum occurs in Iraqi soils at less than 30-cm winter precipitation.[111] A climatic transect in Israel illustrates the presence and depths of carbonate and more soluble salts with mean annual precipitation (Fig. 11-18). Of course, the trends are general and

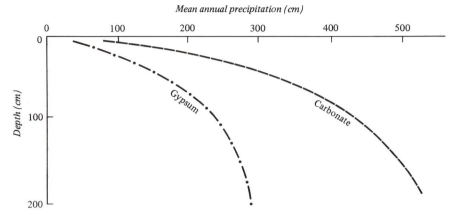

Fig. 11-18 Schematic diagram showing relationship between kind of pedogenic salt, the top of the horizon of salt accumulation with depth, and precipitation, in southern Israel. (Modified from Dan and Yaalon,[31] Fig. 5.)

are affected by soil textures and the kind and solubility of the salts delivered to a particular site. Of interest also is the lack of pedogenic carbonate in areas of extreme aridity where pedogenic gypsum is forming. Apparently the lack of much vegetation results in such low soil CO_2 values[21] that the common Ca salt is the sulphate rather than the carbonate.

Trends in Iron, Aluminum, and Phosphorus

Extracts of Fe, Al, and P display rather predictable trends with climate. Much of this was discussed and referenced in Chapter 8 (Fig. 8-11). Soils in relatively humid environments have an accumulation of oxalate-extractable Fe and Al and a depletion of acid-extractable P that seem to correlate fairly well with precipitation. Arid environments, cold or hot, either do not show trends, or rates of accumulation and depletion are so slow that only the very old soils display trends.

INDICES AS EXAMPLES OF POOR CORRELATION BETWEEN CLIMATE AND SOIL

It should not be assumed that all soil properties will show definite and predictable associations with present-day climate, for some do not. The examples used in the last sections were picked to demonstrate trends. Here several important examples of the lack of such trends will be discussed.

Harden and Taylor[46] have applied the Harden index of profile development (Ch. 1) to chronosequences in four climatic regions and concluded that, depending on how the index is calculated, the change in the index with time can be essentially the same despite differences in climate. The soil-moisture regimes and soils compared are aridic in the Las Cruces, New Mexico area, xeric inland and xeric coastal in California, and udic in Pennsylvania. In one analysis, the profile index was calculated using the four properties that best correlate with age for each chronosequence; these were not always the same property for each chronosequence. Although there is scatter in the plots, the index data from the four soil-moisture regimes plotted against time are all very similar. When eight soil properties are used to calculate the index, the index data for the aridic soil chronosequence plotted at lower values relative to the rest, but those for the other three climatic regions again overlapped.

The general lack of separation of index data for such contrasting climates is surprising and difficult to explain from a climatic point of view. Perhaps a different manipulation of the data will better reflect the climatic differences. However, the fact that all four areas did give such similar plots against time is good news for workers who might use the index to assign broad ages to soils in contrasting climatic regions.

Two Holocene soil chronosequences in the Southern Alps, New Zealand have modified indices of profile anisotropy, mIPA (Ch. 1), that are similar despite fairly large differences in climate. Compared here are soils in the Mt. Cook area, with mean annual values of precipitation and temperature of about 400 cm and 7 to 8 °C, respectively, and the Ben Ohau Range with values of about 100 cm and 4 °C. The data for oxalate-extractable Fe and Al are quite similar, and that for acid-extractable P has higher mIPA values for the area characterized by the lower precipitation and colder temperature, contrary to what many might have predicted (Fig. 11-19). One interpretation of these data is that soils in both climates are characterized by highly leaching conditions, so much so that the differences in climate are not reflected by differences in the mIPAs. To explain this, one could evoke the threshold concept of geomorphologists.[93] As used in geomorphology, the concept states that a gradual change in external factors might not show up as a response in the system. With continued gradual change, however, a threshold might be reached, and the system responds quickly and dramatically. Arroyo cutting is an example of such a threshold. A pedologic threshold might be defined in the same way. In the example here, the pedologic processes have not responded to the differences in climate; perhaps even greater differences are required for a definite and consistent pedologic response. Thus, one could speculate that along particular portions of a climatic gradient, a pedologic gradient is well expressed, but along other portions of a climatic gradient, a pedologic response is not registered. This will continue to be so until a pedologic threshold is crossed.

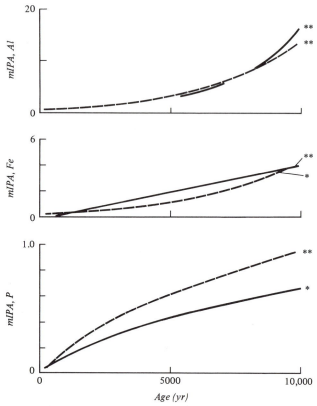

Fig. 11-19 Variation for mIPA for oxalate-extractable aluminum and iron, and acid-extractable P for two chronosequences in the Southern Alps, New Zealand. Solid line is for Mt. Cook area, dashed line for Ben Ohau Range. Significance of correlation coefficient: *, at .05 level; **, at .01 level. (Unpubl. data of author.)

RECONSTRUCTION OF PAST CLIMATES FROM PEDOLOGIC DATA

Morphology and various other properties of soils can be used to infer past climates. This is important to Quaternary research because in many places the soils represent hiatuses in the depositional record and, thus, are the only record left of certain time intervals. Before paleoclimatic interpretations can be made, however, one has to be sure that the observed feature was truly imparted on the soil by a past climate and is not related to other factors of soil formation. It has commonly been recognized, for example, that the effect of a longer interval of soil formation can give the same pedologic result as a climatic change. Moreover, although many workers have used soil features to infer past climates, such studies must be backed up with quantitative soil data. Pawluk[83] correctly urges caution in these interpretations and points out that there may be several genetic pathways for the formation of soil features and that a different former climate may be only one such pathway.

Soil properties vary in their usefulness as tools for paleoclimatic interpretation. If the soil has remained at the surface since the climatic change, the property or properties used to decipher changes in climate must have been resistant enough to persist in the soil, and not be altered entirely during subsequent pedogenesis. Obviously, those soil properties that alter readily with changing environmental conditions, such as pH, cannot be used as indicators of past conditions. The same is true for buried soils. Here, however, the properties imparted to the soil during pedogenesis must be resistant to subsequent diagenetic alteration to be useful in paleoclimatic reconstruction. Morphological features can be ranked by their persistence with changing environmental conditions or with burial (Table 11-5). For example, those horizons or features ranked as relatively persistent or persistent are most liable to carry a legacy of former conditions.

Success in deciphering paleoclimate from soils evidence will depend partly on the field area chosen. The most promising areas are those of rather sharp transition from one climatic region to another. As an illustration, if one is working in the center of a large region of Mollisols that are characterized by a relatively wide range in climate, a past change in climate may have left the area of interest still within an area of Mollisol formation. It is possible that no diagnostic soil feature would record this climatic change. In contrast, if one worked at the margins of the Mollisol area, where they grade into either the Aridisols or the Alfisols, and if a past shift in climate were accompanied by a shift in the geographical position in the soil orders, then marked changes in the soil properties would have accompanied the change in climate, and one could hope to read these properties as due to a past different climate in either surface or buried soils.

Table 11-5
Relative Persistence of Soil Horizons and Features as
Possible Indicators of Former Pedologic Conditions

Easily altered	*Relatively persistent*	*Persistent*
Mollic epipedon	Histic epipedon	Oxic horizon
Ochric epipedon	Umbric epipedon	Placic horizon
Salic horizon	Albic horizon	Argillic horizon
Gypsic horizon	Cambic horizon	Natric horizon
Mottles	Argillic horizon	K horizon
	Spodic horizon	Plinthite
	Calcic horizon	Duripan
	Fragipan	

(Modified from Yaalon,[114] Table 1.)

An example of the above is the work of Janda and Croft[52] on soil formation and Quaternary climates in the northeastern San Joaquin Valley of California. Soils on terraces of all ages are Xeralfs, and the field evidence suggests that they all have formed under a climate conducive to Xeralf formation. Janda and Croft then looked at the west coast distribution of Xeralfs and noted that they presently lie in areas characterized by 16- to 84-cm mean annual precipitation, falling mainly in the winter, and 11.3 to 18.3 °C mean annual temperature. They concluded that this wide range in climatic parameters probably would result in Xeralf formation, and thus, rather large changes in the climate could take place without those changes being registered in soil morphological features. About all they could conclude was that the Quaternary climates in that area probably had not deviated beyond the above precipitation and temperature values.

McFadden[72] came to a similar conclusion as that above in the xeric-to-semiarid-to-arid transect he studied. In the xeric climate, there is little evidence for climatic change in the soils. In contrast, evidence for climatic change is fairly clear in the soils formed in the semiarid and arid climates.

Overall Soil Morphology

Some soils in a region, either at the surface or buried, might have a soil morphology quite different from those currently being formed in the region. If they have, one might be able to reconstruct the past climate by comparing the present-day climate and morphology of soils in the region with the climate and morphology of the soil or soils in question. As a guideline, one might use data like that in Fig. 11-5 to determine how much of a change is suggested by the differences in soil morphologies. This figure also illustrates the importance of working with soils in regions along the boundary between two contrasting soil orders. Again, for soils located in the middle of a large soil-order region, a fair amount of climatic change could result in little morphological change.

Thick loess sections in central Europe contain buried soils with contrasting morphologies that are interpreted as due to climatic change.[61,62] Soils formed during interglacials are mainly leached of carbonate and strongly weathered, with Bt horizons; these formed under forest vegetation. In contrast, soils formed during the glacials under a cold continental climate and steppe vegetation are less leached and weakly developed.

Some areas have experienced large climatic change, as shown by the morphologies of different-aged soils. In southwestern Australia, thick Oxisols lie on old dissected plateau remnants, whereas younger soils in the river valleys are much less leached, and some contain $CaCO_3$ and are alkaline.[77] The area is characterized by less than 50-cm annual precipitation. Clearly, the Oxisols could not form in the present climate because it is doubtful if much water wets as deeply as their thickness. A much wetter former climate, perhaps extending back into the Tertiary, is postulated to explain the presence of the Oxisols. A similar interpretation has been made in California, where the only Oxisol described in the coterminous United States seems to have formed between 45 and 25 my ago.[103] Present mean annual climatic values in the area are 51-cm precipitation and 16 °C, and the interpretation is that a tropical climate is responsible for the thick, highly weathered soil. Oxisols commonly require a udic to perudic soil moisture regime and isohyperthermic temperatures for formation;[34] groundwater, however, is important in the formation of some of them,[67] and it is difficult to put climatic parameters on soils affected by groundwater. A somewhat similar, highly weathered soil, formed from basalt, formed during the Cretaceous in the Negev Desert.[100] In yet another example of dramatic climatic change based on pedologic data, Bown and others[19] report the existence of ancient fragipans in Oligocene terrestrial deposits in the Western Desert of Egypt (Fig. 11-20). They interpret the features to indicate humid conditions with tree vegetation, in contrast to an interpretation by a previous worker that the area at the time was semi-arid and almost treeless.

Organic Matter Distribution and Content

Soil properties associated with the organic matter of the soil are not too useful in reconstructing paleoclimate. Because organic matter properties of soil reach a steady state quite rapidly (Fig. 8-8), with change in environmental conditions, the distribution and content of the organic matter quickly adjust to values in equilibrium with the new conditions. This is especially true for surface soils, and if climatic change is to be deciphered from these properties, the change would have to have taken

Fig. 11-20 Fragipans (f) within Oligocene sediments in the Western Desert of Egypt. The hammer used for scale is circled. (Photograph courtesy of M. J. Kraus.)

place in less time, before the present, than the time necessary to reach the steady state for that environment. An A horizon in a buried soil might be useful, but the organic matter is usually so depleted that the A horizon is seldom recognizable on either color or content of organic matter constituents. A paleoclimatic reconstruction might be possible if some of the organic matter properties could be shown to have been inherited from a former vegetation cover (Ch. 10) and if the change in vegetation was due to a change in climate.

Bt-Horizon Properties

The position of the Bt horizon, its color, and the amount of clay all might contribute to an understanding of past climates. Correct interpretations are hard to make, however.

The position of the Bt horizon should relate to the climate under which the soil formed. It is commonly reported that aridland soils are shallow and that soils in progessively wetter climates are progressively thicker; the Bt-horizon position would parallel these trends. One might be able to relate the position of the Bt horizon to the water-holding capacity of the soil and to water movement within the soil.

Recent work on two soils on the coast of Southern California illustrates the above approach by relating depth of Bt-horizon features to the calculated depth of wetting. Parent material for the younger soil is a sand that is at least several tens of thousands of years old.[109] The soil is characterized by clay bands (lamellae) in the Bt horizon. Band positions are considered to coincide with the soil-moisture wetting front at the time of band formation. Calculations suggest that the deeper bands could not form under the present soil-moisture regime; rather, they require a greater annual or seasonal moisture excess (precipitation minus potential evapotranspiration for those months in which precipitation exceeds the latter). It was estimated that only the shallower bands could form under the present soil-moisture regime, which is characterized by a mean annual precipitation of 26 cm and a temperature of 17 °C. A possible analogue for the proposed wetter climate was suggested by a locality farther north along the coast, where the mean annual climatic parameters are about 80 cm and 14 °C.

Features of a nearby older soil are interpreted to show the effects of the same proposed climatic change.[110] The soil is a Xeralf, with a prominent Bt horizon that extends to 76 cm depth, which is immediately underlain by a duripan to 466 cm depth. The duripan is designated as a Btqm horizon, and is interpreted as the lower part of a Bt horizon that has subsequently been cemented with silica. The above-mentioned climatic change is called upon to explain the Btqm horizon. During the wetter regime, the lower part of the Bt horizon extended to 466 cm. The climate changed subsequently to the present one, which on average, wets to about the top of the duripan. Silica, released by weathering since the climatic change, has been translocated downward to form the duripan and the latter has protected that part of the profile from further weathering and clay translocation.

A similar approach has been used by Reheis[86] to explain the greater depth of pedogenic clay in soils in Montana that are older than the late-Wisconsin glaciation. Soil-moisture conditions calculated for glacial conditions suggest that sufficient moisture was available to translocate clays to the observed depths.

A surface soil that is redder than younger surface soils in the area might be used as an indication of a former warmer climate.[26,27,38,70,94] This can only be done, however, with equivalent parent materials. Moreover, the duration of soil formation has to be known fairly certainly so that the time factor can be ruled out as a cause of the redness. If it can be demonstrated that the red soil formed over a time span similar to that over which the younger non-red soils formed, a temperature change might well have occurred.

The amount of clay present in a soil can be used to indicate past climatic change if the duration of soil formation is precisely known. Here, if the clay content of an old buried soil is greater than that in a soil formed under the present climate, and if both soils formed over a similar time span, it can be concluded that the older soil formed in an environment conducive to a more rapid production of clay. To increase the rate of clay production, one might postulate an increase in precipitation, an increase in temperature, or both. Variation in the rate of eolian dust influx also has to be considered. In order for this approach to really work, we need data on clay production in post-glacial time in many environments, but these data are not too abundant. Furthermore, we need better data on the age of the soils and the duration of their formation, but these data also are difficult to obtain.

The above approach to paleoclimate can be demonstrated with two field examples in the western United States. Hunt and Sokoloff[50] remarked that strongly developed pre-Wisconsin soils of the Rocky Mountain region may have developed in the same time span as the weakly developed post-Wisconsin soils, but under different climatic conditions. This may well be true, but I do not know of any climatic region in the United States in which post-Wisconsin soils are anywhere as strongly developed as the pre-Wisconsin soils they describe. As a second example, Morrison and Frye[76] state that the moderately developed Churchill Geosol in the Lake Lahontan Basin probably formed in not much more than 5000 years, during the mid-Wisconsin. This soil formed from a sand (see soil formed on early-Wisconsin deposit at Mustang, Figs. 11-9 and 12-2). Again in reconnaissance studies in many areas, I have not seen a 5000-year-old soil anywhere as well developed as that one, given a similar parent material. In this case, the suggestion has been made that more time is required to form the Churchill Soil.[14] More recent data bear this out, for the revised Quaternary history of Lake Lahonton suggests that as much as or more than ten times the 5000 years suggested above were available for formation of that soil.[71] This does not negate the approach, however, which I feel is valid. A climate different from the current climate may have been present when the soils formed; this is just very hard to prove by pedologic criteria.

Complex soil morphologies in England are explained by climatic change from interglacials to glacials.[23,27] Red argillic horizons formed during major interglacials, which may have been warmer and wetter than the present climate. Cryoturbation processes under periglacial conditions during the following glacial disrupted the argillic horizon, as shown by both field and thin-section evidence. These processes associated with interglacials and glacials may have been repeated several times in the older soils. Such argillic horizons are termed paleo-argillic, to denote this complex history.

Clay Mineralogy

Clay-mineral formation and transformation in the soil are slow processes in some environments, and therefore clay mineralogy may be a useful tool in assessing past climatic influences on the soil. Here I will simplify the situation and mention only the interpretations that can be made for kaolinite, montmorillonite, and palygorskite.

The stability of a clay mineral in a changing, soil-leaching environment is a function of the clay mineral originally formed and the direction of the climatic change.[11,84,101] As mentioned earlier, kaolinite and halloysite formation is favored by a high leaching environment, and montmorillonite by a relatively low leaching environment. In well-drained parent materials, therefore, the clay minerals formed are generally a function of climate. A change in soil-leaching environment triggered by a change in climate may bring about a change in the clay mineralogy, but the climatic change usually has to be in one direction to be read. For example, many studies have shown kaolinite to be a stable mineral, one that will persist for long intervals of time in a neutral to alkaline environment. Thus, if kaolinite is formed in a soil under a fairly high leaching and relatively acid environment, and the climate becomes arid, kaolinite may persist in the soil and serve as an indicator of a former wetter climate. In places, however, there is a suggestion that montmorillonite can form from kaolinite in the appropriate soil-chemical environment.[113] Montmorillonite, on the other hand, is less stable than kaolinite and will change to the latter if greater leaching conditions obtain. In summary, therefore, a climatic change from humid to arid could be read in the clay mineralogy, but one from arid to humid most likely could not be read. One final point should be made and that is that clay-mineral transformation is a slow process, and therefore the climatic change would have had to persist long enough to allow time for slow reactions to take place. If the climatic change were of short duration, it may not be registered in the clay mineralogy.

A transect from the relatively humid east side of the Sierra Nevada into the dry western Great Basin can serve as an example of the use of clay mineralogy to help define Quaternary climatic change (climatic data are given in the caption for Fig. 11-9).[11] The clay mineralogy of soils developed from tills, outwash deposits, and lacustrine deposits is closely correlated with the present climate. Halloysite is the predominant clay mineral in the humid environment, and montmorillonite is predominant in the arid environment. The change from one mineral assemblage to the other takes place at the present position of the pine-sagebrush vegetation boundary, at about 40-cm mean annual precipitation. The interesting thing about this study is that the clay mineralogy of soils formed on river-terrace deposits of all ages changes at about the same location under similar present-day environmental conditions. This is taken to mean that long-term past climatic change toward a wetter soil-moisture regime in the western Great Basin has not taken place, for if it had, it probably would be seen as predominant kaolinite or halloysite in the older soils far east of their easternmost predominant occurrence in the younger soils. The above conclusions seem to be valid for a large number of sites along the eastern Sierra Nevada.[13] This would be compatible with the cold-dry model for at least part of the glacials.[39] Alternatively, if climatic changes had occurred, the soil-water chemistry did not change sufficiently to alter the equilibrium mineral assemblage (see Fig. 4-5). Perhaps this is another example of a pedologic threshold, or it could be that climatic change may have been significant but not of long enough duration to alter the previously formed clay minerals significantly.

More work could be done on this aspect of paleopedology. Data on clay-mineral stabilities in changing environments are needed, as are data on the time needed to convert original minerals to new species. Furthermore, these data should be looked at in view of independent data on past climates. For example, if the major portion of clay formation takes place in a relatively warm and dry interglacial interval, and the subsequent glacial climate is wetter and cooler, the cooler temperatures may inhibit alteration to a new clay-mineral species if the rate of alteration can be shown to be strongly temperature-dependent. If this is the case, the clay mineralogy of old soils will carry more of the record of the interglacials than of the glacials. Data presented in Chapter 8, however, suggest that alteration can be fairly rapid in relatively cold environments as long as sufficient moisture is present; surely a mean annual precipitation of 100 cm seems sufficient.

The presence of palygorskite and sepiolite in soils has the potential to be an excellent indicator of long-term aridity, as shown by their common occurrence with carbonate horizons,[5] and the calculated stability field[102] for palygorskite with respect to the soil-water chemistry. Calcareous soils in the Texas High Plains contain both of these minerals as a parent-material component.[9] Under the present pedogenic regime, both minerals are being degraded. More regional studies should be undertaken to determine the general climatic conditions under which these minerals are stable and unstable there and elsewhere in the southwestern United States. Only then can statements be made about the paleoclimatic implications based on these mineralogical markers of semiarid and arid conditions.

A final point on using clay-mineral species for paleoclimatic reconstruction is that, like so many other soil properties, it is generally not possible to separate the effects of precipitation from those of temperature. For example, in a particular region, the climatic change may have been toward both higher precipitation and higher temperature. The overall effect on the soil-leaching environment and on the soil-water chemistry may have been nil, and thus one would not expect a change in clay mineralogy to accompany the climatic change. What may happen in this case would be an acceleration in the rate of alteration of primary minerals to clay minerals.

Rather than rely solely on specific clay-mineral species for paleoclimatic interpretation, one can use the depth of pedogenic clay-mineral alteration. The depth of such alteration should be closely related to the depth of soil-moisture penetration, provided sufficient time was available for alteration to be registered in the clay minerals. This approach has been used in the central Yukon Territory, Canada to determine paleoclimates.[37] Soils have formed from glacial and glaciofluvial deposits of late-Wisconsin, early-Wisconsin or pre-Wisconsin, and early-Pleistocene ages. Under the present pedogenic environment, which is subarctic and semiarid, it is calculated that the water penetration is about 20 cm. The soil on the Wisconsin deposits probably formed under that climate, and the pedogenic alteration of clay minerals in glaciofluvial deposits is close to that depth. For the progressively older soils, the depth of pedogenic, clay-mineral alteration is 93 and 190$^+$ cm, respectively; a more humid climate in the past seems to be required for alteration to extend to those depths.

Mineral Etching

In only one study has mineral etching been used to infer past climate. The etching of hornblende grains has been shown to vary with depth and with age of parent material on Baffin Island.[65] It was noted, however, that the rate of etching changed markedly, with some time intervals characterized by low rates and others by high rates. It is suggested that low rates of etching are related to limited soil moisture and high rates to excessive soil moisture, and these are related to climate. A cold, dry climate is inferred for the times of limited soil moisture, and a mild, moist climate for the times of excessive soil moisture.

Position of Horizon of CaCO₃ Accumulation and Morphology of the K Horizon

Because the presence or absence of $CaCO_3$ in a soil is related to soil-water movement, and hence to climate, the position and morphological features of $CaCO_3$ accumulations may provide insight into past climates.

The position of the top of the zone of $CaCO_3$ accumulation is related to the regional climate, as long as the water-holding capacity of the soil is taken into account. Because Arkley[3] found that the tops of many of the Bk or K horizons plot close to the sharp break in the calculated soil-water-movement curves that represents a rapid decrease in water movement with depth (Fig. 11-3), he concluded that climatic change does not have to be called upon to explain the position of the tops of such horizons. This seems to be a reasonable conclusion. However, before one can explain the lack of such correspondence between climate and the top of the Bk or K horizon as being paleoclimatically significant, one would have to investigate thoroughly the possibility of surface erosion or deposition as reasons for the top of such a horizon being either closer to or farther from the surface, respectively, than predicted.

The position of the Bk and/or K horizons is related to that of the Bt horizon; in most places, they lie directly beneath the Bt. If, however, they lie some distance below the base of the Bt, and this position can be shown not to be of groundwater origin, the possibility of a shift to a wetter climate, driving the $CaCO_3$ to greater depth, should be investigated. I have seen this in one locality in Wyoming, where the parent material is a well-drained gravel of a high river terrace. About 20 cm of unweathered Cu material separates the base of the Bt horizon from the top of the Bk horizon. One interpretation is that the top of the Bk was 20 cm higher in the past and that weathering of iron-bearing minerals was inhibited in the carbonate environment. A subsequent climatic change resulted in deeper water penetration to force the top of the Bk horizon to greater depths, but the change has been recent enough so that the silicate grains in the former Bk horizon have not altered to form either a Cox or a Bw horizon. However, if the Bk horizon extends up into the base of the Bt horizon, there may have been a climatic change toward aridity as long as no other reasons for the upward movement of the top of the Bk can be demonstrated. Two reasons for non-climatic upward movement of the top of the Bk can be mentioned. One is that as clay accumulates in the Bt horizon, the water-holding capacity of the horizon increases. The result of this is that more water is held in the Bt than previously was the case. Hence, the water carrying Ca^{2+} and HCO_3^- through the Bt does not move so deeply, and these ions precipitate to form a Btk horizon. The other, non-climatic reason for upward movement of the top of the $CaCO_3$ accumulation layer involves the K horizon. As the K horizon forms, the pores of the soil are progressively plugged so that deep water penetration is inhibited. The downward-moving waters, therefore, are held at the top of the K, and eventually ionic concentrations in the soil solution build up to the point where $CaCO_3$ precipitates out. In this way, the top of the K moves toward the surface, engulfing the base of the B horizon in the process.[43]

Depths of carbonate accumulations have been used to suggest climatic change in some semiarid and arid parts of Southern California.[72] The Holocene soils have a shallow accumulation of carbonate, whereas the accumulation in Pleistocene soils is three to four times as deep. In fact, late Pleistocene soils have two carbonate accumulation maxima with depth—a shallow one related to the present Holocene climate and a deeper one related to the greater effective soil moisture of the late-Pleistocene.

One other aspect of climatic change with respect to carbonate soils is possible variation in accumulation rate with climatic change. For southern New Mexico, Machette and others[68] have evidence that the rate of carbonate accumulation during interpluvials could be twice that of the pluvials. They do not suggest reasons for this, but one obvious one is vegetation type and percent cover. During pluvials, for instance, the source areas may be more restricted relative to those of the interpluvials. To be able to read climatic change on these data, one needs well-dated soils in a chronosequence or buried soils from which accumulation rates over various time intervals can be determined.

Lateral and altitudinal displacements of the calcic–non-calcic soil boundary along climatic transects in soils of different age can also be used as indicators of climatic change. Richmond[88,89] reports such displacements in many parts of the Rocky Mountain region (Fig. 11-21), and they are also seen in deposits along the eastern side of the Sierra Nevada near Reno and in the Colorado Piedmont near Boulder. In all these transects, the calcic–non-calcic soil boundary lies at a higher altitude, and therefore in a wetter present climate, on the older deposits. It is believed that the older soils formed during an interval of time warmer than the present, or possibly, that the combination of precipitation and temperature at that time resulted in $CaCO_3$ accumulation in soils at a higher altitude than

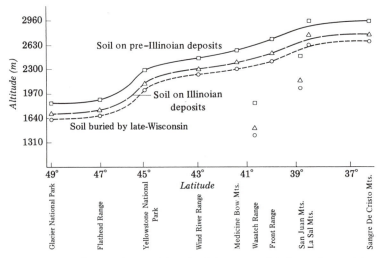

Fig. 11-21 Altitudinal relationship of calcic–non-calcic boundary with age of parent material, Rocky Mountains. (Taken from Richmond,[90] Fig. 2.)

would occur under the present climate. These explanations are plausible, if the position of the calcic–non-calcic soil boundary is not related to the different water-holding capacities of the different-aged soils. For example, the pre-Illinoian soils commonly have Bt horizons of high water-holding capacity, whereas the soils formed from Wisconsin deposits often lack Bt horizons, and thus, their water-holding capacities are less. Under certain climates, it might be possible for Ca^{2+} and HCO_3^- to be leached from the soils with low water-holding capacity, yet be retained to form a Bk horizon in an adjacent older soil with high water-holding capacity.

It is difficult to explain the preservation of $CaCO_3$ in older soils at altitudes at which adjacent younger soils show no accumulation of $CaCO_3$. Richmond[88] offers the following observations. The cooler glacial climates that prevailed subsequent to the formation of the Bk horizons in the older soils should have readily dissolved and removed the $CaCO_3$ because $CaCO_3$ is more soluble at lower temperatures (Fig. 5-8). Other factors, however, may have decreased the rate at which $CaCO_3$ could have been removed from the soil. Richmond includes the limiting chemical activity with a colder glacial climate and the possibility of long seasons during which the ground was frozen as possible explanations for the preservation of the Bk horizons. Another possibility is that perhaps some areas did not experience much of an increase in soil leaching during the glacial climates, a hypothesis that could be supported by the clay-mineral data in the Sierra Nevada-Great Basin transect and references on late Quaternary climatic change, mentioned in the previous section.

Morphological features of K horizons may provide important data on solution and reprecipitation of $CaCO_3$ that perhaps owe their origin to climatic change.[20,22] Some K horizons, for example, have solution pits on their upper surface. If it can be shown that the solution pits are not due to deeper water penetration accompanying ground-surface lowering by erosion, this evidence suggests a change toward a wetter climate. Other soils show evidence of formation of laminated K horizons, brecciation of these horizons, and subsequent cementation of the breccia. Although the processes responsible for this history are not clearly understood, they could be due to climatic change. Perhaps brecciation occurs during intervals of drier climate and cementation of the fragments during intervals of wetter climate.

Position of Gypsic and Salic Horizons, and Associated Weathering Features

The position of salt accumulations could be one of the better indications of climatic change in arid regions. The reason for this is that, because of their high solubility, a change in soil-moisture regime could be reflected in a change in the position of salt accumulation. If the change is toward a drier soil-moisture regime, it might show as an accumulation of salts high in the profile where it is disjunct with respect to the rest of the profile; for example, salts occur in the Bt horizons in some soils in Arizona.[78] In this case, the climatic-change trend is toward diminishing soil moisture with time. In other places, there might be several positions of salt accumulation maxima within a profile, as exemplified by a soil in the central Sinai.[32] Here again, the interpretation could be one of increasing aridity with time.

Because of the high solubility of these salts, there are some problems in using them in paleoclimatic reconstruction. One is that if the younger climate is the more humid one, salts could be leached from the profile and the record of previous aridity eradicated. Another is the effect of wetter-than-normal years that occur every century or even less frequently. These latter might result in complex salt-accumulation patterns that should not be interpreted as due to climatic change. Finally, surface erosion and deposition could influence the depth of salt accumulation, and may result in complex accumulation patterns.

There is one problem with the above interpretation. Reheis[87] cautions workers to consider the hygroscopic qualities of gypsum, since it can absorb and adsorb larger quantities of water than can silicate mineral grains. She suggests, therefore, that horizons of high gypsum content may persist under a climatic change as long as the excess soil moisture is taken up by the gypsum. Perhaps a soil-moisture threshold has to be exceeded for the level of gypsum accumulation to move to greater depth. Laboratory experiments could be performed to test this hypothesis.

Despite the possible ephemeral nature of salt-accumulation horizons with fluctuating climates, associated weathering features may be used to infer former positions of salt accumulation. In the northern Sinai Desert and in the northeastern Bighorn Basin, Wyoming, some horizons of salt accumulation in gravelly parent materials are marked by mechanically cracked clasts[40,87] (Fig. 11-22). The cracking is due to the salt weathering mentioned in Chapter 3 and, presumably, depends both on salt content and time. Levels of cracked stones in soils have the potential for marking former horizons of salt accumulation, which indicate former soil-moisture regimes and therefore former climates; the cracked stones would persist as easily recognized features even if a younger, more humid climate leaches the salts from the horizon or soil.

Negative Evidence for Change in Soil-Moisture Regime

The soil data discussed above might be used to indicate climatic change; conversely, some soil data, including chemical and mineralogical data, might be used to indicate that the soil-moisture regime and perhaps the climate have not changed profoundly. One should first be aware of the association of the above properties with time and climate and determine if they are compatible with the present soil-moisture regime. Thick salt or carbonate accumulations, or abundant palygorskite or sepiolite, indicate long intervals of aridic conditions. The lack of both notable depletion of acid-extractable P and an accumulation of various extracts of Fe or Al

Fig. 11-22   Cracked clasts in the By horizon of an Aridisol, Sinai Desert, southern Israel.

in old soils may also support this interpretation. Marked depletion in the P or accumulation of Fe and Al, or abundant clay minerals indicative of leaching environments, however, could be used to indicate a long-term, excessive soil-moisture regime. A problem will be to determine more precisely any former climate on these data, because many combinations of precipitation and temperature can result in a similar soil-moisture regime.

REFERENCES

1. Allen, B.L., 1977, Mineralogy and soil taxonomy, p. 771–796 *in* J.B. Dixon and S.B. Weed, eds., Minerals in soil environments: Soil Sci. Soc. Amer., Madison, Wisconsin.
2. Arkley, R.J., and Ulrich, R., 1962, The use of calculated actual and potential evapotranspiration for estimating potential plant growth: Hilgardia, v. 32, p. 443–462.
3. Arkley, R.J., 1963, Calculation of carbonate and water movement in soil from climatic data: Soil Sci., v. 96, p. 239–248.
4. Arkley, R.J., 1967, Climates of some Great Soil Groups of the western United States: Soil Sci., v. 103, p. 389–400.
5. Bachman, G.O., and Machette, M.N., 1977, Calcic soils and calcretes in the southwestern United States: U.S. Geol. Surv. Open-File Rept. 77-794, 163 p.
6. Barshad, I., 1966, The effect of a variation in precipitation on the nature of clay mineral formation in soils from acid and basic igneous rocks: 1966 Internat. Clay Conf. (Jerusalem), Proc. v. 1, p. 167–173.
7. Beaven, P.J., and Dumbleton, M.J., 1966, Clay minerals and geomorphology in four Caribbean Islands: Clay Minerals, v. 6, p. 371–382.
8. Bigham, J.M., Golden, D.C., Buol, S.W., Weed, S.B., and Bowen, L.H., 1978, Iron oxide mineralogy of well-drained Ultisols and Oxisols: II. Influence on color, surface area, and phosphate retention: Soil Sci. Soc. Amer. Jour., v. 42, p. 825–830.
9. Bigham, J.M., Jaynes, W.F., and Allen, B.L., 1980, Pedogenic degradation of sepiolite and palygorskite in the Texas High Plains: Soil Sci. Soc. Amer. Jour., v. 44, p. 159–167.
10. Birkeland, P.W., 1968, Correlation of Quaternary stratigraphy of the Sierra Nevada with that of the Lake Lahontan area, p. 469–500 *in* R.B. Morrison and H.E. Wright, Jr., eds., Means of correlation of Quaternary successions: Internat. Assoc. Quaternary Res., VII Cong., Proc. v. 8, 631 p.
11. ———, 1969, Quaternary paleoclimatic implications of soil clay mineral distribution in a Sierra Nevada–Great Basin transect: Jour. Geol., v. 77, p. 289–302.
12. ———, 1978, Soil development as an indication of relative age of Quaternary deposits, Baffin Island, N.W.T., Canada: Arctic Alp. Res., v. 10, p. 733–747.
13. Birkeland, P.W., and Janda, R.J., 1971, Clay mineralogy of soils developed from Quaternary deposits of the eastern Sierra Nevada, California: Geol. Soc. Amer. Bull., v. 82, p. 2495–2514.
14. Birkeland, P.W., and Shroba, R.R., 1974, The status of the concept of Quaternary soil-forming intervals in the western United States, p. 241–276 *in* W.C. Mahaney, ed., Quaternary environments: York Univ. Geog. Mono. No. 5, Toronto, Canada.
15. Birkeland, P.W., Burke, R.M., and Walker, A.L., 1980, Soils and subsurface rock-weathering features of Sherwin and pre-Sherwin glacial deposits, eastern Sierra Nevada, California: Geol. Soc. Amer. Bull., v. 91, p. 238–244.
16. Bockheim, J.G., 1979, Properties and relative age of soils of southwestern

Cumberland Peninsula, Baffin Island, N.W.T., Canada: Arctic Alp. Res., v. 11, p. 289–306.

17. ———, 1980, Properties and classification of some desert soils in coarse-textured glacial drift in the Arctic and Antarctic: Geoderma, v. 24, p. 45–69.

18. ———, 1982, Properties of a chronosequence of ultraxerous soils in the Trans-Antarctic Mountains: Geoderma, v. 28, p. 239–255.

19. Bown, T.M., Kraus, M.J., Wing, S.L., Fleagle, J.G., Tiffney, B.H., Simons, E.L., and Vondra, C.F., 1982, The Fayum primate forest revisited: Jour. Human Evol., v. 11, p. 603–632.

20. Bretz, J.H., and Horberg, L., 1949, Caliche in southeastern New Mexico: Jour. Geol., v. 57, p. 491–511.

21. Brook, G.A., Folkoff, M.E., and Box, E.O., 1983, A world model of soil carbon dioxide: Earth Surface Proc. Landforms, v. 8, p. 79–88.

22. Bryan, K., and Albritton, C.C., Jr., 1943, Soil phenomena as evidence of climatic changes: Amer. Jour. Sci., v. 241, p. 469–490.

23. Bullock, P., and Murphy, C.P., 1979, Evolution of a paleo-argillic brown earth (Paleudalf) from Oxfordshire, England: Geoderma, v. 22, p. 225–252.

24. Burke, R.M., 1979, Multiparameter relative dating (RD) techniques applied to morainal sequences along the eastern Sierra Nevada, California and Wallowa Lake area, Oregon: Ph.D. thesis, Univ. of Colorado, Boulder, 166 p.

25. Burke, R.M., and Birkeland, P.W., 1979, Reevaluation of multiparameter relative dating techniques and their application to the glacial sequence along the eastern escarpment of the Sierra Nevada, California: Quaternary Res., v. 11, p. 21–51.

26. Catt, J.A., 1979, Soils and Quaternary geology in Britain: Jour. Soil Sci., v. 30, p. 607–642.

27. Chartres, C.J., 1980, A Quaternary soil sequence in the Kennet Valley, central southern England: Geoderma, v. 23, p. 125–146.

28. Claridge, G.G.C., and Campbell, I.B., 1982, A comparison between hot and cold desert soils and soil processes: Catena, Suppl. 1, p. 1–28.

29. Crandell, D.R., 1965, The glacial history of western Washington and Oregon, p. 341–353 *in* H.E. Wright, Jr., and D.G. Frey, eds., The Quaternary of the United States: Princeton University Press, Princeton.

30. ———, 1969, Surficial geology of Mount Rainier National Park, Washington: U.S. Geol. Surv. Bull. 1288, 41 p.

31. Dan, J., and Yaalon, D.H., 1982, Automorphic saline soils in Israel: Catena, Suppl. 1, p. 103–115.

32. Dan, J., Yaalon, D.H., Moshe, R., and Nissim, S., 1982, Evolution of Reg soils in southern Israel and Sinai: Geoderma, v. 28, p. 173–202.

33. Eswaran, H., and Sys, C., 1970, An evaluation of the free iron in tropical basaltic soils: Pedologie, v. 20, p. 62–85.

34. Eswaran, H., and Tavernier, R., 1980, Classification and genesis of Oxisols, p. 427–442 *in* B.K.G. Theng, ed., Soils with variable charge: New Zealand Society of Soil Science, Lower Hutt.

35. Evans, L.J., and Cameron, B.H., 1979, A chronosequence of soils developed from granitic morainal material, Baffin Island, N.W.T.: Canad. Jour. Soil Sci., v. 59, p. 203–210.

36. Forsyth, J.L., 1965, Age of the buried soil in the Sidney, Ohio, area: Amer. Jour. Sci., v. 263, p. 571–597.
37. Foscolos, A.E., Rutter, N.W., and Hughes, O.L., 1977, The use of pedological studies in interpreting the Quaternary history of central Yukon Territory: Geol. Surv. Canada Bull. 271, 48 p.
38. Frye, J.C., and Leonard, A.B., 1967, Buried soils, fossil mollusks, and late Cenozoic paleoenvironments, p. 429–444 *in* C. Teichert and E.L. Yochelson, eds., Essays in paleontology and stratigraphy: University of Kansas Press, Lawrence.
39. Galloway, R.W., 1983, Full-glacial southwestern United States: Mild and wet or cold and dry?: Quaternary Res., v. 19, p. 236–248.
40. Gerson, R., 1981, Geomorphic aspects of the Elat Mountains, p. 279–296 *in* J. Dan, R. Gerson, H. Koyumdjisky, and D.H. Yaalon, eds., Aridic soils of Israel: Agri. Res. Organization (Israel) Spec. Publication No. 190.
41. Gile, L.H., 1975, Holocene soils and soil-geomorphic relations in an arid region of southern New Mexico: Quaternary Res., v. 5, p. 321–360.
42. ———, 1977, Holocene soils and soil-geomorphic relations in a semiarid region of southern New Mexico: Quaternary Res., v. 7, p. 112–132.
43. Gile, L.H., Peterson, F.F., and Grossman, R.B., 1966, Morphological and genetic sequences of carbonate accumulation in desert soils: Soil Sci., v. 101, p. 347–360.
44. Gile, L.H., Hawley, J.W., and Grossman, R.B., 1981, Soils and geomorphology in the Basin and Range area of southern New Mexico—Guidebook to the desert project: New Mexico Bur. Mines and Mineral Resources Mem. 39, 222 p.
45. Gradusov, B.P., 1974, A tentative study of clay mineral distribution in soils of the world: Geoderma, v. 12, p. 49–55.
46. Harden, J.W., and Taylor, E.M., in press, A quantitative comparison of soil development in four climatic regimes: Quaternary Res.
47. Harradine, F., and Jenny, H., 1958, Influence of parent material and climate on texture and nitrogen and carbon contents of virgin California soils. I. Texture and nitrogen contents of soils: Soil Sci., v. 85, p. 235–243.
48. Hay, R.L., and Jones, B.F., 1972, Weathering of basaltic tephra on the island of Hawaii: Geol. Soc. Amer. Bull., v. 83, p. 317–332.
49. Herbillon, A.J., 1980, Mineralogy of Oxisols and oxic materials, p. 109–126 *in* B.K.G. Theng, ed., Soils with variable charge: New Zealand Society of Soil Science, Lower Hutt.
50. Hunt, C.B., and Sokoloff, V.P., 1950, Pre-Wisconsin soil in the Rocky Mountain Region, a progress report: U.S. Geol. Surv. Prof. Pap. 221-G, p. 109–123.
51. Hurst, V.J., 1977, Visual estimation of iron in saprolite: Geol. Soc. Amer. Bull., v. 88, p. 174–176.
52. Janda, R.J., and Croft, M.G., 1967, The stratigraphic significance of a sequence of Noncalcic Brown soils formed on the Quaternary alluvium of the northeastern San Joaquin Valley, California, p. 157–190 *in* R.B. Morrison and H.E. Wright, Jr., eds., Quaternary soils: Internat. Assoc. Quaternary Res., VII Cong., Proc. v. 9, 338 p.
53. Jenny, H., 1935, The clay content of the soil as related to climatic factors, particularly temperature: Soil Sci., v. 40, p. 111–128.

54. ———, 1941, Factors of soil formation: McGraw-Hill, New York, 281 p.
55. ———, 1941, Calcium in the soil: III. Pedologic relations: Soil Sci. Soc. Amer. Proc., v. 6, p. 27–37.
56. ———, 1950, Causes of the high nitrogen and organic matter content of certain tropical forest soils: Soil Sci., v. 69, p. 63–69.
57. ———, 1961, Comparison of soil nitrogen and carbon in tropical and temperate regions: Missouri Agri. Exp. Sta. Res. Bull. 765, p. 5–31.
58. ———, 1980, The soil resource: Springer-Verlag, New York, 377 p.
59. Jenny, H., Gessel, S.P., and Bingham, F.T., 1949, Comparative study of decomposition rates of organic matter in temperate and tropical regions: Soil Sci., v. 68, p. 419–432.
60. Kämpf, N., and Schwertmann, U., 1983, Goethite and hematite in a climosequence in southern Brazil and their application in classification of kaolinite soils: Geoderma, v. 29, p. 27–39.
61. Kukla, J., 1970, Correlation between loesses and deep-sea sediments: Geologiska Föreningens i Stockholm Förhandlingar, v. 92, p. 148–180.
62. ———, 1977, Pleistocene land-sea correlations. 1. Europe: Earth-Sci. Rev., v. 13, p. 307–374.
63. Lea, P.D., 1983, Glacial history of the southern margin of the Puget Lowland, Washington: M.S. thesis, Univ. of Washington, Seattle.
64. Levine, E.L., and Ciolkosz, E.J., 1983, Soil development in till of various ages in northeastern Pennsylvania: Quaternary Res., v. 19, p. 85–99.
65. Locke, W.W., III, 1979, Etching of hornblende grains in arctic soils: An indicator of relative age and paleoclimate: Quaternary Res., v. 11, p. 197–212.
66. Loughnan, F.C., 1969, Chemical weathering of the silicate minerals: American Elsevier Publ. Co., New York, 154 p.
67. MacFarlane, M.J., 1976, Laterite and landscape: Academic Press, New York, 151 p.
68. Machette, M.N., Harper-Tervet, J., and Timbel, N.R., in press, Surficial deposits of the southwestern United States: Geol. Soc. Amer. Spec. Pap.
69. Marshall, C.E., 1977, The physical chemistry and mineralogy of soils. Vol. II: Soils in place: John Wiley and Sons, New York, 313 p.
70. Matsui, T., 1967, On the relic red soils of Japan, p. 221–224 *in* R.B. Morrison and H.E. Wright, Jr., eds., Quaternary soils: Internat. Assoc. Quaternary Res., VII Cong., Proc. v. 9.
71. McCoy, W.D., 1981, Quaternary aminostratigraphy of the Bonneville and Lahontan basins, western United States, with paleoclimatic interpretations: Ph.D. thesis, Univ. of Colorado, Boulder, 603 p.
72. McFadden, L.D., 1982, The impacts of temporal and spatial climatic changes on alluvial soils genesis in Southern California: Ph.D. thesis, Univ. of Arizona, Tucson, 430 p.
73. Mears, B., Jr., 1981, Periglacial wedges and the late Pleistocene environment of Wyoming's intermontane basins: Quaternary Res., v. 15, p. 171–198.
74. Mohr, E.C.J., van Baren, F.A., and van Schuylenborgh, J., 1973, Tropical soils: Mouton-Ichtiar Baru-van Hoeve, The Hague, Netherlands, 481 p.
75. Moore, T.R., 1978, Soil development in arctic and subarctic areas of Quebec and Baffin Island, p. 379–411 *in* W.C. Mahaney, ed., Quaternary soils: Geo Abstracts, Norwich, England.

76. Morrison, R.B., and Frye, J.C., 1965, Correlation of the middle and late Quaternary successions of the Lake Lahontan, Lake Bonneville, Rocky Mountain (Wasatch Range), southern Great Plains, and eastern mid-west areas: Nevada Bur. Mines Rept. 9, 45 p.

77. Mulcahy, M.J., 1967, Landscapes, laterites, and soils in southwestern Australia, p. 211–230 *in* J.N. Jennings and J.A. Mabbutt, eds., Landform Studies from Australia and New Guinea: Australian National University Press, Canberra.

78. Nettleton, W.D., Witty, J.E., Nelson, R.E., and Hawley, J.W., 1975, Genesis of argillic horizons in soils of desert areas of the southwestern United States: Soil Sci. Soc. Amer. Proc., v. 39, p. 919–926.

79. Nettleton, W.D., Nelson, R.E., Brasher, B.R., and Derr, P.S., 1982, Gypsiferous soils in the western United States, p. 147–168 *in* J.A. Kittrick, D.S. Fanning, and L.R. Hossner, eds., Acid sulfate weathering: Soil Sci. Soc. Amer. Spec. Publ. 10, Madison, Wisconsin.

80. Page, W.D., 1972, The geological setting of the archaeological site at Oued El Akarit and the paleoclimatic significance of gypsum soils, southern Tunisia: Ph.D. thesis, Univ. of Colorado, Boulder, 111 p.

81. Paquet, H., and Millot, G., 1973, Geochemical evolution of clay minerals in the weathered products of soils of Mediterranean climates: Internat. Clay Conf., 1972, Soc. Espanola de Arcillas-Assoc. Internat. pour l'Etude des Argiles, Madrid, v. 1, p. 255–261.

82. Pastor, J., and Bockheim, J.G., 1980, Soil development on moraines of Taylor Glacier, lower Taylor Valley, Antarctica: Soil Sci. Soc. Amer. Jour., v. 44, p. 341–348.

83. Pawluk, S., 1978, The pedogenic profile in the stratigraphic section, p. 61–76 *in* W.C. Mahaney, ed., Quaternary soils: Geo Abstracts Ltd., Norwich, England.

84. Pedro, G., Jamagne, M., and Bejon, J.C., 1969, Mineral interactions and transformations in relation to pedogenesis during the Quaternary: Soil Sci., v. 107, p. 462–469.

85. Porter, S.C., 1969, Pleistocene geology of the east-central Cascade Range, Washington: Guidebook for 3rd Pacific Coast Friends of the Pleistocene field conference, 54 p.

86. Reheis, M.C., in press, Three soil chronosequences on the Rock Creek terraces, south-central Montana, and implications for paleoclimate: U.S. Geol. Surv.

87. ———, in press, The soil chronosequence on the Kane fans, Bighorn Basin, Wyoming: U.S. Geol. Surv.

88. Richmond, G.M., 1962, Quaternary Stratigraphy of the La Sal Mountains, Utah: U.S. Geol. Surv. Prof. Pap. 324, 135 p.

89. ———, 1965, Glaciation of the Rocky Mountains, p. 217–230 *in* H.E. Wright, Jr., and D.G. Frey, eds., The Quaternary of the United States: Princeton Univ. Press, Princeton, 922 p.

90. ———, 1972, Appraisal of the future climate of the Holocene in the Rocky Mountains: Quaternary Res., v. 2, p. 315–322.

91. Rodin, L.E., and Bazilevič, N.I., 1966, The biological productivity of the main vegetation types in the Northern Hemisphere of the Old World: Forestry Absts., v. 27, p. 369–372.

92. Ross, G.J., 1980, The mineralogy of Spodosols, p. 127–143 *in* B.K.G. Theng, ed., Soils with variable charge: New Zealand Soc. Soil Sci., Lower Hutt.

93. Schumm, S.A., 1977, The fluvial system: John Wiley and Sons, New York, 338 p.

94. Schwertmann, U., Murad, E., and Schulze, D.G., 1982, Is there Holocene reddening (hematite formation) in soils of axeric temperate areas?: Geoderma, v. 27, p. 209–223.

95. Scott, W.E., 1977, Quaternary glaciation and volcanism, Metolius River area, Oregon: Geol. Soc. Amer. Bull., v. 88, p. 113–124.

96. Sherman, G.D., 1952, The genesis and morphology of the alumina-rich laterite clays, p. 154–161 *in* Problems of clay and laterite genesis: Amer. Inst. Mining and Metallurgical Engr., New York.

97. Shroba, R.R., 1977, Soil development in Quaternary tills, rock-glacier deposits, and taluses, southern and central Rocky Mountains: Ph.D. thesis, Univ. of Colorado, Boulder, 424 p.

98. Shroba, R.R., and Birkeland, P.W., 1983, Trends in late Quaternary soil development in the Rocky Mountains and Sierra Nevada, western United States, p. 145–156 *in* S.C. Porter, ed., Late-Quaternary environments of the United States, vol. 1. The late Pleistocene: University of Minnesota Press, Minneapolis.

99. Singer, A., 1966, The mineralogy of the clay fractions from basaltic soils in the Galilee, Israel; Jour. Soil Sci., v. 17, p. 136–147.

100. ———, 1975, A Cretaceous laterite in the Negev Desert, southern Israel: Geol. Mag., v. 112, p. 151–162.

101. ———, 1979/1980, The paleoclimatic interpretation of clay minerals in soils and weathering profiles: Earth-Sci. Rev., v. 15, p. 303–326.

102. Singer, A., and Norrish, K., 1974, Pedogenic palygorskite occurrences in Australia: Amer. Miner., v. 59, p. 508–517.

103. Singer, M.J., and Nkedi-Kizza, P., 1980, Properties and history of an exhumed Tertiary Oxisol in California: Soil Sci. Soc. Amer. Jour., v. 44, p. 587–590.

104. Strakhov, N.M., 1967, Principles of lithogenesis, v. 1: Oliver and Boyd Ltd., Edinburgh, 245 p.

105. Tardy, Y., 1971, Characterization of the principal weathering types by the geochemistry of waters from some European and African crystalline massifs: Chem. Geol., v. 7, p. 253–271.

106. Tedrow, J.C.F., 1977, Soils of the polar landscapes: Rutgers University Press, New Brunswick, N.J., 638 p.

107. The Times of London, 1967, The Times atlas of the world: Houghton Mifflin Co., Boston.

108. Thornthwaite, C.W., and Mather, J.R., 1957, Instructions and tables for computing potential evapotranspiration and the water balance: Drexel Inst. Techn., Lab. of Climatology, Publs. in Climatology, v. 10, no. 3, 311 p.

109. Torrent, J., Nettleton, W.D., and Borst, G., 1980, Clay illuviation and lamella formation in a Psammentic Haploxeralf in Southern California: Soil Sci. Soc. Amer. Jour., v. 44, p. 363–369.

110. Torrent, J., Nettleton, W.D., and Borst, G., 1980, Genesis of a Typic Durixeralf of Southern California: Soil Sci. Soc. Amer. Jour., v. 44, p. 575–582.

111. Tucker, M.E., 1978, Gypsum crusts (gypcrete) and patterned ground from northern Iraq: Zeitschrift für Geomorph., v. 22, p. 89–100.
112. Uehara, G., and Gillman, G., 1981, The mineralogy, chemistry, and physics of tropical soils with variable charge clays: Westview Tropical Agri. Series, No. 4, Westview Press, Inc., Boulder, Colorado, 170 p.
113. Weaver, R.M., Jackson, M.L., and Syers, J.K., 1971, Magnesium and silicon activities in matrix solutions of montmorillonite-containing soils in relation to clay mineral stability: Soil Sci. Soc. Amer. Proc., v. 35, p. 823–830.
114. Yaalon, D.H., 1971, Soil-forming processes in time and space, p. 29–39 *in* D.H. Yaalon, ed., Paleopedology: Israel University Press, Jerusalem.

12

Application of soils
to geomorphological studies

There are many applications of soils to geomorphological studies. One common one is to help decipher or put limits on possible climatic and/or vegetation change during the Quaternary and older periods of geological time, topics covered in Chapters 10 and 11. Another is to aid in the subdivision and correlation of unconsolidated sediments, as these are of great interest to geologists, geographers, and archeologists. Finally, soils are useful in neotectonic studies, for they can be displaced by faults and overlie faults, used to help date deposits or surfaces that have been deformed, or related to tectonics in other ways.

USE OF SOILS IN QUATERNARY STRATIGRAPHIC STUDIES

Soils are important to the subdivisions of Quaternary sediments, whether the soils are at the surface or buried. They are used primarily to aid in the subdivision of a local succession of deposits, to provide data on the lengths of time that separate periods of deposition, and to facilitate short- and long-range correlation. This rather special field of study is called soil stratigraphy, and its history and methods are reviewed quite thoroughly by Morrison.[59,62]

The use of soils for stratigraphic purposes demands that the investigator have a thorough background in pedology. It is of prime importance to be able to distinguish those features of the soil profile that are mainly geological in origin from those that are distinctly pedologic. A knowledge of processes responsible for profile development is one working tool of the soil stratigrapher, as is a good understanding of the influence of soil-forming factors on soil genesis. As an illustration, one of the first tasks of the investigator in soil-stratigraphic research is to reconstruct the landscape and delineate the factors that prevailed when the soil in question formed.

In any stratigraphic study, soils should not be used solely for correlation, to the exclusion of other field criteria. The best answers come from the integration of data from several disciplines. Thus, the soil evidence must be reconciled with evidence concerning geological events and processes, and knowledge of past climatic trends, as well as palynological data. Furthermore, one should integrate all relevant data in correlation studies, such as paleomagnetic, volcanic ash stratigraphy, and other data mentioned by Morrison.[60]

Stratigraphic studies of soils also are useful in pedologic studies, since only by a thorough knowledge of the soil in its stratigraphic setting can one understand all the factors responsible for the development of a certain profile. Indeed, as more of the many factors in soil-profile development are seen to interact, it becomes more difficult to isolate the effects of any one factor. It is still possible in many places, however, to distinguish between the main soil-forming factors and those of relatively minor importance.

Soil Stratigraphy and Soil-Stratigraphic Units (Geosols and Pedoderms)

Stratigraphy is that branch of geology dealing with the sequence, age, and correlation of rock bodies. Soil stratigraphy involves the use of soils in defining a local stratigraphic succession, estimating the ages of units associated with the soils, and suggesting short- or long-range correlations. Soil-stratigraphic units can be formally or informally named, and examples of both follow.

Soils used formally in some soil-stratigraphic studies are called pedostratigraphic units, according to the new stratigraphic code for North America.[64] Because the term "soil" has different connotations among geologists, engineers, and soil scientists, the fundamental pedostratigraphic unit is the geosol, a term proposed long ago by Morrison.[59] Geosols are recognized on pedologic criteria, they are restricted to buried soils, and they must have a known stratigraphic position and be traceable laterally in the field. The top of the unit is placed at the top of the uppermost pedologic horizon, commonly an A horizon or a truncated B horizon, and the lower boundary is the base of the lowest pedologic horizon, commonly the Cox/Cu boundary. Names for formally named geosols are taken from the name of some nearby geographical place; an example is the Churchill Geosol. If one cares to keep the nomenclature informal, one can just use the term soil in combination with the underlying and overlying units. An example would be a post-Tahoe–pre-Tioga soil of the Sierra Nevada. Geosols and informal pedostratigraphic soils should be described well at the type locality, and morphological variation laterally should be indicated. It is hoped that some laboratory characterization will accompany the description. Although not specified in the code, laterally dissimilar morphologies of a geosol are called facies,[59,62] as is done in geological studies (e.g., sedimentary facies). Thus, because of variation in pedogenic environments, a calcic facies of a particular geosol might grade laterally to an argillic-horizon facies and that to a gleyed facies characterized by properties induced by a reducing pedogenic environment.

The International Association for Quaternary Research (INQUA) has a Commission on Paleopedology, which has circulated an unpublished soil-stratigraphic guide (R.B. Parsons, written commun., 1981). Pedoderm is the formal pedo-stratigraphic unit they advocate, and it is defined much like a geosol. A pedoderm should be mappable, and should have significant areal extent, just as a geosol should have. By this is meant that these should be significant features of the stratigraphic succession of an area, not just isolated occurrences. One advantage "pedoderm" has over "geosol" is that relict and exhumed soils are included. Hence, soils associated with surface deposits of different ages (soils of post-incisive chronosequences) can be given formal status. This is important because surface soils surely are more commonly used in stratigraphic studies than are the few known buried soils, at least in many regions.

One thing to be aware of is that, although formal stratigraphic names are nice, one does not need them to do high-quality soil-stratigraphic work involving soils. We should not get into the pitfall of formally naming all kinds of soils. The focus of such studies is to use the soil data as a guide to the age of a particular deposit, and this work can be done just as well if the soils are used informally as formally. Such informal names could be linked to the name of the deposit from which the soil formed. Thus, the strongly developed soil formed from the Rocky Flats alluvium in the Colorado Piedmont[73] would be called the post-Rocky Flats soil. Many soils in the Piedmont have been named for interglacials of the midcontinent, and although this was very useful in the days of the earlier studies, such nomenclature is best avoided now. The post-Rocky Flats soil, even though it has regional significance, cannot be called a geosol because it is not buried. It could be called a pedoderm, such as the Louisville Pedoderm. Proliferation of names, however, are not always in the best interests of people trying to understand the stratigraphic succession in an area not too familiar to them.

Soil-Stratigraphic Units as an Aid in Defining Stratigraphic Units

The North American Stratigraphic Code[64] has many advantages over the old code,[1] one of which is that the difficulties of adequately defining the more common Quaternary stratigraphic units has been rectified. Geological-climatic units are no longer used, and soils are no longer used to define lithostratigraphic units. Chronostratigraphic units can still be defined on soils. The main use of soils, however, is to define allostratigraphic units, a new unit that should prove to be useful in Quaternary stratigraphic studies.

A lithostratigraphic unit is a body of rocks distinguished mainly on lithologic characteristics. In practice, therefore, two rock bodies (formations) would be separated on physical features that ideally are recognizable in the field as, for example, the separation of a green shale from a red sandstone. Although either a constructional or an erosional surface form may help in the recognition of a lithostratigraphic unit, surface form should not be a primary factor in the definition of such a unit. In many areas in which Quaternary deposits have been mapped, the deposits of all ages (e.g., tills) commonly have identical lithologies, and there is no practical way to separate the deposits in the field by the properties of the deposits themselves. Soils have been used to classify some of these units as formations,[56,70] but this is no longer advocated.

A chronostratigraphic unit is a rock body whose boundaries depend on geological time; that is, the upper and lower boundaries are isochronous over large areas. Soils can be used in defining such boundaries, as long as the boundaries of the soil-stratigraphic units are nearly the same age over large areas. Richmond,[70] Morrison,[56] and Morrison and Frye[61] give examples of chronostratigraphic units defined on soils, but all of this work was completed before the new stratigraphic code was established.

In many areas, the Quaternary deposits one commonly studies can be classified as allostratigraphic units.[64] These are mappable stratified bodies of sedimentary rock or sediments whose boundaries are laterally traceable disconformities. The upper boundary can be a geomorphic surface and associated surface soil. Although not specified in the new code, I see no reason to exclude lichenometric or rock-weathering criteria in identifying such units.[9] Indeed, the estimation of age and correlation are strengthened using more criteria than just soils. The lower boundary of the unit could be a buried soil. Unlike lithostratigraphic formations, allostratigraphic units are not identified on lithology; commonly in a particular area, all units would have a similar lithology. The fundamental unit is the alloformation. Thus, referring back to the example mentioned before, the early Quaternary(?) Rocky Flats alluvium could be formally designated the Rocky Flats Alloformation. In the Sierra Nevada, tills could be defined as alloformations, such as the Tahoe Alloformation, and the latter would be distinguished on a multitude of relative-dating criteria, only one of which is soils.[17]

DETERMINING THE TIME OF SOIL FORMATION

The time of formation for a particular soil is determined by its stratigraphic position relative to adjacent deposits and soils.[4,56,59,70] This time interval can be approximated for both buried and surface soils.

The time of soil formation is best dated with buried soils. In Fig. 12-1, B, for example, buried soil "a" formed between the deposition of units I and II, and buried soil "b" between units II and III. Moreover, soil "a" is similar whether buried or at the surface, and so is soil "b"; hence, one

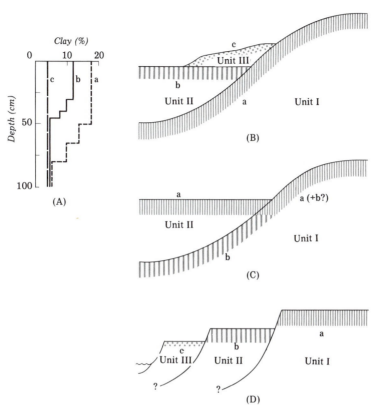

Fig. 12-1 Hypothetical relationships between soils and deposition units (I, old to III, young). Approximate clay contents for the soils in (B), (C), and (D) are shown in (A).

interpretation is that the major properties of soils "a" and "b" were imparted to the soils during the time intervals between depositional episodes, whether the soil remained at the surface or was buried. Furthermore, because soil "a" is more strongly developed than soil "b," soil formation during soil-forming interval "b" had little effect on soil "a" where it remained at the surface. Figure 12-1, C indicates that the strongest soil formed after the deposition of unit II; hence, soil "b" is recognizable as such only where buried. Although soil "b" was present on the landscape prior to the formation of soil "a," soil "a" is strong enough in development to mask most of the features of soil "b" in surface positions. Therefore, the major soil-forming interval here is that that formed soil "a."

It is more difficult to determine the timing of formation of surface soils where buried counterparts do not exist. In Fig. 12-1, D, river terraces of various ages are depicted. In this case, soil-forming intervals cannot be proved. The maximum time available for the formation of each soil is that from the time deposition ceased to the present. Thus, if soil formation proceeded at a relatively uniform rate since deposition, soil "a" may have had only about one-half its present amount of clay before soil "b" began to form, but soil "b" should have had most of its clay formed before soil "c" began to form. The soils could have formed in less time, but there is no evidence on the minimum time for formation.

Determining the ages of soils on slopes is even more difficult. Soil-stratigraphic relationships at the base of the slope may help put limits on the age of the slope and, thus, on the soils there. Hence, the entire slope catena has to be studied; examples in Chapter 9 provide data on what can be expected. In the example in Iowa, it was shown that the deposits at the base of the slope were of Holocene age, indicating that the slopes were eroding during much of the earlier Holocene, and have been relatively stable only for the last 3000 years. In the example in the Sierra Nevada, the thick, strongly developed soils in the moraine footslope deposits were used to argue for a long interval of slope stability. Finally, multiple buried soils in Australia suggest periodic erosion and stability of slopes. Such K cycles also are recognized in New Zealand.[45] It is still difficult to predict soil development with slope position, no matter how well the deposits and soils at the base of the slope have been studied. Localized areas of stable slopes can persist for long periods of time, and be marked by well-developed soils.

CONCEPT AND EXPLANATION OF THE SOIL-FORMING INTERVAL

The concept of the soil-forming interval was one outgrowth of soil-stratigraphic studies in the western United States.[59,62] Where soil profiles are nearly identical in both buried and surface occurrences (Fig. 12-1, B), the soil can be said to have formed during a discrete soil-forming interval bracketed by the ages of adjacent depositional units; the soil is younger than the unit from which it has formed, and it is older than the unit that buries it. A good example of a soil-forming interval is the Churchill Geosol of the Great Basin.[56] This soil has formed on what was considered to be early-Wisconsin deposits and is now found either at the surface, where it has been exposed to soil formation since deposition ceased, or buried beneath late-Wisconsin deposits. Obviously, the buried soil had much less time to form relative to the surface soil, yet both are nearly identical in profile morphology and on laboratory analysis (Fig. 12-2). In addition, the clay mineralogy of both relict and surface occurrences of the soil is identical, as is the degree of etching of the pyroxene grains (J.G. LaFleur, written report, 1972). Soil-forming intervals can be documented in many other localities in the United States and in the world.[59]

There are several ways to explain the fact that buried and surface occurrences of the same soil-stratigraphic units are nearly identical, although each has been exposed to soil-forming processes a different

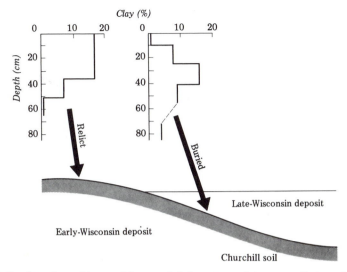

Fig. 12-2 Stratigraphic position and laboratory data on relict and buried occurrences of the Churchill soil (geosol where buried) in the Great Basin. (Data from Morrison,[56] Table II, and Birkeland,[5] Fig. 3.) See text for discussion that the early-Wisconsin deposit is Illinoian.

length of time. Morrison[56,59,62] and Morrison and Frye[61] proposed that the soil-forming intervals were characterized by unique climatic conditions that resulted in accelerated soil formation over relatively short time spans; in constrast, the climates prevailing between the soil-forming intervals were not conducive to soil formation. In the Lake Lahontan area of Nevada, for example, the soil-forming intervals are thought to have occurred over short intervals of time during which both precipitation and temperature were above their present values (Fig. 12-3). Clay-mineral data are amenable to this interpretation, although other interpretations of the clay-mineral data are possible.[7]

Soils forming during an interval of fluctuating climate, as visualized above, probably would form in a stepwise fashion, in which an accelerated rate of development coincided with optimum climatic conditions, and declining rate of development coincided with marginal climatic conditions. Curve A in Fig. 12-4 depicts this and can be used to help explain the field evidence for a soil-forming interval. Assume that soil formation went on at all sites in question until time "a." At this time, a climatic change took place, and part of the soil was buried by sediments. Because of the direction of the climatic change, the development of the soil that remained at the surface would have slowed down considerably, as shown by the flattened curve between times "a" and "b." If we examine both soils at time "b," the surface soil that formed in the time interval from 0 to "b" would be similar to that exposed to pedogenesis from 0 to "a," but buried from "a" to "b." In places, it might be possible to recognize the buried soil

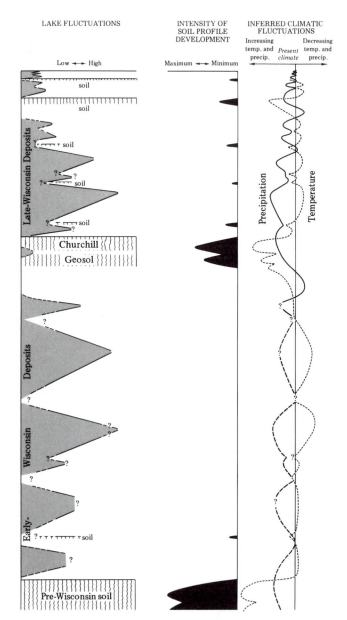

Fig. 12-3 Stratigraphy and lake fluctuations of Pleistocene Lake Lahontan, Nevada, and the stratigraphic position of the major soils. (Taken from Morrison and Frye,[61] Fig. 2.) Inferred climates during all events, relative to that of the present, are depicted, as are relative intensities of soil-profile development during the soil-forming intervals. See text for discussion that early-Wisconsin deposit is Illinoian.

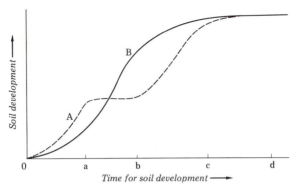

Fig. 12-4   Hypothetical curves of soil development with time under (A), a fluctuating climate, and (B), a constant climate.

formed during interval "a"–"b," and it should be weakly developed. Continuing in time, if the surface soil remained at the surface from times "b" to "c," a soil-forming interval, then theoretically it should be more strongly developed than the buried soil that formed between 0 and "a."

One other way of explaining the similarity in soils, buried and at the surface, is that they may have reached a steady state in development so that further change is limited. For example, assume a constant climate over a long interval of time. The development of the soil probably would proceed as depicted in Fig. 12-4, curve B.[4,40,84] As indicated, once a soil has attained fairly strong development, a greater length of time adds only a little to further development. If the soil had reached the development indicated by time "c," and part of it became buried and part remained at the surface until time "d," both would look similar if examined at time "d." This is common for some strongly developed profiles, but some data in Chapter 8 challenged the steady-state concept.

It is theoretically possible, however, to have accelerated soil development without a change in climate. Janda and Croft[39] point out that if the curve for soil development under a constant climate is like that of Fig. 12-4, curve B, soil development, especially with sandy parent materials, could accelerate between times "a" and "b" with no climate change. The reason for this is that as organic matter and clay accumulate in the soil, the water-holding capacity increases, and this should promote more clay formation from the weathering of non-clay minerals. A period of rapid clay formation continues until the curve begins to flatten as the more weatherable minerals are depleted.

Various problems raised by the concept of the soil-forming interval may have answers other than those given above.[8] One is that the intervals are envisaged as being of short duration, such as 5000 years for the formation of the Churchill Geosol (Figs. 12-2 and 12-3).[61] The problem is that it is difficult to document soil formation that has been that rapid in the past 5000 years in any environment in the western United States. This means that it is difficult to locate a climatic analogue for the formation of the Churchill Geosol. However, recent stratigraphic work in the Lake Lahonton sediments indicates that much more time is available to form that soil,[53,74] and I suspect that this is true for many other soils for which formation intervals of short duration are advocated. Another problem is the focus on temperature as the main factor in accelerating pedogenesis (Fig. 12-3), yet it was shown in Chapter 8 that soils in some alpine areas can form quite rapidly in spite of low temperatures. Alpine soils in the Colorado Front Range are an example of such rapid formation,[10] yet they are frozen for about one-half the year. This raises questions of what happens to soils that remain at the surface during glacial climates, supposedly times when soil formation is inhibited. Finally, one should also consider surface erosion as one means of explaining the similarity of a particular surface soil with a buried soil.

Part of the reason for the skepticism on the soil-forming intervals is that few soils for which such intervals have been advocated have been quantitatively analyzed in the laboratory. In one outcrop in the Antelope Valley of California, a stratigraphic situation much like that in Figure 12-2 exists, but the deposits are older. D.J. Ponti (1982 written report) has analyzed the soils in detail and finds that the surface soil has more clay than the buried soil. In other words, clay accumulation continued at the surface, but the buried equivalent soil has not gained any clay since burial. In general, this is what one would expect.

USING SOIL-STRATIGRAPHIC UNITS FOR SUBDIVISION AND CORRELATION OF QUATERNARY DEPOSITS

Soils can be quite useful for subdivision and correlation of Quaternary successions.[18,24,47,60,62,63] In this section, examples of both short- and long-range correlation, using soils, will be presented. For the most part, correlations based solely on soils do not come easily. This is because soils change laterally in their properties, because the factors responsible for soil morphology commonly change laterally. Thus, one is faced with comparing soils of about the same age, or duration of formation, in many different environmental settings. Spodosols are compared with Aridisols, thick soils with thin soils, high-clay soils with low-clay soils, because these encompass some of the wide variety of soils and soil properties that can be expected at any one stratigraphic horizon.

Soils have long been an important stratigraphic tool in subdividing and correlating Quaternary deposits in the midcontinent; few students of Quaternary studies, for example, have not heard of the Sangamon soil.[27] In Illinois in particular, soils have had a major role in the stratigraphic subdivision,[81] and voluminous data have been published on the soils and sediments. Iowa is another area in which soils have played a key role in the development of the Quaternary stratigraphic succession.[31,72]

The La Sal Mountains (Utah) study of Richmond[70] provides an excellent example of the use of soils to differentiate and correlate deposits and soils in a localized area. Combining the pedologic and geological record, Richmond was able to recognize soils at many stratigraphic intervals (Fig. 12-5). Based partly on their development, it was demonstrated that some soils mark major interglacials, whereas others mark fairly short intervals of time during which deposition ceased. The area is one of rapid change in environmental factors over short distances. The mountains reach to about 4170 m, rising about 2100 m above the surrounding Colorado Plateau. Vegetation varies with altitude and climate. The highest summits are in the alpine zone, and with decreasing altitude, one goes through a sub-alpine zone of spruce and fir; a montane zone of aspen; a foothills zone of scrub oak, mountain mahogany, ponderosa pine, juniper, and pinyon; and a sagebrush-grass zone. What makes this study so important to Quaternary stratigraphic studies is that for each formation (alloformation under the new stratigraphic code[64]), many depositional facies were recognized, and for each soil-stratigraphic unit, several pedologic facies were recognized. Soils formed on deposits of all ages vary in many of their properties, and this is mainly a function of the climate as related to altitude (Fig. 12-6; see also Fig. 11-17). Non-calcic soils formed at the higher altitudes, calcic soils at the lower. The problem of age assignment and correlation in the field, however, is confused by the fact that the altitude at which soils of a particular age change from non-calcic to calcic varies with age of the parent material (Fig. 12-7). Thus, in some places one has to compare the development of the non-calcic facies of a young soil with that of the calcic facies of a nearby older soil to work out the stratigraphic succession. Richmond[70] used these data to suggest correlations with other glaciated areas in the Rocky Mountains and with the midcontinent glacial succession using soils as one major tool.

Morrison[56,57,58] has done some exceptionally detailed work on the subdivision of the deposits and soils of the Lake Bonneville and Lake Lahontan areas in the western United States. Soils are found both at the surface and buried and can be traced rather continuously in both areas. They are key markers to the mapping because in many places what are now allostratigraphic units are separable only on the basis of their position relative to the stratigraphically diagnostic soils. The stratigraphic succession in both areas seems to be rather similar, once one has matched the relative development of the soils in both areas. Recent work, however, by Scott and others[74] in the Lake Bonneville deposits suggests that alterations should be made in both the stratigraphic placement of some of the soils recognized by Morrison and in their duration of formation.

The local subdivision of a stratigraphic succession using soils is fairly straightforward, but correlation using soils is more difficult. The methods of correlation using soils were set forth by Richmond[70] and Morrison.[56,59] They suggested that the first task in any one area is to rank the soils in

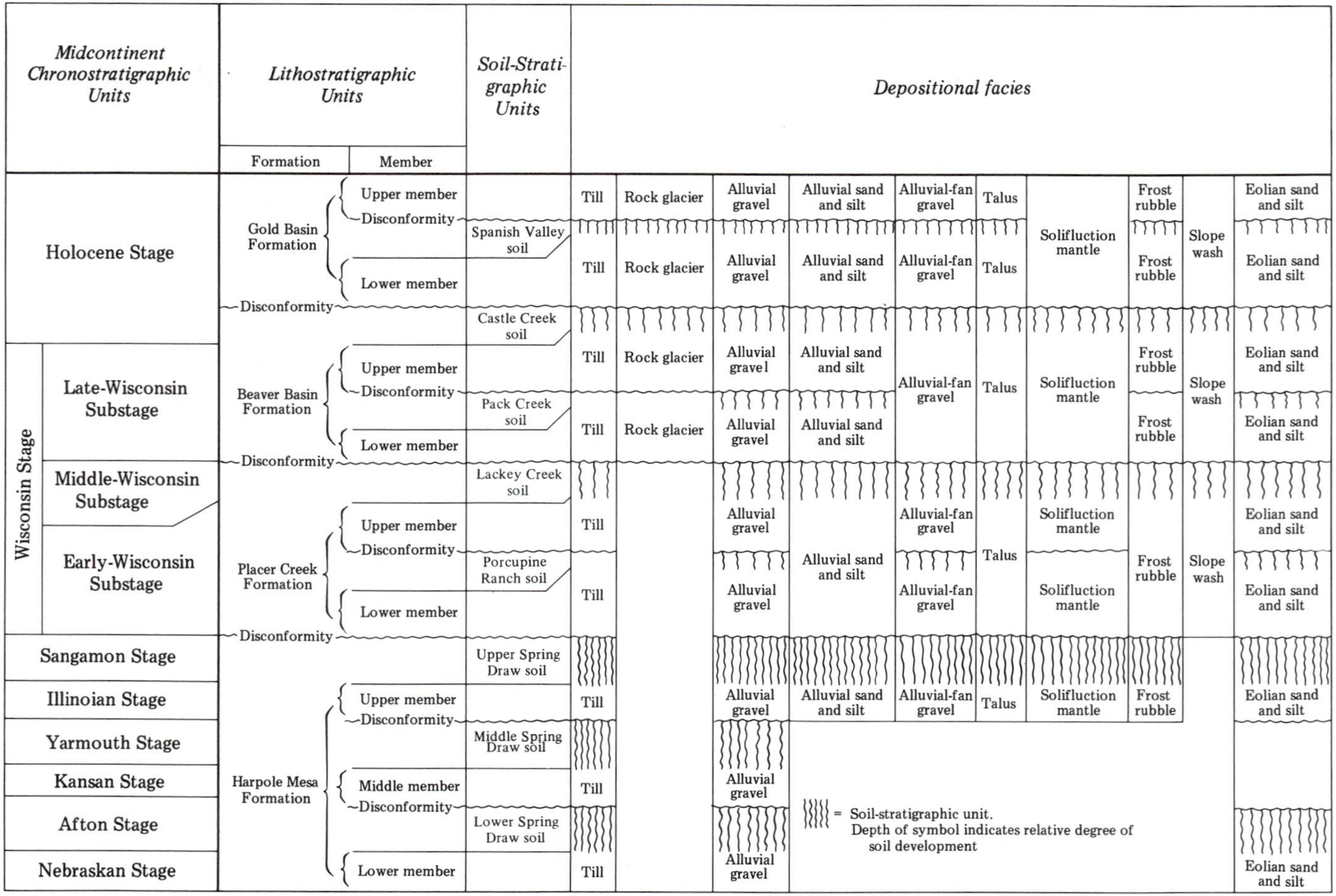

Fig. 12-5 Quaternary stratigraphic units in the La Sal Mountains, Utah, and tentative correlation with chronostratigraphic units of the midcontinent. (Taken from Richmond,[79] Fig. 9 and Table 11.) Formations would be reclassified as alloformations under the new strati-

Midcontinent Chronostratigraphic Units	Lithostratigraphic Units — Formation	Lithostratigraphic Units — Member	Soil-Stratigraphic Units	Depositional facies — Till	Rock glacier	Alluvial gravel	Alluvial sand and silt	Alluvial-fan gravel	Talus	Solifluction mantle	Frost rubble	Slope wash	Eolian sand and silt
Holocene Stage	Gold Basin Formation	Upper member / Disconformity / Lower member	Spanish Valley soil; Castle Creek soil	Till	Rock glacier	Alluvial gravel	Alluvial sand and silt	Alluvial-fan gravel	Talus	Solifluction mantle	Frost rubble	Slope wash	Eolian sand and silt
Wisconsin Stage — Late-Wisconsin Substage	Beaver Basin Formation	Upper member / Disconformity / Lower member	Pack Creek soil; Lackey Creek soil	Till	Rock glacier	Alluvial gravel	Alluvial sand and silt	Alluvial-fan gravel	Talus	Solifluction mantle	Frost rubble	Slope wash	Eolian sand and silt
Wisconsin Stage — Middle-Wisconsin Substage				Till	Rock glacier	Alluvial gravel	Alluvial sand and silt	Alluvial-fan gravel	Talus	Solifluction mantle	Frost rubble	Slope wash	Eolian sand and silt
Wisconsin Stage — Early-Wisconsin Substage	Placer Creek Formation	Upper member / Disconformity / Lower member	Porcupine Ranch soil	Till		Alluvial gravel	Alluvial sand and silt	Alluvial-fan gravel	Talus	Solifluction mantle	Frost rubble	Slope wash	Eolian sand and silt
Sangamon Stage			Upper Spring Draw soil			Alluvial gravel	Alluvial sand and silt	Alluvial-fan gravel		Solifluction mantle			Eolian sand and silt
Illinoian Stage		Upper member / Disconformity		Till		Alluvial gravel	Alluvial sand and silt	Alluvial-fan gravel	Talus	Solifluction mantle	Frost rubble		Eolian sand and silt
Yarmouth Stage			Middle Spring Draw soil										
Kansan Stage	Harpole Mesa Formation	Middle member / Disconformity		Till									
Afton Stage			Lower Spring Draw soil										
Nebraskan Stage		Lower member		Till		Alluvial gravel	Alluvial sand and silt	Alluvial-fan gravel	Talus	Solifluction mantle	Frost rubble		Eolian sand and silt

≈ = Soil-stratigraphic unit. Depth of symbol indicates relative degree of soil development

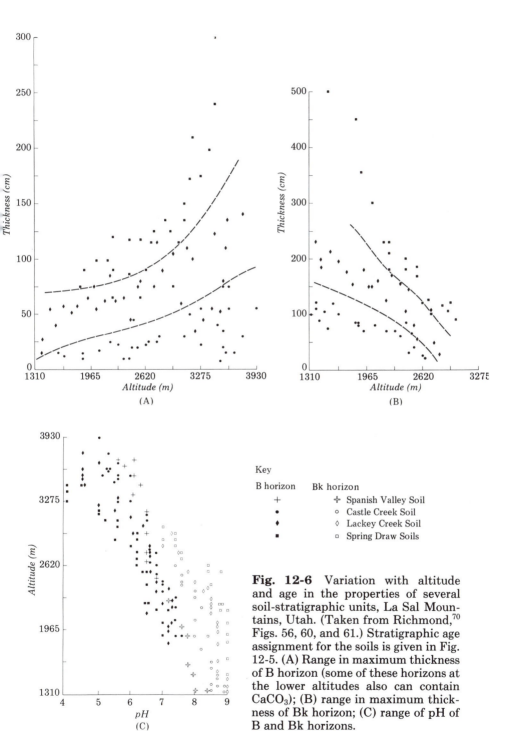

Key

B horizon   Bk horizon

+   ⊹ Spanish Valley Soil

●   ○ Castle Creek Soil

◆   ◇ Lackey Creek Soil

■   □ Spring Draw Soils

Fig. 12-6 Variation with altitude and age in the properties of several soil-stratigraphic units, La Sal Mountains, Utah. (Taken from Richmond,[70] Figs. 56, 60, and 61.) Stratigraphic age assignment for the soils is given in Fig. 12-5. (A) Range in maximum thickness of B horizon (some of these horizons at the lower altitudes also can contain $CaCO_3$); (B) range in maximum thickness of Bk horizon; (C) range of pH of B and Bk horizons.

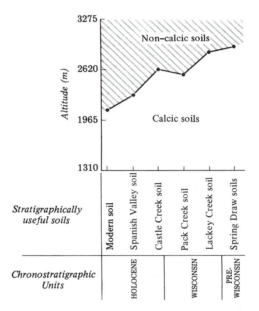

Fig. 12-7 Average altitude of the boundary between non-calcic and calcic soils of a different age, La Sal Mountains, Utah. (Taken from Richmond,[70] Fig. 19.) The Lackey Creek soil may be older than depicted here.

the local succession on the basis of their relative degree of profile development; some will be weakly developed, others moderately developed, and still others strongly developed. Ranking on development will not always be greater with age of the deposits, because some weakly developed soils can form and be preserved as buried soils. In the other area to which correlation is to be made, the soils are similarly ranked, again with respect to that local succession. Because the environment of each local succession will not be identical, the time-equivalent soils in one succession may bear little resemblance to those in another. As an example, a 12-cm-thick, weakly developed Aridisol in one environment may be correlated with a 50-cm-thick, moderately developed Alfisol in another environment. The soils and deposits in both areas are then matched. Commonly, the basic framework here is provided by the youngest strongly developed soil in each area. If the timing of the periods of deposition and soil formation are approximately the same, and deposition occurred during the glaciations, the youngest strongly developed soil often is the soil that formed mainly during the last major interglacial in the midcontinent, the Sangamon. All other strongly developed soils of pre-Sangamon age also are matched and, because they, too, may mark major interglacials, they are important to the correlation. The weakly and moderately developed soils that lie stratigraphically between the strongly developed ones can then be matched.

There are several problems with the above proposal for using soils in correlation. One is that the ranking on development is qualitative and relative within a sequence of soils in a particular environment; hence, one soil in one environment might be ranked as weakly developed, but the same soil in another environment could be strongly developed. I would prefer to use a quantitative ranking of soil development, such as the Harden index.[32] In addition, quantitative soil data should be used in correlations. For example, recent work by Scott and others[74] on the Lake Bonneville sediments and soils in Utah suggests that a major soil formed between about 100,000 and 25,000 years ago (equivalent to the Churchill soil and geosol of Figs. 12-2 and 12-3). The soil is correlated around the area on the total grams of pedogenic $CaCO_3$, and the results are quite encouraging. Machette[51] uses similar data to assign ages to surficial deposits in southcentral Utah. Once the local successions are worked out, correlation can proceed. Thus, it could be that a soil with 50 g $CaCO_3/cm^2/$ profile will correlate with one elsewhere with only one-half as much carbonate. This same approximate magnitude of difference might hold for all the soils being correlated between the two areas. Other soil properties could be used for correlation in the same fashion; a logical one would be grams of pedogenic clay.

The need for assembling the geological and the pedologic data for the most meaningful correlation cannot be overemphasized. One has to have a thorough knowledge of the geological history in the area, including the kind and rate of geological processes that have been operative. Soils have to be studied in the field for their stratigraphic position and for their morphology and history, and they should be analyzed quantitatively for their critical properties. Only then can all the data be assembled to give the best reasonable correlations. One should avoid merely counting back in time and fitting all the stratigraphic units into the so-called known standard sequences, because our information on these is always changing. It seems fairly well accepted now, for example, that the Placer Creek Formation (Fig. 12-5) probably correlates with the Illinoian of the midcontinent, as do the early-Wisconsin deposits of Figures 12-2 and 12-3. Regional correlation charts, using soils as one correlation criterion, have been prepared,[6,34] but parts of these are now out of date and should be revised using more current information.

An example of long-range correlation, using soils as one criterion, is that for Quaternary loesses and the deep-sea record as discussed by Kukla.[43,44] Detailed study was made of loesses and their associated fossil content and buried soils in Central Europe. A marked glacial-interglacial cyclicity is apparent, and there are many similarities in the paleoenvironment deduced from both soils and the paleontological record. These cycles are then shown to correlate well with those of the marine oxygen-isotope record to demonstrate a correlation between the marine and terrestrial records. Kukla argues that, of the land records, the loess-soil record is one of the best, since it contains the detail necessary for such correlation.

USING SOILS TO DATE FAULTING

Faults cut Quaternary deposits in many areas of active faulting, and in places, soils must be used to decipher the age of faulting. The impetus for many recent studies is to determine the seismic hazard of known faults near nuclear power plant sites.[75] The reason for relying on soils for dating is that material for radiocarbon dating either is not present or is not present in the most useful stratigraphic position. Recent studies have been conducted in many environments.[12,13,22,42,52,78] Commonly, the work is done in artificial exposures, such as backhoe trenches (see examples in McCalpin[52]). First, detailed maps are made of the cut walls showing the relationships of the deposits and soils to the fault(s). Sediments and soils overlying the fault, but not displaced by it, are younger than the fault, whereas those cut by the fault predate one or more episodes of faulting. Using soils in these studies is no easy task. First, soils have to be recognized as such, for many of them are buried. Then, an age has to be estimated, keeping in mind all the factors that can affect the morphology and other properties of soils. In this way, the period of faulting can be given approximate age brackets.

A detailed study of Quaternary faulting, using calcic soils west of Albuquerque, New Mexico indicates recurrent fault movement over the past 0.5 million years.[50] Analyses were made both on the percent and amount (g) of carbonate for each soil profile. Because of recurrent normal downfaulting, several buried soils are present just east of the fault (Fig. 12-8, loc. c, d, and e); materials shed from the fault scarp buried each soil soon after movement. Farther east, the buried soils merge to form one surface soil (loc. f); west of the fault, the soils also merge to form a single composite profile (loc. a). Because some of the carbonate is primary, it was subtracted from the total carbonate to calculate the pedogenic carbonate in each profile. Geological relationships suggest that the geomorphic surface cut by the fault is about 0.5 my old. Dividing the total pedogenic carbonate, as represented by the sum of carbonate of the buried soils, by 0.5 my gives a long-term average rate of accumulation of pedogenic carbonate. Knowing the amount of pedogenic carbonate in each buried soil, one can estimate the duration of soil formation between recurrent episodes of fault movement. Four faulting events are recognized over the past 0.5 my, the most recent of which probably took place about 20,000 years ago (Fig. 12-9). Individual fault displacements ranged from about 2 to 8 m, and the recurrence intervals range from 90,000 to 190,000 years. Two fairly reasonable assumptions were used in this study. One is that the rate of carbonate accumulation has been constant. The other is that the amount of time encompassed by erosion of the fault scarp and burial of the soil on the down-thrown side is negligible relative to the amount of time necessary for surface stability and soil formation.

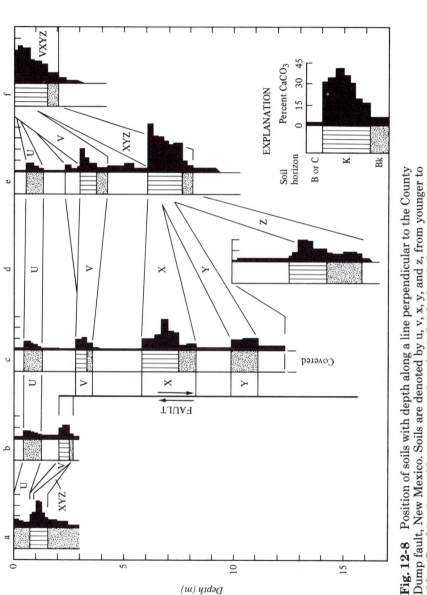

Fig. 12-8 Position of soils with depth along a line perpendicular to the County Dump fault, New Mexico. Soils are denoted by u, v, x, y, and z, from younger to older. Combinations of letters mean that the individual soils have merged laterally to form one composite soil; its age and properties are denoted by letter combinations (e.g., xyz). Lower-case letters at the top of the figure (a, b, c, d, and e) denote the position of the soils in the field, shown to scale in the lower diagram of Figure 12-9. (Modified from Machette,[50] Fig. 6.)

WEST   EAST

a. 400,000 Years B.P. Prior to fault event 1. Soil Z (100,000 yr old) has formed, and then fault event 1 took place at the indicated position

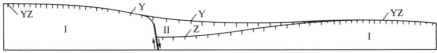

b. 310,000 Years B.P. Prior to fault event 2. Soil Z buried by unit II following fault event 1, surface restabilized, and soil Y (90,000 yr old) has formed

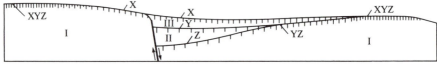

c. 120,000 Years B.P. Prior to fault event 3. Soil Y buried by unit III following fault event 2, surface restabilized, and soil X (190,000 yr old) has formed

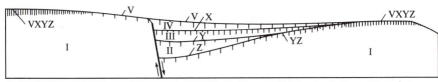

d. 20,000 Years B.P. Prior to fault event 4. Soil X buried by unit IV following fault event 3, surface restabilized, and soil V (100,000 yr old) has formed

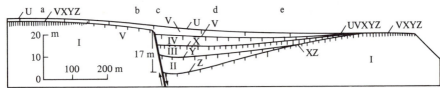

e. Present Day Soil V buried by unit V following fault event 4, surface restabilized, and soil U (20,000 yr old) has formed

Fig. 12-9 Schematic cross sections, showing the sequence of events over the last 0.5 my in the vicinity of the County Dump fault (B.P, before present; depositional units, I, old to V, young). Soils are designated u, v, x, y, and z; the horizontal spacing of the vertical hachure pattern is proportional to the duration of soil formation. Localities for soils shown in Figure 12-8 are designated a, b, c, d, e, and f in the lowest figure. (Taken from Machette,[50] Fig. 9.)

These studies could be extended to other parts of the western United States where calcic soils exist. Crude dating could be done knowing the grams of $CaCO_3/cm^2$/profile for the soils of interest and either knowing or estimating the regional rate of accumulation of pedogenic carbonate (Fig. 8–14). Colman[20] did this in assigning a 0.25 my age to a calcic soil that caps deformed sediments in southeastern Utah, and this helped date tectonics related to flowage of a bedrock salt unit. Salt beds in the region are being studied as possible disposal sites of nuclear wastes, and this study, although not at a proposed site, helps provide information on the more recent tectonics of salt-cored anticlines.

Although pedogenic $CaCO_3$ is useful in the above example because it is ubiquitous in the western United States, other properties could be used to date faults. Examples are the absolute amounts (g) of pedogenic clay or salts more soluble than $CaCO_3$. Regional rates of accumulation have to be known for these materials, however, before the age can be estimated. Or, stratigraphically significant soils can be used. For example, soils are one major relative dating criterion used to subdivide and estimate the ages of glacial and glaciofluvial deposits in the western United States. Using such buried diagnostic soils as one criterion, McCalpin[52] found that in the Rio Grande Rift in Colorado, the vertical displacements range from 0.8 to 2.9 m/fault event, and the estimated recurrence intervals range from 10,000 to 100,000 years. Finally, one could try using the Harden index[32] to obtain broad age estimates of fault-related soils, keeping in mind that the index may provide reasonable estimates despite differences in climate between areas.[33]

Soils can also be used to date geomorphic surfaces that have been tectonically deformed. In this case, it may not be necessary to dig artificial trenches to obtain broad ages for the structural events. River terraces serve as a good example of this. Terraces might converge or diverge downstream for a variety of reasons, only one of which might be tectonic. If one can be well assured that the pattern is tectonic,[5,41,52,55,66,77] broad age ranges can be determined for the time of deformation. Ages can be estimated either from the soil formed in the uppermost terrace materials, or if loess or ash deposition is interspersed with soil formation on the top of the terraces, the degree of complexity of the soils and deposits may increase with the older units as long as erosion has not eradicated the record. Or, it might be that the soils on top of the terrace are eroded and that the soils and deposits that form the colluvial wedge on the terrace scarp may display a complexity with time that relates to the ages of the associated surfaces. Study of the soils and geomorphic relations might indicate that the 100,000-year-old terrace has been warped, but the 30,000-year-old terrace has not. Deformation, therefore, is bracketed between the latter two ages. Strike-slip faults that cut across terraces[48] or other landforms also can be dated using soils.

Still another geomorphic setting in which soils can be used to help date tectonic features is the alluvial-fan environment. Alluvial fans are ubiquitous throughout the Basin and Range Province of the western United States, and elsewhere. Commonly, the history of the fan has been an alternation between downcutting and alluviation, so that the age relationship of deposits across the fan are complex[21,38]; in addition, some fans may be segmented in the down-fan direction.[15,16] Several interpretations can be offered to explain these features. Surely climatic change is one, as it is fairly well accepted that there have been late-Quaternary climatic changes in the region. Some features, however, are related to recent tectonics, and dating the deposits and surfaces using soils can help put age limits on faulting.

USING SOILS IN ARCHEOLOGICAL STUDIES

Soils can be of use to archeologists in the interpretation of stratified sites. They can be used to indicate the ages of certain layers, past moisture conditions, or perhaps past vegetation. I do not think their potential has yet been realized in archeological studies. In addition, archeological excavations, because of their extreme detail, yield much information that is useful to soils studies.

The initial major problem with stratified archeological sites is to be able to differentiate between geological and pedologic materials, as well as materials altered by humans. Throughout this book, there are examples by which one can identify pedologic materials, whether at the surface or buried. Farrand[25] has been especially active in sediment analysis of archeological sites and how they might be used to infer paleoenvironment. Useful analytical techniques are particle size, shape and sorting, bulk density, thin section, mineralogy of all size fractions, and the usual soil-laboratory analyses.[19] These must be used in the context of the field setting for the best geological-pedologic interpretation. Thus, it is important, if not mandatory, that the person responsible for the pedologic interpretation study the site rather than just be handed bags of materials for analysis.

One major use of soils at archeological sites is to identify depositional hiatuses, for if materials are not being deposited, soils can form. Haynes[36] has been a leader in these studies, and his regional review, although it requires updating,[34] is an example of how to incorporate soils into the study of archeologically important, late-Quaternary alluvial deposits. However, few detailed soil analyses accompanied these early studies.

Holliday[37] has set a model for soil-archeolgical studies in his work at the Lubbock Lake State and National Archaeological Landmark site, Texas, in which field relationships of deposits and soils are combined with quantitative soil data. The geology of the area is complex, since the following contrasting depositional facies are compressed into a small area: lacustrine, fluvial under both poorly and well-drained conditions, hillslope colluvial, and eolian. Soils have formed from deposits of the above facies. About 70 radiocarbon dates provide age control for the sediments and soils.

The soils vary in their parent materials, in the duration of time they formed, and in pedogenic environment and so contrast markedly (Figs. 1-12c and 12-10). The Firstview soil formed in 2000 years; those along the former valley axis have high clay content and are gleyed due to reducing conditions that accompanied a high water table. In contrast, the valley-

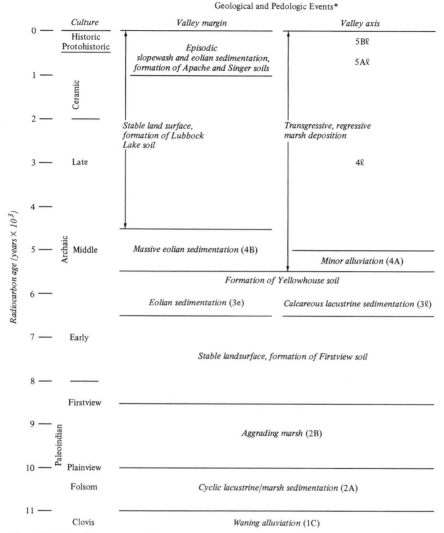

Fig. 12-10 Summary of late-Quaternary geology, soils, and culture at the Lubbock Lake Landmark, Texas. The main depositional units are numbered from oldest (1) to youngest (5) their subdivisions are indicated by capital letters (2A, 2B). Lacustrine and eolian units are designated by l and e, respectively. (Taken from Holliday,[37] Fig. 11.)

margin soil facies is a sandy, calcareous A/C profile formed in eolian sand. Near the middle Holocene, the weakly developed Yellowhouse soil formed over a period of about 500 years. This soil is formed in the same two depositional facies as that above. The Lubbock Lake soil, the most distinctive one at the site, formed mainly in eolian sediments; the duration of soil formation varies from about 3500 years, where buried, to 4500 years, where it occurs at the surface. The A horizon is so distinctive that it was given a geological designation and used as a marker bed in the early studies at the site. This emphasizes the necessity to be certain that pedologic units are distinguished from geological ones. Where best developed, the Lubbock Lake soil is an A/Bt/Bk profile with stage I to II carbonate morphology. Because this soil occurs in several topographic settings, catena relationships (Ch. 9) are commonly well expressed. The Apache soil represents from 350 to 600 years of pedogenesis and has formed mainly in eolian materials. Both Bw and Bt horizons are present, and stage I carbonate morphology in Bk horizons is common. Finally, the Singer soil has formed in the Historic period and usually is an A/weak Bw or C/Bk profile with weak stage I carbonate morphology. Extensive laboratory and thin-section data are used to characterize these soils and derive chrono-functions.

Stratigraphy, quantitative data on soils, and some radiocarbon dates are used to suggest correlations of the above units with sediments and soils at other well-known archeological sites in the High Plains of Texas.[37] These are the Blackwater Draw locality 1 and the Plainview site. Thus, it appears that the patterns are regional, and this enhances the usefulness of soils in broad correlation schemes.

Other workers are using soils to help decipher environmental history of archeological sites. A site in an arroyo environment in Wyoming has four soils (three buried) formed in sediments that date from the late Pleistocene to the present.[67] The older soils predate the local altithermal, are the best developed of the four, and have carbonate superimposed on noncalcic profiles; presumably this happened during the altithermal time of relative aridity. The two younger soils postdate the altithermal and are weakly developed. Not enough radiocarbon dates are available to determine how well this site correlates with the sites in Texas. Another study that has used soils is that of the late Holocene sediments at Delaware Canyon, Oklahoma.[26] If correlation of soils could be proved, it would give greater credence to soil-stratigraphic correlations on a rather small time scale over large distances.

Benedict[2,3] is known for his detailed geological-archeological work in alpine areas of the Colorado Front Range, and although soils are not the basic tool used in age assignment, soil horizons are commonly related to the geological and cultural materials. Most of the cultural materials are found at shallow depth in the upper soil horizons, in contrast to arroyo sites in which the deposits can be thick and the soils and cultural materials widely separated vertically. Till, alluvium, slopewash, colluvium, and loess, generally modified to some extent by frost-sorting processes, form the complex parent materials; a loess unit commonly is the uppermost unit throughout the area. Soils are formed through all materials and, in places, buried soils are present. Hearths usually predate the youngest loess, and the latter commonly is the parent material for the A horizon and, in places, part of the B horizon. The detailed archeological work points out that the uppermost loess, although thin, was deposited at different times and contains artifacts.

Soils have also been used either to help date or to evaluate the paleoenvironment of much older archeological sites. Although not essential in dating the hominid remains and archeological materials, calcic soils occur within the deposits at Olduvai Gorge, Africa and are used to help date some intervals of deposition and to verify the fact that a semiarid climate existed throughout the Pleistocene.[35] The Calico archeological site in the Mojave Desert is controversial because workers are not unanimous as to whether the materials are human artifacts or not. At any rate, the soil formed in the uppermost part of the deposit containing "artifacts" is a strongly developed A/Bt/Bk profile with stage II carbonate morphology.[11] Comparison with other calcic soils in the region suggests an age of at least 80,000 years, and this is compatible with uranium-series dates of around 200,000 years on groundwater carbonate near the base of the deposit.

The above examples should alert archeologists to the potential use of soils in their research. They also can help in the search for early man sites. For example, early man is considered to be no older than latest Pleistocene in the United States, and soil morphology might indicate that deposits beneath a particular soil are too old to yield meaningful results.

Finally, it should be pointed out that some chemical tests might be useful in determining the impact of humans on present or buried land surfaces. Phosphorus determinations have been found to be useful, because human remains and debris are high in this element.[23,65,80,83] Some of these interpretations might be difficult to make, however, and workers first should be aware of the pedologically distinctive extracts and trends (Chs. 5 and 8).

OTHER APPLICATIONS AND POTENTIAL MEANS OF DATING SOILS

Several other applications of soils can be listed, but will not be discussed in detail. One is in the reconstruction of ancient environments, in combination with paleontological and sedimentalogical studies.[14,68,69] For example, it is now recognized by some workers that soils can add information about some ancient environments that is not available through the usual paleontological research. Another is the influence of soil horizons on landscapes, since some horizons can be so indurated that they have a profound influence on landscape evolution.[30,54] Still other applications have to do with determining if the formation of some features is basically pedogenic or geological. An example of the latter is the presence of a material called silcrete. This is a highly indurated, quartz-rich material cemented with crystalline and amorphous silica, which is especially widespread in parts of Australia.[46]

Although the dating of deposits or landscapes by morphological criteria is emphasized throughout the book, other methods might be used. In Chapter 5, examples are given of the dating of carbonate horizons by radiocarbon and uranium-series methods. Several less common ones will be mentioned here, and most need more work to prove their usefulness. Amino acids have been used to help date fossils[79] and identify buried soils,[29,49] and their variation in content, concentration, and ratios have the potential to help date soils.[28] Rosholt[71] has devised a complicated, uranium-trend method for the dating of sediments. In one test of this method for dating till in the Rocky Mountains, an age of 130,000 years for a Bull Lake till (-Illinoian?) is close to what was predicted from the soil properties, and arguments about correlations with other deposits in the region and with the marine oxygen-isotope record.[76] Finally, thermoluminescence techniques[82] might be used to date sediments associated with soils.

REFERENCES

1. American Commission on Stratigraphic Nomenclature, 1961, Code of stratigraphic nomenclature: Amer. Assoc. Petrol. Geol. Bull., v. 45, p. 645–665.
2. Benedict, J.B., 1981, The Fourth of July Valley: Center for Mountain Archaeology Rept. No. 2, Ward, Colorado, 139 p.
3. Benedict, J.B., and Olson, B.L., 1978, The Mount Albion Complex: Center for Mountain Archaeology Rept. No. 1, Ward, Colorado, 213 p.
4. Birkeland, P.W., 1967, Correlation of soils of stratigraphic importance in western Nevada and California, and their relative rates of profile development, p. 71–91 *in* R.B. Morrison and H.E. Wright, Jr., eds., Quaternary soils: Internat. Assoc. Quaternary Res., VII Cong., Proc. v. 9.
5. ———, 1968, Correlation of Quaternary stratigraphy of the Sierra Nevada with that of the Lake Lahontan area, p. 469–500 *in* R.B. Morrison and H.E. Wright, Jr., eds., Means of correlation of Quaternary successions: Internat. Assoc. Quaternary Res., VII Cong., Proc. v. 8.
6. Birkeland, P.W., Crandell, D.R., and Richmond, G.M., 1971, Status of correlation of Quaternary stratigraphic units in the western coterminous United States: Quaternary Res., v. 1, p. 208–227.
7. Birkeland, P.W., and Janda, R.J., 1971, Clay mineralogy of soils developed from Quaternary deposits of the eastern Sierra Nevada, California: Geol. Soc. Amer. Bull., v. 82, p. 2495–2514.
8. Birkeland, P.W., and Shroba, R.R., 1974, The status of the concept of Quaternary soil-forming intervals in the western United States, p. 241–276 *in* W.C. Mahaney, ed., Quaternary environments: York University (Toronto, Canada) Geogr. Mono. No. 5.
9. Birkeland, P.W., Colman, S.M., Burke, R.M., Shroba, R.R., and Meierding, T.C., 1979, Nomenclature of alpine glacial deposits, or, what's in a name?: Geology, v. 7, p. 532–536.
10. Birkeland, P.W., Burke, R.M., and 13 soil students, 1982, Quantitative data for an alpine soil chronosequence, Colorado Front Range: Amer. Quaternary Assoc., Prog. and absts., 7th Biennial Conf., p. 70.

11. Bischoff, J.L., Shlemon, R.J., Ku, T.L., Simpson, R.D., Rosenbauer, R.J., and Budinger, F.E., Jr., 1981, Uranium-series and soil-geomorphic dating of the Calico archaeological site, California: Geology, v. 9, p. 576–582.

12. Borchardt, G., Rice, S., and Taylor, G., 1980, Paleosols overlying the Foothills fault system near Auburn, California: California Division of Mines and Geology Special Report 149, 38 p.

13. Borchardt, G., Taylor, G., and Rice, S., 1980, Fault features in soils of the Mehrten Formation, Auburn damsite, California: California Division of Mines and Geology Special Report 141, 45 p.

14. Bown, T.M., and Kraus, M.J., 1981, Lower Eocene alluvial paleosols (Willwood Formation, northwest Wyoming, U.S.A.) and their significance for paleoecology, paleoclimatology, and basin analysis: Palaeogeography, Palaeoclimatology, Palaeoecology, v. 34, p. 1–30.

15. Bull, W.B., 1964, Geomorphology of segmented alluvial fans in western Fresno County, California: U.S. Geol. Surv., Prof. Pap. 352-E, 129 p.

16. ———, The alluvial-fan environment: Prog. Phys. Geogr., v. 1, p. 222–270.

17. Burke, R.M., and Birkeland, P.W., 1979, Re-evaluation of multiparameter relative dating techniques and their application to the glacial sequence along the eastern escarpment of the Sierra Nevada, California: Quaternary Res., v. 11, p. 21–51.

18. Catt, J.A., 1979, Soils and Quaternary geology in Britain: Jour. Soil Sci., v. 30, p. 607–642.

19. Catt, J.A., and Weir, A.H., 1976, The study of archeologically important sediments by petrographic techniques, p. 65–91 *in* D.A. Davidson and M.L. Shackley, eds., Geoarchaeology: G. Duckworth and Co., Ltd., London.

20. Colman, S.M., 1983, Influence of the Onion Creek salt diapir on the late Cenozoic history of Fisher Valley, southeastern Utah: Geology, v. 11, p. 240–243.

21. Denny, C.S., 1965, Alluvial fans in the Death Valley region, California and Nevada: U.S. Geol. Surv. Prof. Pap. 466, 62 p.

22. Douglas, L.A., 1980, The use of soils in estimating the time of last movement of faults: Soil Science, v. 129, p. 345–352.

23. Eidt, R.C., 1977, Detection and examination of anthrosols by phosphate analysis: Science, v. 197, p. 1327–1333.

24. Evans, L.J., 1982, Dating methods of Pleistocene deposits and their problems: VII. Paleosols: Geoscience Canada, v. 9, p. 155–160.

25. Farrand, W.R., 1975, Sediment analysis of a prehistoric rockshelter: The Abri Pataud: Quaternary Res., v. 5, p. 1–26.

26. Ferring, C.R., ed., 1982, The late Holocene prehistory of Delaware Canyon, Oklahoma: Contrib. in Arch. No. 1, Inst. Appl. Sci., North Texas State University, Denton, 331 p.

27. Follmer, L.R., 1982, The geomorphology of the Sangamon surface; its spatial and temporal attributes, p. 117–146 *in* C.E. Thorn, ed., Space and time in geomorphology: George Allen and Unwin, London.

28. Forman, S.L., Brigham, J.K., and Miller, G.H., 1982, Amino acids in soil: Concentrations, isoleucine epimerization and geological applications: Geol. Soc. Amer. Absts. with Programs, v. 14, p. 490–491.

29. Goh, K.M., 1972, Amino acid levels as indicators of paleosols in New Zealand soil profiles: Geoderma, v. 7, p. 33–47.

30. Goudie, A., 1973, Duricrusts in tropical and subtropical landscapes: Oxford University Press, London, 174 p.

31. Hallberg, G.R., 1980, Pleistocene stratigraphy in east-central Iowa: Iowa Geol. Surv. Tech. Info. Ser. No. 10, 168 p.

32. Harden, J.W., 1982, A quantitative index of soil development from field descriptions: Examples from a chronosequence in central California: Geoderma, v. 28, p. 1–28.

33. Harden, J.W., and Taylor, E.M., in press, A quantitative comparison of soil development in four climatic regimes: Quaternary Res.

34. Hawley, J.W., Bachman, G.O., and Manley, K., 1976, Quaternary stratigraphy in the Basin and Range and Great Plains Provinces, New Mexico and western Texas, p. 235–274 *in* W.C. Mahaney, ed., Quaternary stratigraphy of North America: Dowden, Hutchinson, and Ross, Inc., Stroudsberg, Penn.

35. Hay, R.L., 1976, Geology of the Olduvai Gorge: University of California Press, Berkeley, 203 p.

36. Haynes, C.V., Jr., 1968, Chronology of late-Quaternary alluvium, p. 591–631 *in* R.B. Morrison and H.E. Wright, Jr., eds., Means of correlation of Quaternary successions: University of Utah Press, Salt Lake City.

37. Holliday, V.T., 1982, Morphological and chemical trends in Holocene soils at the Lubbock Lake Archeological Site, Texas: Ph.D. thesis, Univ. of Colorado, Boulder, 285 p.

38. Hooke, R.L., 1972, Geomorphic evidence for late-Wisconsin and Holocene tectonic deformation, Death Valley, California: Geol. Soc. Amer. Bull., v. 83, p. 2073–2098.

39. Janda, R.J., and Croft, M.G., 1967, The stratigraphic significance of a sequence of Noncalcic Brown soils formed on the Quaternary alluvium of the northeastern San Joaquin Valley, California, p. 157–190 *in* R.B. Morrison and H.E. Wright, Jr., eds., Quaternary soils: Internat. Assoc. Quaternary Res., VII Cong., Proc. v. 9.

40. Jenny, H., 1980, The soil resource: Springer-Verlag, New York, 377 p.

41. Keller, E.A., Johnson, D.L., Rockwell, T.K., Clark, M.N., and Dembroff, G.R., 1981, Quaternary stratigraphy, soil geomorphology, chronology and tectonics of the Ventura, Ojai, and San Paula areas, western Transverse Ranges, California: unpubl. Friends of the Pleistocene field trip guidebook, 125 p.

42. Kirkham, R.M., 1977, Quaternary movements on the Golden fault, Colorado: Geology, v. 5, p. 689–692.

43. Kukla, J., 1970, Correlation between loesses and deep-sea sediments: Geologiska Föreningess i Stolkholm Förhandlingar, v. 92, 148–180.

44. ———, 1977, Pleistocene land-sea correlations. 1. Europe: Earth-Sci. Rev., v. 13, p. 307–374.

45. Laffan, M.D., and Cutler, E.J.B., 1977, Landscapes, soils, and erosion of a small catchment in the Wither Hills, Marlborough: 1. Landscape periodicity, slope deposits, and soil pattern: New Zealand Jour. Sci., v. 20, 37–48.

46. Langford-Smith, T., ed., 1978, Silcrete in Australia: Dept. Geogr., Univ. of New England, Australia, 304 p.

47. Leamy, M.L., Milne, J.D.G., Pullar, W.A., and Bruce, J.G., 1973, Paleopedology and soil stratigraphy in the New Zealand Quaternary succession: New Zealand Jour. Geol. Geophys., v. 16, p. 723–744.

48. Lensen, G.J., 1968, Analysis of progressive fault displacement during downcutting at the Branch River terraces, South Island, New Zealand: Geol. Soc. Amer. Bull., v. 79, p. 545–556.

49. Limmer, A.W., and Wilson, A.T., 1980, Amino acids in buried soils: Jour. Soil Sci., v. 31, p. 147–153.

50. Machette, M.N., 1978, Dating Quaternary faults in the southwestern United States by using buried calcic paleosols: U.S. Geol. Surv. Jour. Res., v. 6, p. 369–381.

51. ———, 1982, Guidebook to the late Cenozoic geology of the Beaver Basin, south-central Utah: U.S. Geol. Surv. Open-File Report 82–850, 42 p.

52. McCalpin, J.P., 1982, Quaternary geology and neotectonics of the west flank of the northern Sangre de Cristo Mountains, south-central Colorado: Colo. School Mines Quarterly, v. 77, no. 3, 97 p.

53. McCoy, W.D., 1981, Quaternary aminostratigraphy of the Bonneville and Lahontan basins, western United States, with paleoclimatic implications: Ph.D. thesis, Univ. of Colorado, Boulder, 603 p.

54. McFarlane, M.J., 1976, Laterite and landscape: Academic Press, New York, 151 p.

55. Milne, J.D.G., 1973, Map and sections of river terraces in the Rangitikei Basin, North Island, New Zealand: New Zealand Soil Surv. Report 4.

56. Morrison, R.B., 1964, Lake Lahontan: Geology of the southern Carson Desert, Nevada: U.S. Geol. Surv. Prof. Pap. 401, 156 p.

57. ———, 1965, Lake Bonneville: Quaternary stratigraphy of eastern Jordan Valley, south of Salt Lake City, Utah: U.S. Geol. Surv. Prof. Pap. 477, 80 p.

58. ———, 1966, Predecessors of Great Salt Lake, p. 77–104 *in* W.L. Stokes, ed., The Great Salt Lake: Guidebook to the geology of Utah, no. 20.

59. ———, 1967, Principles of Quaternary stratigraphy, p. 1–69 *in* R.B. Morrison and H.E. Wright, Jr., eds., Quaternary soils: Internat. Assoc. Quaternary Res., VII Cong., Proc. v. 9.

60. ———, 1968, Means of time-stratigraphic division and long-distance correlation of Quaternary successions, p. 1–113 *in* R.B. Morrison and H.E. Wright, Jr., eds., Means of correlation of Quaternary successions: Internat. Assoc. Quaternary Res., VII Cong., Proc. v. 8.

61. Morrison, R.B., and Frye, J.C., 1965, Correlation of the middle and late Quaternary successions of the Lake Lahontan, Lake Bonneville, Rocky Mountain (Wasatch Range), southern Great Plains, and eastern midwest areas: Nevada Bur. Mines Report 9, 45 p.

62. Morrison, R.B., 1978, Quaternary soil stratigraphy—concepts, methods, and problems, p. 77–108 *in* W.C. Mahaney, ed., Quaternary soils: Geo Abstracts, Norwich, England.

63. Mulcahy, M.J., and Churchward, H.M., 1973, Quaternary environments and soils in Australia: Soil Sci., v. 116, p. 156–169.

64. North American Commission on Stratigraphic Nomenclature, 1983, North American stratigraphic code: Amer. Assoc. Petrol. Geol., v. 67, p. 841–875.

65. Proudfoot, B., 1976, The analysis and interpretation of soil phosphorus in archaeological contexts, p. 93–113 *in* D.A. Davidson and M.L. Shackley, eds., Geoarchaeology: G. Duckworth and Co., Ltd., London.

66. Putnam, W.C., 1942, Geomorphology of the Ventura region, California: Geol. Soc. Amer. Bull., v. 53, p. 691–754.

67. Reider, R.G., 1980, Late Pleistocene and Holocene soils of the Carter/Kerr-McGee archeological site, Powder River Basin, Wyoming: Catena, v. 7, p. 301–315.

68. Retallack, G., 1981a, Fossil soils: Indicators of ancient terrestrial environments, p. 55–102 in K.J. Niklas, ed., Paleobotany, Paleoecology and evolution, v. 1: Praeger Publishers, New York.

69. ———, 1981b, Two new approaches for reconstructing fossil vegetation with examples from the Triassic of eastern Australia, p. 271–295 in J. Gray, A.J. Boucot, and W.B.N. Berry, eds., Communities of the past: Hutchinson Ross Publishing Co., Stroudsberg, Penn.

70. Richmond, G.M., 1962, Quaternary stratigraphy of the La Sal Mountains, Utah: U.S. Geol. Surv. Prof. Pap. 324, 135 p.

71. Rosholt, J.N., 1980, Uranium-trend dating of Quaternary sediments: U.S. Geol. Surv. Open-File Rept. 80–1087.

72. Ruhe, R.V., 1969, Quaternary landscapes in Iowa: Iowa State Univ. Press, Ames, 255 p.

73. Scott, G.R., 1963, Quaternary geology and geomorphic history of the Kassler Quadrangle, Colorado: U.S. Geol. Surv. Prof. Pap. 421-A, 70 p.

74. Scott, W.E., McCoy, W.D., Shroba, R.R., and Rubin, M., in press, Reinterpretation of the exposed record of the last two cycles of Lake Bonneville, western United States: Quaternary Res.

75. Shlemon, R.J., 1978, Late Quaternary evolution of the Camp Pendleton-San Onofre State Beach coastal area, northwestern San Diego County, California: Unpubl. report for Southern California Edison Co. and San Diego Gas and Electric Co., 123 p.

76. Shroba, R.R., Rosholt, J.N., and Madole, R.F., 1983, Uranium-trend dating and soil B horizon properties of till of Bull Lake age, North St. Vrain drainage, Front Range, Colorado: Geol. Soc. Amer. Absts. with Programs, v. 15, p. 431.

77. Suggate, R.P., 1965, Late Pleistocene geology of the northern part of the South Island, New Zealand: New Zealand Geol. Surv. Bull. 77, 91 p.

78. Swan, F.H., III, Schwartz, D.P., and Cluff, L.S., 1980, Recurrence of moderate to large magnitude earthquakes produced by surface faulting on the Wasatch fault zone, Utah: Bull. Seis. Soc. Amer., v. 70, p. 1431–1462.

79. Wehmiller, J.F., 1982, A review of amino acid racemization studies in Quaternary mollusks: Stratigraphic and chronologic applications in coastal and interglacial sites. Pacific and Atlantic coasts, United States, United Kingdom, Baffin Island and tropical islands: Quaternary Sci. Rev., v. 1, p. 83–120.

80. White, E.M., 1978, Cautionary note on soil phosphate data interpretation for archaeology: Amer. Antiquity, v. 43, p. 507–508.

81. Willman, H.B., and Frye, J.C., 1970, Pleistocene stratigraphy of Illinois: Illinois State Geol. Surv. Bull. 94, 204 p.

82. Wintle, A.G., and Huntley, D.J., 1982, Thermoluminescence dating of sediments: Quaternary Sci. Rev., v. 1, p. 31–53.

83. Woods, W.I., 1977, The quantitative analysis of soil phosphate: Amer. Antiquity, v. 42, p. 248–252.

84. Yaalon, D.H., 1971, Soil-forming processes in time and space, p. 29–39 in D.H. Yaalon, ed., Paleopedology: Israel Univ. Press, Jerusalem, 350 p.

Data necessary for describing a soil profile

HORIZON NOMENCLATURE

It should be emphasized, first, that horizon nomenclature commonly depends not only on the properties of the particular horizon, but also on those of the overlying and underlying horizons and the parent material or the presumed parent material. Hence, it is not uncommon early in the soil description to change horizon designation because what was one set of horizons identified down from the surface may change after more is learned of the adjacent horizons and the parent material.

As mentioned in Chapter 1, there are certain rules to follow when more than one lower-case letter is used with the master horizon designation. In most cases, more than one lower-case letter can be used. These letters are always written first: a, e, i, h, r, s, t, v, and w. Further, none of the latter letters is used in combination, except for Bhs and Crt. If more than one lower-case letter is used and the horizon is not buried, these, if used, are written last: c, f, g, m, and x. If used, b commonly is last, unless field or laboratory data indicate horizon features that are post-burial. For example, if carbonate accumulates in a Bt horizon after burial, the designation is Btbk. For B-horizon designation, t has precedence over w, s, and h, and the latter are not used in combination with t. If other lower-case letters are used with t, the t comes first (e.g. Bto).

Because most of the soil literature in the United States uses the old soil-horizon nomenclature, Table A-1 is provided to convert the old symbols to the new ones.

It is important that the terminology developed by soil scientists is used to describe soils.[8] Examples of a variety of soil-profile descriptions with accompanying laboratory analyses are given by the Soil Survey Staff.[9] The following properties should be recorded:

Depth The top of the uppermost mineral horizon (A or E) is taken as zero depth. The 0-horizon thickness is measured up from that point (2 to 0 cm), and all other horizons down from that point (0 to 8 cm).

Table A-1
Soil-horizon Symbol Conversions

MASTER HORIZONS		SUBORDINATE DEPARTURES		
Old	*New*	*Old*	*New*	*Description*
O	O	—	a	Highly decomposed organic matter
O1	Oi, Oe	b	b	Buried soil horizon
O2	Oa, Oe	cn	c	Concretions or nodules
A	A	—	e	Intermediately decomposed organic matter
A1	A	f	f	Frozen soil
A2	E	g	g	Strong gleying
A3	AB or EB	h	h	Illuvial accumulation of organic matter
AB	—	—	i	Slightly decomposed organic matter
A&B	E/B	ca	k	Accumulation of carbonates
AC	AC	m	m	Strong cementation
B	B	sa	n	Accumulation of sodium
B1	BA or BE	—	o	Residual accumulation of sesquioxides
B&A	B/E	p	p	Plowing or other disturbance
B2	B or Bw	si	q	Accumulation of silica
B3	BC or CB	r	r	Weathered or soft bedrock
C	C	ir	s	Illuvial accumulation of sesquioxides
R	R	t	t	Accumulation of clay
		—	v	Plinthite
		—	w	Color or structural B
		x	x	Fragipan character
		cs	y	Accumulation of gypsum
		sa	z	Accumulation of salts

(From Guthrie and Witty.[5])

Color List dominant color and size and color variation of prominent mottles. Use the Munsell Soil Color Chart (Munsell Color Co., Inc., Baltimore) or other suitable charts that use the Munsell color notation. List the moisture state when taken.

Consistence This is a measure of the adherence of the soil particles to the fingers, the cohesion of soil particles to one another, and the resistance of the soil mass to deformation. Because this property varies with moisture content, it is taken when the soil is dry, moist, and wet. The wet consistence (natural wetness or artifical wetness) is useful in determining texture classes in the field and is composed of two quantities, stickiness and plasticity. Stickiness is measured by compressing the soil between thumb and forefinger and noting the adherence of the soil to either upon release of pressure. The classes recognized are *non-sticky*—no adherence when pressure is released; *slightly sticky*—soil adheres slightly upon release of pressure and stretches only slightly before being pulled apart; *sticky*—soil adheres on release of pressure and stretches before being pulled apart; *very sticky*—soil adheres strongly and will sustain a fair amount of stretching before rupture.

Plasticity is measured by rolling the soil between thumb and forefinger in an attempt to form a thin rod. Several classes are recognized: *nonplastic*—no rod forms; *slightly plastic*—a weak rod forms that is easily deformed and broken; *plastic*—a rod forms that will resist moderate deformation and breakage during moderate handling; *very plastic*—a rod forms and is readily bent and otherwise manipulated before breakage. Wet consistence is very important in determining a change in soil texture with depth and textural class; it is a major field clue to textural change if several adjacent soil horizons in a profile lie within the same textural class.

Texture Determine the textural class of the less than 2-mm fraction, by noting the grittiness and wet consistence. Broad guidelines are given in the rating chart (Fig. A-1), but for more accuracy, one should determine the limits for himself, using samples with known particle-size distribution. Greater than 2-mm particles should be described according to size, and volume percent of the soil they occupy. Weight percent can be determined in the field with a screen and hand-held portable scale. Be watchful for shape and lithologic changes, as they may indicate parent materials of more than one origin.

Structure Describe type (Table 1-4), size, and grade of structure. Size classes vary with type of structure as shown in Table A-2. Grade is a classification of structural development: *single grain*—no bonding between particles; *massive*—no ped formation, but there is enough interparticle bonding for the soil to stand in a vertically cut face; *weak*—a few peds are barely observable, and much material is unaggregated; *moderate*—peds are easily observable in place and most material is aggregated; *strong*—mass consists entirely of distinctly visible peds. In general, structural grade is stronger with increasing amounts of clay-size particles.

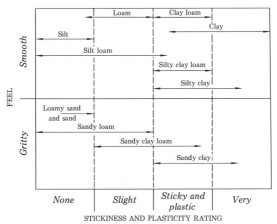

Fig. A-1 Approximate relations between texture class, grittiness, and wet consistence.

Table A-2
Variation in Size Classes with Structure

Size class	Granule diameter (mm)	Plate thickness (mm)	Block diameter (mm)	Prism diameter (mm)
vf (very fine)	<1	<1	<5	<10
f (fine)	1–2	1–2	5–10	10–20
m (medium)	2–5	2–5	10–20	20–50
c (coarse)	5–10	5–10	20–50	50–100
vc (very coarse)	>10	>10	>50	>100

Clay films Record their occurrence, frequency, and thickness. Films occur as colloidal stains on grains, as bridges between adjacent grains, or aligned along pores or ped faces. Frequency classification is based on the percent of the ped faces and/or pores that contain films: *very few*—less than 5%; *few*—5–25%; *common*—25–50%; *many*—50–90%; and *continuous*—90–100%. The thickness of films is determined with a hand lens: *thin*—film is so thin that very fine sand grains stand out; *moderately thick*—very fine sand grains are so enveloped by the film that the grain outlines are indistinct, yet the grains impart microrelief to film; and *thick*—very fine and fine sand grains are enveloped by clay, forming a film with a smooth appearance; films are visible without magnification. A hand lens is required for the study of films.

Silt caps In places, one can detect translocation of silt and clay by the presence of these materials on the tops of clasts in the soil. Burns[3] has noted that the caps become thicker and cover a greater proportion of the clast surface with time, and he and S.L. Forman (personal commun., 1983) define stages of development from work in alpine and arctic regions (Table A-3).

Table A-3
Stages of Development of Silt Caps

Stage	Thickness (mm); cover (%)	Morphology
1	Thickness: < 1 Cover: 1–25	Stages 1–4 are recognized as coatings on the tops of clasts
2	Thickness: 1–4 Cover: 25–75	
3	Thickness: 4–7 Cover: 75–90	
4	Thickness: 5–10 Cover: 90–100	
5	Thickness: 10–20 Cover: 100	Caps are connected between clasts to form bridges
6	Thickness: 10–30 Cover: 100	Void spaces are filled with silt and clay

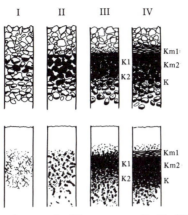

Fig. A-2 Sketch of carbonate buildup stages (I, II, III, and IV) for gravelly (top) and nongravelly (bottom) parent materials. (Taken from Gile and others,[4] © 1966, The Williams & Wilkins Co., Baltimore.)

pH Record field value, using a field test kit.

Carbonates Note the distribution of carbonate, estimate the volume percent, and classify on stage of development[4] (Fig. A-2 for the first four stages, and Table A-4 for all stages and descriptions).

In places, there may be not stage II morphology in a sequence of soils; rather, filaments of stage I become so common that the horizon meets the percent requirements for stage II. Holliday[6] suggests that these latter occurrences be termed IIf to point to their filamentous morphology.

I want to inject a word of caution on the recognition of carbonate morphological stages. In places, carbonate can be deposited on vertical faces by laterally seeping waters and thereby mask the pedogenic carbonate morphology.[7] Hence, to study the morphology of pedogenic carbonate and avoid surficial cementation, one may have to dig back about 1 m.

Indicate effervescence with dilute (~1 N) HCl: *very slight*—few bubbles; *slight*—bubbles readily observed; *strong*—bubbles form a low foam; *violent*— foam is thick and has a "boiling" appearance.

Salts For salts more soluble than carbonates, Bockheim[2] has devised a six-stage scheme based on the morphology of the salts and the degree of induration of the horizon (Table A-5).

Cementation Record the kind of cementing agent, whether it is continuous or discontinuous, estimate the volume percent it occupies, and estimate how strongly the horizon is cemented: *weak*—material is brittle and can be broken with the hands; *strong*—material is brittle and broken easily with a hammer; *indurated*—material is brittle and broken only with a sharp hammer blow.

Table A-4
Stages of Carbonate Morphology

STAGE	GRAVELLY PARENT MATERIAL	NONGRAVELLY PARENT MATERIAL
I	Thin discontinuous clast coatings; some filaments; matrix can be calcareous next to stones; about 4% $CaCO_3$	Few filaments or coatings on sand grains; <10% $CaCO_3$
I+	Many or all clast coatings are thin and continuous	Filaments are common
II	Continuous clast coatings; local cementation of few to several clasts; matrix is loose and calcareous enough to give somewhat whitened appearance	Few to common nodules; matrix between nodules is slightly whitened by carbonate (15–50% by area), and the latter occurs in veinlets and as filaments; some matrix can be noncalcareous; about 10–15% $CaCO_3$ in whole sample, 15–75% in nodules
II+	Same as stage II, except carbonate in matrix is more pervasive	Common nodules; 50–90% of matrix is whitened; about 15% $CaCO_3$ in whole sample

Continuity of fabric high in carbonate

III Horizon has 50–90% K fabric with carbonate forming an essentially continuous medium; color mostly white; carbonate-rich layers more common in upper part; about 20–25% $CaCO_3$

Many nodules, and carbonate coats so many grains that over 90% of horizon is white; carbonate-rich layers more common in upper part; about 20% $CaCO_3$

III+ Most clasts have thick carbonate coats; matrix particles continuously coated with carbonate or pores plugged by carbonate; cementation more or less continuous; >40% $CaCO_3$

Most grains coated with carbonate; most pores plugged; >40% $CaCO_3$

Partly or entirely cemented

IV Upper part of K horizon is nearly pure cemented carbonate (75–90% $CaCO_3$) and has a weak platy structure due to the weakly expressed laminar depositional layers of carbonate; the rest of the horizon is plugged with carbonate (50–75% $CaCO_3$)

V Laminar layer and platy structure are strongly expressed; incipient brecciation and pisolith (thin, multiple layers of carbonate surrounding particles) formation

VI Brecciation and recementation, as well as pisoliths, are common

(Taken from Gile and others,[4] Bachman and Machette,[1] with further modification by R.R. Shroba, written commun., 1982.)

Table A-5
Morphological Stages of Salt Accumulation

Salt stage	Maximum salt morphology	Electroconductivity of salt-enriched horizon (mmho cm^{-1})*
0	None	<0.6
I	Coating on stone bottoms	0.6–5
II	Few flecks (<20% of surface area of cut has accumulations that are about 1 to 2 mm in diameter)	5–18
III	Many flecks (>20% of surface area has flecks as above)	18–25
IV	Weak cementation**	25–40
V	Strong cementation**	40–60
VI	Indurated horizon**	60–100$^+$

*A measure of salt content.
**These terms are defined in the next section.
(Taken from Bockheim.[2])

Horizon boundaries Record width of transition zone from the overlying to the underlying horizon (distinctness) and the topography of the zone. Distinctness classes are *very abrupt*—no greather than 1 mm; *abrupt*—1 mm–2.5 cm; *clear*—2.5–6 cm; *gradual*—6–12.5 cm; *diffuse*—> 12.5 cm. Topography descriptions are *smooth*—boundaries are parallel to ground surface; *wavy*—boundary undulates and depressions are wider than they are deep; *irregular*—boundary undulates and depressions are deeper than they are wide; *broken*—parts of horizon are disconnected laterally.

Percentage estimate It is important to estimate the percent by volume of various soil features, such as gravel or carbonate content or extent of mottling. Figure A-3 is provided to aid with such estimates.

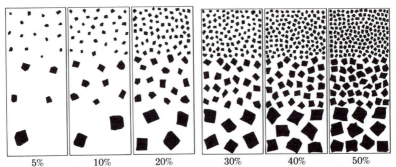

Fig. A-3 Charts for estimating percent, by volume, of various soil features. (Taken from Yaalon,[10] © 1966, The Williams & Wilkins Co., Baltimore.)

REFERENCES

1. Bachman, G.O., and Machette, M.N., 1977, Calcic soils and calcretes in the southwestern United States: U.S. Geol. Surv. Open-File Rept. 77–794, 163 p.
2. Bockheim, J.G., 1981, Soil development in cold deserts of Antarctica, p. 16–17 *in* D.H. Yaalon, ed., International conference on aridic soils (Abstracts): Jerusalem, Israel.
3. Burns, S.F., 1980, Alpine soil distribution and development, Indian Peaks, Colorado Front Range: Unpubl. Ph.D. thesis, Univ. of Colorado, Boulder, 360 p.
4. Gile, L.H., Peterson, F.F., and Grossman, R.B., 1966, Morphological and genetic sequences of carbonate accumulation in desert soils: Soil Sci., v. 101, p. 347–360.
5. Guthrie, R.L., and Witty, J.E., 1982, New designations for soil horizons and layers and the new *Soil Survey Manual:* Soil Sci. Soc. Amer. Jour., v. 46, p. 443–444.
6. Holliday, V.T., 1982, Morphological and chemical trends in Holocene soils at the Lubbock Lake archeological site, Texas: Unpubl. Ph.D. thesis, Univ. of Colorado, Boulder, 285 p.
7. Lattman, L.H., 1973, Calcium carbonate cementation of alluvial fans in southern Nevada: Geol. Soc. Amer. Bull., v. 84, p. 3013–3028.
8. Soil Survey Staff, 1951, Soil survey manual: U.S. Dept. Agri. Handbook no. 18, 503 p.
9. ———, 1975, Soil taxonomy: U.S. Dept. Agri. Handbook no. 436, 754 p.
10. Yaalon, D.H., 1966, Chart for the quantitative estimation of mottling and of nodules in soil profiles: Soil Sci., v. 102, p. 212–213.

APPENDIX 2

Climatic conditions in the United States

Throughout the text, data are given for soils located in different parts of the United States. Rather than give the climatic data for each of these places, the following maps are presented to give the reader a general picture of the climate at the various places mentioned in the text. (Taken from the Oxford World Atlas, Saul B. Cohen, ed. © 1973, Oxford University Press, New York, p. 80–81.)

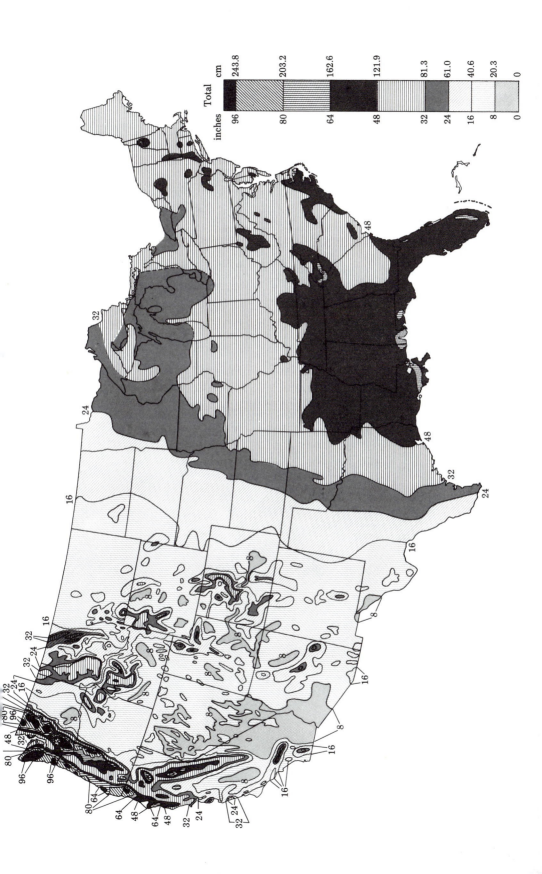

inches		cm
96		243.8
		203.2
80		
		162.6
64		
		121.9
48		
		81.3
32		
24		61.0
16		40.6
8		20.3
0		0

Total

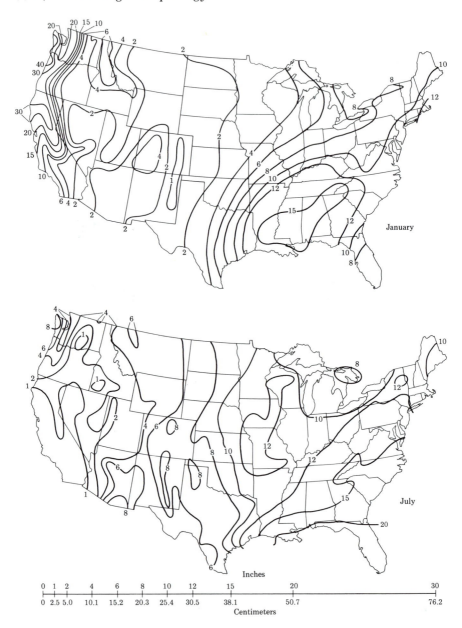

Precipitation

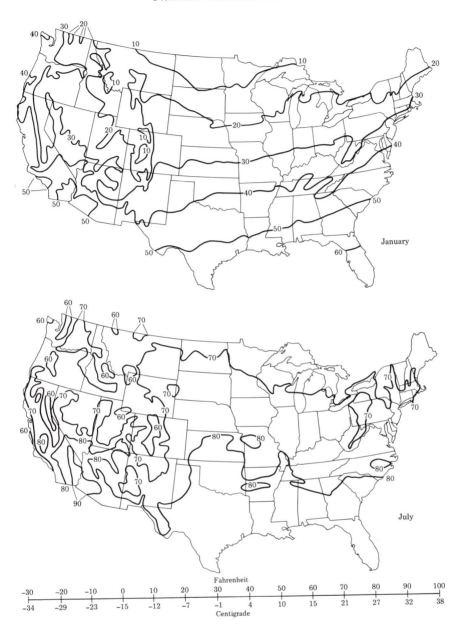

Temperature

Index